CW01672123

MST124

Essential mathematics 1

Book C

This publication forms part of an Open University module. Details of this and other Open University modules can be obtained from the Student Registration and Enquiry Service, The Open University, PO Box 197, Milton Keynes MK7 6BJ, United Kingdom (tel. +44 (0)845 300 6090; email general-enquiries@open.ac.uk).

Alternatively, you may visit the Open University website at www.open.ac.uk where you can learn more about the wide range of modules and packs offered at all levels by The Open University.

To purchase a selection of Open University materials visit www.ouw.co.uk, or contact Open University Worldwide, Walton Hall, Milton Keynes MK7 6AA, United Kingdom for a brochure (tel. +44 (0)1908 858779; fax +44 (0)1908 858787; email ouw-customer-services@open.ac.uk).

The Open University, Walton Hall, Milton Keynes, MK7 6AA.

First published 2014.

Edited, designed and typeset by The Open University, using the Open University TEX System.

Printed in the United Kingdom by Page Bros, Norwich.

ISBN 978 1 7800 7359 0

1.1

Woodland Carbon CO₂de

The Open University has had Woodland Carbon Code Pending Issuance Units assigned from Doddington North forest creation project (IHS ID103/26819) that will, as the trees grow, compensate for the greenhouse gas emissions from the manufacture of the paper in MST124 Essential mathematics 1: book C. More information can be found at https://www.woodlandcarboncode.org.uk/

Contents

Unit 7

Differentiation methods and integration

Introduction

The first half of this unit continues with the topic of Unit 6, *differential calculus*. You'll meet the formulas for the derivatives of five standard functions, namely sin, cos, tan, exp and ln. You'll also learn some further ways of differentiating complicated functions by combining the formulas for the derivatives of standard functions.

By the end of this part of the unit you should have acquired skills in differentiating a wide variety of functions. You'll need these skills in the later parts of the module, and in any further modules that you study that involve calculus. So it's important that you become proficient in them. The only way to do that is to practise, so you should aim to do all the activities in this unit, and as many further activities in the practice quizzes and exercise booklets as you have time for, until you feel confident with your skills.

In the first half of the unit you'll also learn how to use your computer to find and work with derivatives, and you'll see some further applications of differentiation.

In the second half of the unit you'll start to learn about *integral calculus*, which can be thought of as the reverse of differential calculus. You've seen that in differential calculus you start off knowing the values taken by a continuously changing quantity throughout a period of change, and you use this information to find the values taken by the rate of change of the quantity throughout the same period. In integral calculus, you do the opposite: you start off knowing the values taken by the rate of change of a quantity throughout a period of change, and you use this information to find the values taken by the quantity throughout the same period. This process is called *integration*. The topic of integral calculus continues in Unit 8.

1 Derivatives of further standard functions

In this section you'll meet the formulas for the derivatives of the five standard functions sin, cos, tan, exp and ln.

1.1 Derivatives of sin, cos and tan

In this first subsection we'll obtain formulas for the derivatives of the trigonometric functions $f(x) = \sin x$, $f(x) = \cos x$ and $f(x) = \tan x$.

When you're working with these functions, it's essential to remember that the input variable x is the size of an angle *measured in radians*. This is indicated by the formulas for the functions, because x represents a number, and when the size of an angle is given just as a number, with no units, then the units are assumed to be radians. You met this convention in Unit 4. For example, $\sin 10$ denotes the sine of 10 radians, whereas $\sin 10^\circ$ denotes the sine of 10 degrees.

When we're dealing with the derivatives of trigonometric functions, we always measure angles in radians. This is because it makes the formulas for the derivatives of trigonometric functions simpler. If you'd like to know more about this, then, once you've completed Sections 1 and 2 of this unit, read the document *Why do we use radians?* on the module website.

Derivative of sin

Our first aim in this subsection is to obtain a formula for the derivative of the sine function, $f(x) = \sin x$.

As you know, the derivative of a function is itself a function, and therefore the derivative has a graph. So, to get a rough idea of what the derivative of the sine function might be like, let's try to roughly sketch its graph, by estimating the gradient of the graph of the sine function at various points.

Figure 1 shows the graph of the sine function, drawn on axes with equal scales, and some tangents to the graph.

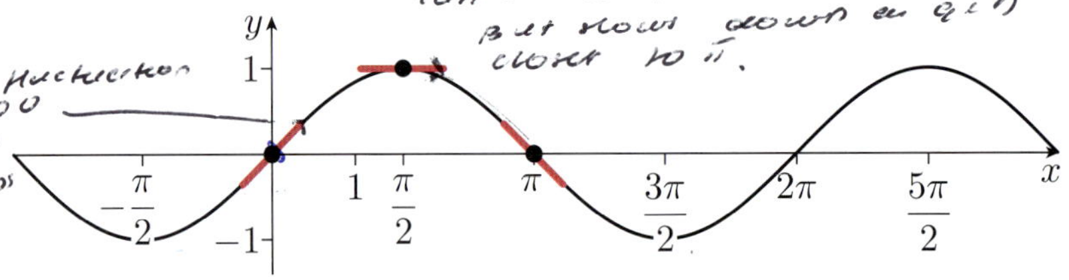

Figure 1 The graph of $f(x) = \sin x$

When $x = 0$, the gradient of the graph of the sine function seems to be about the same as the gradient of the line $y = x$; that is, it seems to be about 1. As x increases, the graph gradually gets less steep – that is, its gradient gradually decreases, until $x = \pi/2$, when the gradient seems to be zero. The gradient decreases slowly when x is only a little larger than 0, but decreases more rapidly as x gets closer to $\pi/2$.

In other words, for x between 0 and $\pi/2$ the value of the derivative seems to decrease from 1 to zero, slowly at first, but then more rapidly. So its graph must look something like the sketch in Figure 2.

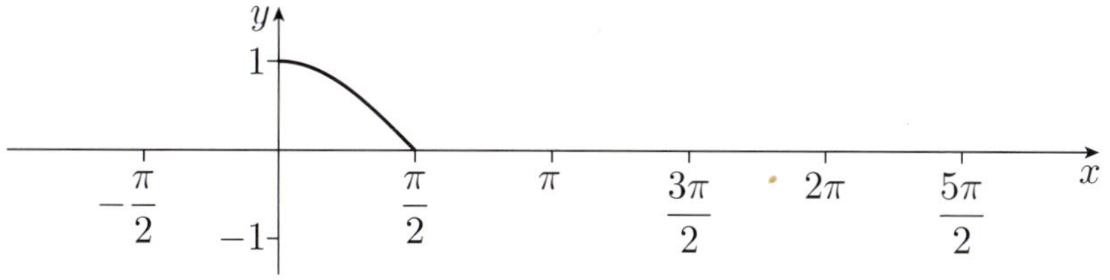

Figure 2 A sketch of the graph of the derivative of $f(x) = \sin x$, for x between 0 and $\pi/2$

Now look at the graph of the sine function for x between $\pi/2$ and π. The gradient starts at zero, then it becomes negative. At first the graph is not very steep – that is, the gradient has small negative values – but it becomes steeper and steeper – that is, the gradient takes negative values of greater and greater magnitude. Eventually, when $x = \pi$, the gradient seems to be about the same as the gradient of the line $y = -x$; that is, it seems to be about -1. The gradient decreases fairly rapidly when x is only a little larger than $\pi/2$, but decreases more slowly as x gets closer to π.

So, for x between $\pi/2$ and π, the value of the derivative seems to decrease from 0 to -1, fairly rapidly at first, but then more slowly. So the graph of the derivative must look something like the sketch in Figure 3.

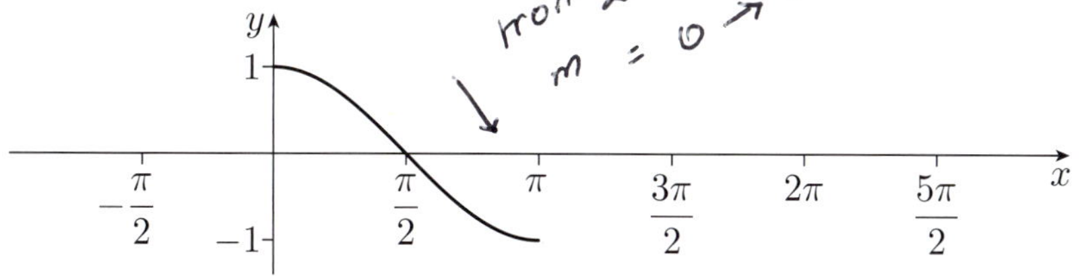

Figure 3 A sketch of the graph of the derivative of $f(x) = \sin x$, for x between 0 and π

In the next activity you're asked to extend the sketch in Figure 3 to cover more values of x.

Activity 1 *Extending the sketch of the graph of the derivative of sin*

By considering the gradients of the graph of the sine function (shown in Figure 1) for x between π and $3\pi/2$, and then for x between $3\pi/2$ and 2π, work out what the graph of the derivative of the sine function must look like for these values of x. Hence extend the sketch of the derivative of the sine function in Figure 3 to cover all values of x between 0 and 2π.

In Activity 1 you should have worked out that the graph of the derivative of the sine function for x between 0 and 2π must look something like the sketch in Figure 4.

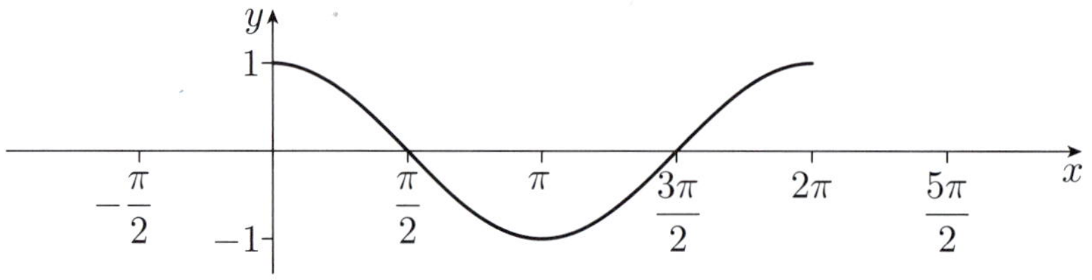

Figure 4 A sketch of the graph of the derivative of $f(x) = \sin x$, for x between 0 and 2π

Since the graph of the sine function repeats every 2π units, its gradients also repeat every 2π units. So you can now extend the sketch of the graph of the derivative to cover any interval of values of x that you wish. For example, Figure 5 shows a sketch of the graph for values of x between $-\pi$ and 3π.

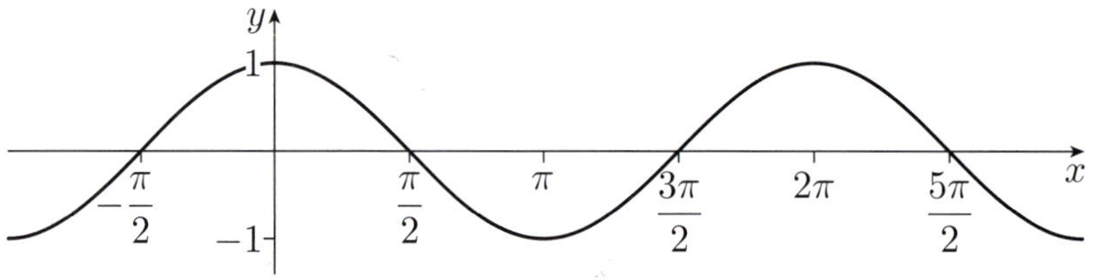

Figure 5 A sketch of the graph of the derivative of $f(x) = \sin x$, for x between $-\pi$ and 3π

The graph in Figure 5 should look familiar to you – it looks very like the graph of the cosine function! So it seems possible that the derivative of the sine function is the cosine function – in other words, that

$$\text{if } f(x) = \sin x, \text{ then } f'(x) = \cos x.$$

This is indeed true. It can be confirmed by using differentiation from first principles, but this is quite tricky to do. If you'd like to see some details about how it's done, then have a look at the document *The derivative of the sine function* on the module website.

Derivative of cos

Let's now look at obtaining a formula for the derivative of the cosine function, $f(x) = \cos x$. You could get a rough idea of what this derivative must be like by using a similar procedure to the one that you've just seen for the sine function. Figures 6 and 7 show the graph of $f(x) = \cos x$, and the sketch of its derivative that's obtained from this procedure.

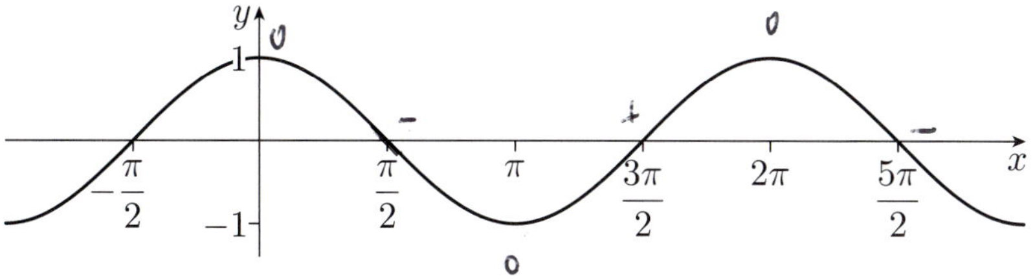

Figure 6 The graph of $f(x) = \cos x$

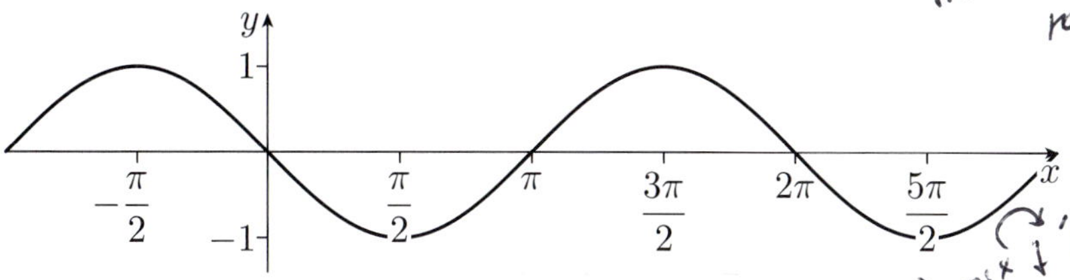

Figure 7 A sketch of the graph of the derivative of $f(x) = \cos x$

The graph in Figure 7 looks like the graph of the sine function, reflected in the x-axis. That is, it looks like the graph of $y = -\sin x$. So it seems possible that the derivative of the cosine function is the negative of the sine function – in other words, that

$$\text{if } f(x) = \cos x, \text{ then } f'(x) = -\sin x.$$

Again, this is indeed true. One way to confirm it is to use differentiation from first principles, but, as with the sine function, this is a tricky process.

A simpler way to confirm it is to deduce it from the fact that the derivative of the sine function is the cosine function. As you know, you can obtain the graph of the cosine function by translating the graph of the sine function by $\pi/2$ to the left. It follows that you can obtain the graph of the *derivative* of the cosine function by translating the graph of the *derivative* of the sine function by $\pi/2$ to the left. (This is because the two derivatives give the gradients of the two graphs.)

The derivative of the sine function is the cosine function, whose graph is shown in Figure 8. Translating this graph by $\pi/2$ to the left gives the graph in Figure 9. Because of the symmetries of the graphs of sine and cosine, this graph is the graph of $y = -\sin x$. So the derivative of $f(x) = \cos x$ is indeed $f'(x) = -\sin x$.

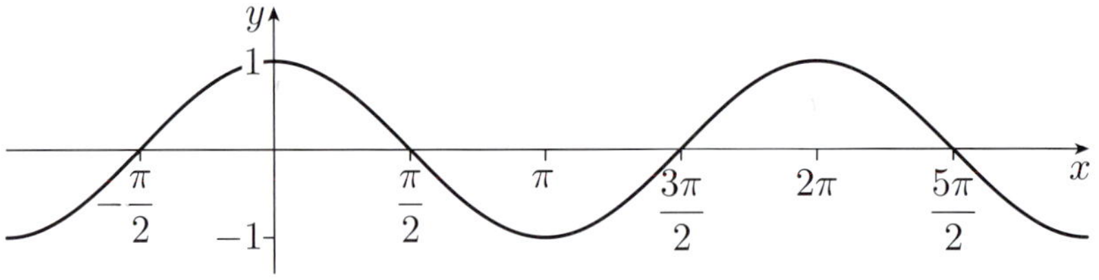

Figure 8 The graph of $y = \cos x$

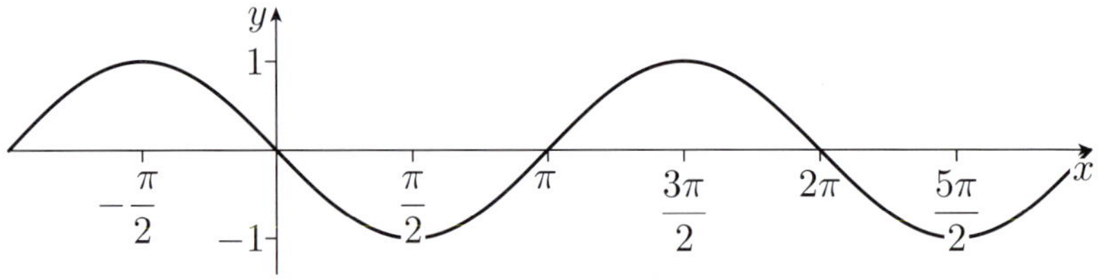

Figure 9 The graph of $y = \cos x$ translated by $\pi/2$ to the left

Derivative of tan

The final derivative of a trigonometric function that you'll meet in this subsection is the derivative of the tangent function. In fact,

$$\text{if } f(x) = \tan x, \text{ then } f'(x) = \sec^2 x.$$

Remember that $\sec^2 x$ means $(\sec x)^2$, and also that $\sec x$ means $1/(\cos x)$. So another way to write the derivative of $f(x) = \tan x$ is

$$f'(x) = \frac{1}{(\cos x)^2}.$$

However, we usually use the form $f'(x) = \sec^2 x$, for conciseness.

This formula for the derivative of the tangent function can be worked out by using the fact that $\tan x = (\sin x)/(\cos x)$, together with the formulas for the derivatives of $f(x) = \sin x$ and $f(x) = \cos x$. The process involves a rule for combining derivatives that you'll meet later in the unit, and you'll see the justification of the formula there.

For now, you might like to compare the graphs of $y = \tan x$ and $y = \sec^2 x$, shown in Figure 10, and convince yourself that the graph of $y = \sec^2 x$ does seem, at least roughly, to give the gradients of the graph of $y = \tan x$.

In particular, notice that the gradients of the graph of $y = \tan x$ are always positive and that the graph of $y = \tan x$ is the least steep whenever x is a multiple of π.

when y = tan x close to multiples of π → the graph becomes less steep.

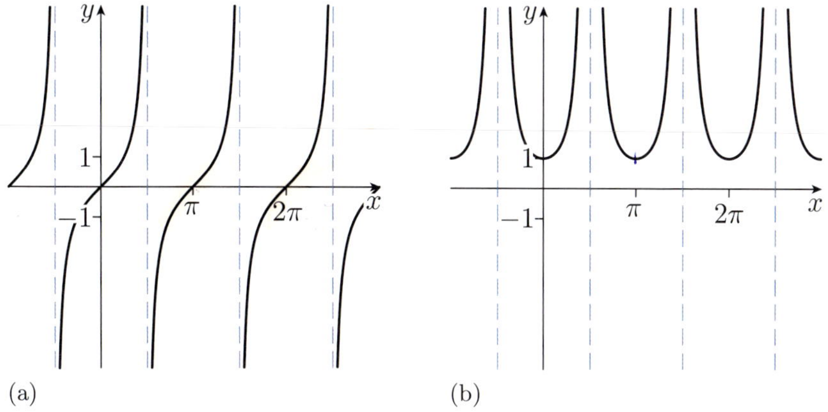

Figure 10 The graphs of (a) $y = \tan x$ (b) $y = \sec^2 x$

Here's a summary of the three derivatives that you've met in this subsection.

Derivatives of sin, cos and tan

$$\frac{d}{dx}(\sin x) = \cos x \qquad \frac{d}{dx}(\cos x) = -\sin x \qquad \frac{d}{dx}(\tan x) = \sec^2 x$$

In the next activity you're asked to use these formulas, together with the sum and constant multiple rules, to find the derivatives of some more functions. In this activity, and in general, remember that if a function is specified using function notation, then you should usually use Lagrange notation for its derivative, whereas if it's specified by an equation that expresses one variable in terms of another, then you should usually use Leibniz notation.

Activity 2 *Differentiating functions involving sin, cos and tan*

Write down the derivatives of the following functions.

(a) $f(x) = \sin x + \cos x$ (b) $g(u) = u^2 - \cos u$ (c) $P = 6\tan\theta$

(d) $r = -2(1 + \sin\phi)$

1.2 Derivatives of exp and ln

In this subsection you'll meet the formulas for the derivatives of the exponential function, $f(x) = e^x$, and its inverse function, the natural logarithm function $f(x) = \ln x$. Remember that e is a special irrational constant, whose value is about 2.718.

Derivative of exp

Let's look first at the exponential function, $f(x) = e^x$. Its graph is shown in Figure 11.

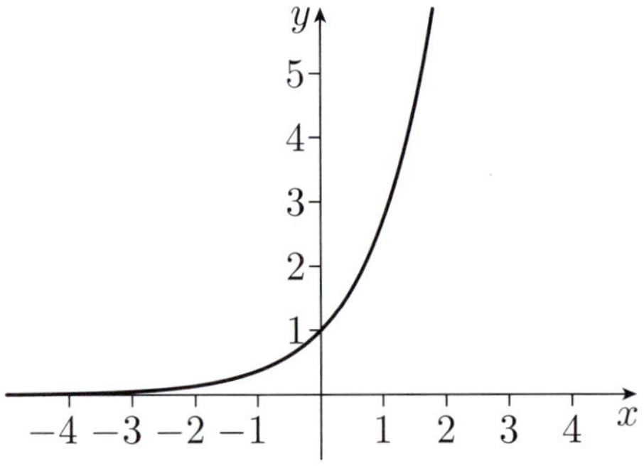

Figure 11 The graph of $y = e^x$

You saw in Unit 3 that the constant e has the special property that the graph of $f(x) = e^x$ has gradient exactly 1 at the point with x-coordinate 0, as shown in Figure 12. So you already know one value of $f'(x)$ for this function f: you know that $f'(0) = 1$.

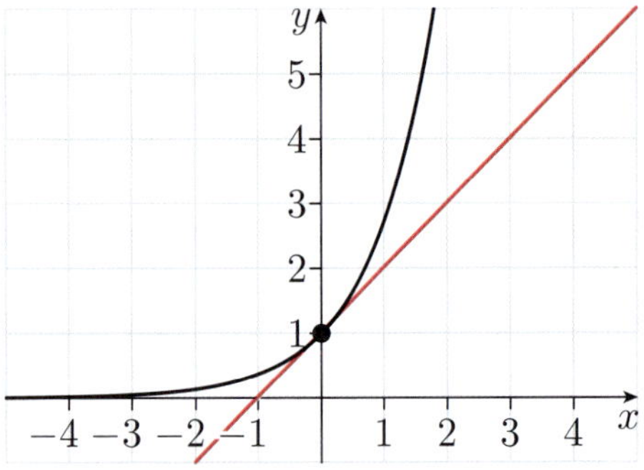

Figure 12 The graph of $y = e^x$ has gradient 1 when $x = 0$

You can find the general formula for the derivative $f'(x)$ of the function $f(x) = e^x$ by using differentiation from first principles. The difference quotient for $f(x) = e^x$ is

$$\frac{f(x+h) - f(x)}{h} = \frac{e^{x+h} - e^x}{h}.$$

As usual, to find a formula for $f'(x)$, you have to work out what happens to the value of the difference quotient as h gets closer and closer to zero. It's not immediately obvious how to do this, but surprisingly you can do it by using the fact that you already know what happens in the particular case when $x = 0$, since you know that $f'(0) = 1$.

When $x = 0$, the difference quotient is

$$\frac{e^{0+h} - e^0}{h} = \frac{e^h - 1}{h}.$$

So you already know that, as h gets closer and closer to zero, the value of this expression gets closer and closer to 1.

Now look again at the difference quotient for a general value of x. You can rearrange it as follows:

$$\frac{e^{x+h} - e^x}{h} = \frac{e^x e^h - e^x}{h} = e^x \left(\frac{e^h - 1}{h}\right).$$

The expression in the brackets is the difference quotient in the case when $x = 0$. So as h gets closer and closer to zero, the expression in the brackets gets closer and closer to 1, and hence the value of the whole expression gets closer and closer to e^x. That is, the formula for the derivative is

$$f'(x) = e^x.$$

So the derivative of the exponential function is the exponential function!

To see that this makes sense, look at the graph of the exponential function in Figure 11. As x increases, the gradient of the graph increases, and it increases more and more rapidly. So as x increases, the value of the derivative of the exponential function increases, and it increases more and more rapidly. The exponential function itself has the property that as x increases, the value of the function increases, and it increases more and more rapidly.

The exponential function, and its constant multiples such as the function $f(x) = 3e^x$, are the only functions that are equal to their own derivatives (except, of course, for functions obtained by restricting the domains of these functions). It's this property that makes the constant e such a special number.

Derivative of ln

Now let's look at the natural logarithm function, $f(x) = \ln x$. Since this function is the inverse of the exponential function $f(x) = e^x$, its graph is the reflection of the graph of $f(x) = e^x$ in the line $y = x$, as shown in Figure 13.

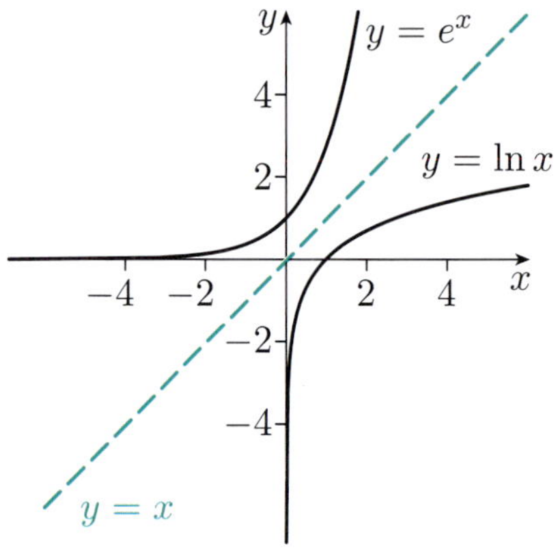

Figure 13 The graphs of $y = \ln x$ and $y = e^x$

So the gradients of the two graphs are related, and hence you can use the formula for the derivative of $f(x) = e^x$ to work out a formula for the derivative of $f(x) = \ln x$. If you'd like to see this done, then have a look at the document *The derivative of the natural logarithm function* on the module website.

However, a quicker way to work out the derivative of the natural logarithm function is to use a general rule, called the *inverse function rule*, that you'll meet later in the unit. This rule tells you how to use the formula for the derivative of *any* invertible function to work out a formula for the derivative of its inverse function. When you meet this rule, you'll see that the formula for the derivative of $f(x) = \ln x$ turns out to be

$$f'(x) = \frac{1}{x}.$$

For now, you might like to compare the graphs of $y = \ln x$ and $y = 1/x$, shown in Figure 14, and convince yourself that the function $y = 1/x$ does seem to give the gradients of the graph of the function $y = \ln x$. Only the right-hand half of the graph of $y = 1/x$ is shown. The left-hand half doesn't give gradients of the graph of $f(x) = \ln x$, because $\ln x$ is defined only for $x > 0$.

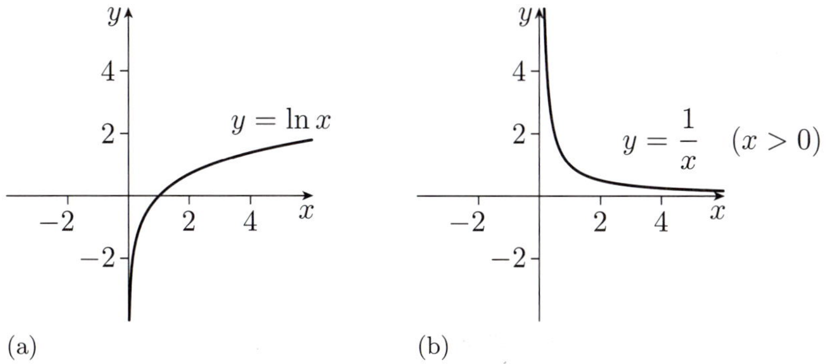

Figure 14 The graphs of (a) $y = \ln x$ (b) $y = 1/x$ $(x > 0)$

The two derivatives that you've met in this subsection are summarised in the box below.

Derivatives of exp and ln

$$\frac{d}{dx}(e^x) = e^x \qquad \frac{d}{dx}(\ln x) = \frac{1}{x}$$

In the next activity you're asked to use these formulas, together with the sum and constant multiple rules and the formulas for the derivatives of trigonometric functions, to find the derivatives of some more functions.

Activity 3 *Differentiating functions involving exp and log*

Find the derivatives of the following functions.

(a) $f(x) = e^x + \ln x$ (b) $h(r) = r - \cos r - 3\ln r$

(c) $v = \dfrac{1}{t} + \ln t$ (d) $w = 5 - 3e^u$ (e) $k = 4(\ln v - \tan v)$

The formulas for the derivatives of standard functions that you've met in this section are all included in the *Handbook*. However, it's worth memorising the formulas for the derivatives of sin, cos, exp and ln, at least, as they occur frequently.

formulas available in Handbook
– but memorise
sin
cos
tan
e^x
ln x
} as they are common.

2 Finding derivatives of more complicated functions

In this section you'll learn some more ways in which you can combine formulas for the derivatives of functions to obtain formulas for the derivatives of other, more complicated functions.

2.1 Product rule

Suppose that you know the formulas for the derivatives of two functions, and you want to know the formula for the derivative of their product. (Remember that the *product* of two functions is obtained by 'multiplying them together'.) For example, you already know the formulas for the derivatives of the functions $f(x) = x^2$ and $g(x) = \sin x$, but suppose that you want to know the formula for the derivative of the function $k(x) = x^2 \sin x$.

You've seen that to obtain a formula for the derivative of the *sum* of two functions, you just add the formulas for the derivatives of the two individual functions – this is the *sum rule*. So you might hope that to obtain a formula for the derivative of the *product* of two functions, you just multiply the formulas for the derivatives of the two individual functions.

Unfortunately, this doesn't work! For example, consider the two power functions $f(x) = x^2$ and $g(x) = x^3$. The derivatives of x^2 and x^3 are $2x$ and $3x^2$, respectively, and multiplying these two derivatives gives $6x^3$. However, the product of the functions $f(x) = x^2$ and $g(x) = x^3$ is the function $k(x) = x^5$, and you already know that the derivative of x^5 is $5x^4$. So multiplying the formulas for the derivatives of the two individual functions doesn't give the right answer for the derivative of their product.

In fact, to find a formula for the derivative of the product of two functions, you need to use the rule in the box below. It's given in Lagrange notation, but you'll see it in Leibniz notation shortly.

Product rule (Lagrange notation)

If $k(x) = f(x)g(x)$, then

$$k'(x) = f(x)g'(x) + g(x)f'(x),$$

for all values of x at which both f and g are differentiable.

The product rule can be proved by using the idea of differentiation from first principles, and you'll see this done at the end of this subsection.

Your main task in this subsection is to learn how to use the product rule. The next example demonstrates how it can be used to differentiate the function $k(x) = x^2 \sin x$, which was mentioned at the beginning of this section.

Example 1 *Using the product rule*

Differentiate the function

$$k(x) = x^2 \sin x.$$

Solution

Identify two functions f and g such that $k(x) = f(x)g(x)$.

Here $k(x) = f(x)g(x)$ where $f(x) = x^2$ and $g(x) = \sin x$.

Find their derivatives.

Then $f'(x) = 2x$ and $g'(x) = \cos x$.

Use the product rule.

By the product rule,

$$\begin{aligned} k'(x) &= f(x)g'(x) + g(x)f'(x) \\ &= x^2 \cos x + (\sin x) \times 2x \end{aligned}$$

Simplify the answer.

$$= x^2 \cos x + 2x \sin x.$$

An alternative way to write the final answer in Example 1 is as $k'(x) = x(x \cos x + 2 \sin x)$. Either form is acceptable.

Here are some similar examples for you to try.

Activity 4 *Using the product rule*

Use the product rule to differentiate the following functions.

(a) $k(x) = x^3 \tan x$ (b) $k(x) = xe^x$

Activity 5 *Using the product rule again*

Check that if you find the derivative of the function $k(x) = x^5$ by writing $x^5 = x^2 \times x^3$ and using the product rule, then you obtain the same answer as you obtain by differentiating it directly.

The product rule can be written in Leibniz notation as follows.

Product rule (Leibniz notation)

If $y = uv$, where u and v are functions of x, then

$$\frac{\mathrm{d}y}{\mathrm{d}x} = u\frac{\mathrm{d}v}{\mathrm{d}x} + v\frac{\mathrm{d}u}{\mathrm{d}x},$$

for all values of x at which both u and v are differentiable.

(The phrase 'u and v are differentiable' in the box above is a condensed way of saying that if we write $u = f(x)$ and $v = g(x)$ then f and g are differentiable at x.)

However it's written, in essence the product rule tells you the following:

$$\begin{pmatrix}\text{derivative}\\ \text{of product}\\ \text{function}\end{pmatrix} = \begin{pmatrix}\text{first}\\ \text{function}\end{pmatrix} \times \begin{pmatrix}\text{derivative}\\ \text{of second}\\ \text{function}\end{pmatrix} + \begin{pmatrix}\text{second}\\ \text{function}\end{pmatrix} \times \begin{pmatrix}\text{derivative}\\ \text{of first}\\ \text{function}\end{pmatrix}.$$

For most people, the best way to remember and use it is as the abbreviated informal version below.

Product rule (informal)

$$\begin{pmatrix}\text{derivative}\\ \text{of product}\end{pmatrix} = (\text{first}) \times \begin{pmatrix}\text{derivative}\\ \text{of second}\end{pmatrix} + (\text{second}) \times \begin{pmatrix}\text{derivative}\\ \text{of first}\end{pmatrix}.$$

When you use this informal version of the product rule, you can usually skip the steps of writing down the first and second functions and their derivatives, and just go straight to applying the rule. To do this, you recite the informal version of the rule in your head: as you think 'first', you write down the formula for the first function, then as you think 'derivative of second', you differentiate the second function in your head and write down the result, and so on. This is illustrated in the next example.

Example 2 *Using the product rule efficiently*

Differentiate the function

$$k(x) = (2x + 1)\ln x.$$

Solution

Identify the first and second functions: here first $= 2x + 1$ and second $= \ln x$. Use the rule

$$\begin{pmatrix}\text{derivative}\\ \text{of product}\end{pmatrix} = (\text{first}) \times \begin{pmatrix}\text{derivative}\\ \text{of second}\end{pmatrix} + (\text{second}) \times \begin{pmatrix}\text{derivative}\\ \text{of first}\end{pmatrix}.$$

By the product rule,

$$k'(x) = (2x+1) \times \frac{1}{x} + (\ln x) \times 2$$

Simplify the answer.

$$= \frac{2x+1}{x} + 2\ln x$$

$$= \frac{2x+1}{x} + \frac{2x\ln x}{x}$$

$$= \frac{2x+1+2x\ln x}{x}.$$

There are various ways to write the final answer in Example 2. For example, the following are all acceptable:

$$2 + \frac{1}{x} + 2\ln x, \quad \frac{1}{x} + 2(1 + \ln x) \quad \text{and} \quad \frac{2x(1+\ln x)+1}{x}.$$

You can practise using the informal version of the product rule in the next activity. Alternatively, you might prefer to continue using the formal version for a little longer, and move to the informal version when you feel more confident with it. You can use the exercises on the product rule in the Unit 7 practice quiz and exercise booklet for further practice.

Whichever version of the product rule you use, remember to simplify your final answers, where possible. You should always do this when you differentiate functions.

Activity 6 *Using the product rule efficiently*

Use the product rule to differentiate the following functions.

(a) $k(x) = (3x^2 + 2x + 1)e^x$ (b) $k(x) = \sin x \cos x$

(c) $k(z) = z \sin z$ (d) $v = (2t^2 - 1)\cos t$ (e) $m = (u^2 + 3)\ln u$

(f) $y = \sqrt{x}\sin x$

Hint: In part (f), convert the square root symbol to index notation before you apply the product rule.

Activity 7 *Using the product rule to find a gradient*

Find dy/dx when $y = x^2e^x$. What is the gradient of the graph of $y = x^2e^x$ at the point with x-coordinate -1?

To finish this subsection, here's a proof of the product rule, using differentiation from first principles. It uses the Lagrange notation form of the product rule, which is repeated below.

Product rule (Lagrange notation)

If $k(x) = f(x)g(x)$, then

$$k'(x) = f(x)g'(x) + g(x)f'(x),$$

for all values of x at which both f and g are differentiable.

A proof of the product rule

Suppose that f and g are functions, and that the function k is given by $k(x) = f(x)g(x)$. Let x denote any value at which both f and g are differentiable. To determine whether $k'(x)$ exists, and find it if it does exist, you have to consider what happens to the difference quotient for k at x, which is

$$\frac{k(x+h) - k(x)}{h}$$

(where h can be either positive or negative, but not zero), as h gets closer and closer to zero. Since $k(x) = f(x)g(x)$, this difference quotient is equal to

$$\frac{f(x+h)g(x+h) - f(x)g(x)}{h}.$$

You can rearrange this expression so that it includes the difference quotients for f and g at x. To do this, you first subtract and add $f(x)g(x+h)$ in the numerator. Then the expression becomes

$$\frac{f(x+h)g(x+h) - f(x)g(x+h) + f(x)g(x+h) - f(x)g(x)}{h},$$

which is equal to

$$\left(\frac{f(x+h) - f(x)}{h}\right) g(x+h) + f(x)\left(\frac{g(x+h) - g(x)}{h}\right).$$

The expression in the first pair of large brackets is the difference quotient for f at x, and the expression in the second pair of large brackets is the difference quotient for g at x. So as h gets closer and closer to zero, the values of these two expressions get closer and closer to $f'(x)$ and $g'(x)$, respectively. Also, the value of $g(x+h)$ gets closer and closer to $g(x)$.

Hence the whole expression gets closer and closer to $f'(x)g(x) + f(x)g'(x)$; that is, $f(x)g'(x) + g(x)f'(x)$. So

$$k'(x) = f(x)g'(x) + g(x)f'(x),$$

which is the product rule.

2.2 Quotient rule

Suppose now that you know the formulas for the derivatives of two functions, and you want to know the formula for the derivative of their quotient. (Remember that a *quotient* of two functions is obtained by 'dividing one by the other'.) For example, you already know the formulas for the derivatives of the functions $f(x) = x^2$ and $g(x) = \sin x$, but suppose that you want to know the formula for the derivative of the function $k(x) = x^2/\sin x$.

You can find the derivative of the quotient of two functions by using the rule in the box below. It's given in Lagrange notation, but you'll see a version in Leibniz notation shortly.

Quotient rule (Lagrange notation)

If $k(x) = f(x)/g(x)$, then

$$k'(x) = \frac{g(x)f'(x) - f(x)g'(x)}{(g(x))^2},$$

for all values of x at which both f and g are differentiable and $g(x) \neq 0$.

You'll see a justification of the quotient rule in Subsection 2.5. Here's an example that illustrates its use.

Example 3 *Using the quotient rule*

Differentiate the function

$$k(x) = \frac{x^2}{\sin x}.$$

Solution

Identify two functions f and g such that $k(x) = f(x)/g(x)$.

Here $k(x) = f(x)/g(x)$ where $f(x) = x^2$ and $g(x) = \sin x$.

Find their derivatives.

Then $f'(x) = 2x$ and $g'(x) = \cos x$.

Use the quotient rule.

By the quotient rule,

$$k'(x) = \frac{g(x)f'(x) - f(x)g'(x)}{(g(x))^2}$$

$$= \frac{(\sin x) \times 2x - x^2 \cos x}{(\sin x)^2}$$

Simplify the answer.

$$= \frac{2x \sin x - x^2 \cos x}{\sin^2 x}.$$

There are various ways to write the final answer in Example 3. For example, the following are all acceptable:

$$\frac{x(2\sin x - x\cos x)}{\sin^2 x}, \quad \frac{2x}{\sin x} - \frac{x^2 \cos x}{\sin^2 x}, \quad \frac{2x - x^2 \cot x}{\sin x}.$$

Here are two similar examples for you to try.

Activity 8 *Using the quotient rule*

Use the quotient rule to differentiate the following functions.

(a) $k(x) = \dfrac{e^x}{x}$ (b) $k(x) = \dfrac{x^3}{2x+1}$

The quotient rule can be written in Leibniz notation as follows.

Quotient rule (Leibniz notation)

If $y = u/v$, where u and v are functions of x, then

$$\frac{\mathrm{d}y}{\mathrm{d}x} = \frac{v\dfrac{\mathrm{d}u}{\mathrm{d}x} - u\dfrac{\mathrm{d}v}{\mathrm{d}x}}{v^2},$$

for all values of x at which both u and v are differentiable and $v \neq 0$.

For most people the best way to remember and use the quotient rule is as the informal version below.

Quotient rule (informal)

$$\begin{pmatrix}\text{derivative}\\ \text{of quotient}\end{pmatrix} = \frac{(\text{bottom}) \times \begin{pmatrix}\text{derivative}\\ \text{of top}\end{pmatrix} - (\text{top}) \times \begin{pmatrix}\text{derivative}\\ \text{of bottom}\end{pmatrix}}{(\text{bottom})^2}.$$

You can use this informal version of the rule in a similar way to the informal version of the product rule. To differentiate a quotient of two functions by using the quotient rule, you recite the informal version of the rule in your head: as you think 'bottom', you write down the formula for the bottom function, then as you think 'derivative of top' you differentiate the top function in your head and write down the result, and so on. (Make sure that you remember that the quotient rule contains a minus sign rather than a plus sign.) This is illustrated in the next example.

Example 4 *Using the quotient rule efficiently*

Differentiate the function

$$k(x) = \frac{2x+1}{\ln x}.$$

Solution

Identify the top and bottom functions: here top $= 2x + 1$ and bottom $= \ln x$. Use the rule

$$\begin{pmatrix}\text{derivative}\\ \text{of quotient}\end{pmatrix} = \frac{(\text{bottom}) \times \begin{pmatrix}\text{derivative}\\ \text{of top}\end{pmatrix} - (\text{top}) \times \begin{pmatrix}\text{derivative}\\ \text{of bottom}\end{pmatrix}}{(\text{bottom})^2}.$$

By the quotient rule,

$$k'(x) = \frac{(\ln x) \times 2 - (2x+1) \times \dfrac{1}{x}}{(\ln x)^2}$$

Simplify the answer. To clear the fraction in the numerator, multiply both the numerator and the denominator by x.

$$= \frac{2x\ln x - (2x+1) \times \dfrac{1}{x} \times x}{x(\ln x)^2}$$

$$= \frac{2x\ln x - 2x - 1}{x(\ln x)^2}.$$

You can practise using the informal version of the quotient rule in the next activity. Alternatively, as with the product rule, you might prefer to continue using the formal version for a little longer, and move to the informal version when you feel more confident with it.

Activity 9 *Using the quotient rule*

Use the quotient rule to differentiate the following functions.

(a) $f(x) = \dfrac{e^x}{x^3}$ (b) $g(u) = \dfrac{u-1}{e^u}$ (c) $z = \dfrac{r+1}{r-1}$

(d) $m = \dfrac{\theta}{\cos\theta}$ (e) $p(t) = \dfrac{\ln t}{t^2}$

In Subsection 1.1 you met the fact that the derivative of $\tan x$ is $\sec^2 x$, but this result wasn't justified there. It can be confirmed by using the quotient rule, as shown in the next example. (Note, however, that this justification is incomplete, as you haven't yet seen a proof of the quotient rule!)

Example 5 *Using the quotient rule to find the derivative of tan*

Show that the derivative of $\tan x$ is $\sec^2 x$.

Solution

Use the fact that $\tan x = (\sin x)/(\cos x)$, and apply the quotient rule.

By the quotient rule,

$$\begin{aligned}\frac{d}{dx}(\tan x) &= \frac{d}{dx}\left(\frac{\sin x}{\cos x}\right)\\ &= \frac{(\cos x)(\cos x) - (\sin x)(-\sin x)}{(\cos x)^2}\\ &= \frac{\cos^2 x + \sin^2 x}{\cos^2 x}\end{aligned}$$

Simplify this expression by using the facts that $\cos^2 x + \sin^2 x = 1$ and $1/\cos x = \sec x$.

$$\begin{aligned}&= \frac{1}{\cos^2 x}\\ &= \sec^2 x,\end{aligned}$$

as required.

You can also use the quotient rule to find the derivatives of the trigonometric functions cosec, sec and cot. These derivatives are given in the box below, and you're asked to confirm them in the next activity.

Derivatives of cosec, sec and cot

$$\frac{d}{dx}(\operatorname{cosec} x) = -\operatorname{cosec} x \cot x$$

$$\frac{d}{dx}(\sec x) = \sec x \tan x$$

$$\frac{d}{dx}(\cot x) = -\operatorname{cosec}^2 x$$

Activity 10 *Finding the derivatives of further trigonometric functions*

Use the quotient rule to confirm the following formulas for derivatives.

(a) $\dfrac{d}{dx}(\cot x) = -\operatorname{cosec}^2 x$ (b) $\dfrac{d}{dx}(\sec x) = \sec x \tan x$

(c) $\dfrac{d}{dx}(\operatorname{cosec} x) = -\operatorname{cosec} x \cot x$

Hint: Use the facts that $\cot x = (\cos x)/(\sin x)$, $\sec x = 1/(\cos x)$ and $\operatorname{cosec} x = 1/(\sin x)$.

The derivatives of cosec, sec and cot can now be included in the list of derivatives of standard functions that you have available to work with. The *Handbook* includes a table of all the standard derivatives given in this module.

2.3 Chain rule

Now that you know the product rule and the quotient rule, you can differentiate a much wider range of functions than you could before. However, there are still some fairly simple functions that you can't differentiate using any of the rules that you've met so far.

For example, consider the function $k(x) = \sin(x^2)$. This function isn't a sum, a product or a quotient of functions that you can differentiate. However, it's still made up of functions that you can differentiate, just combined in a different way. To work out $k(x)$ for a particular value of x, you have to first square x, then find the sine of the result. So the function k is a *composite* of the functions $f(x) = x^2$ and $g(x) = \sin x$, as shown in Figure 15. You met the idea of a composite of two functions in Subsection 3.2 of Unit 3.

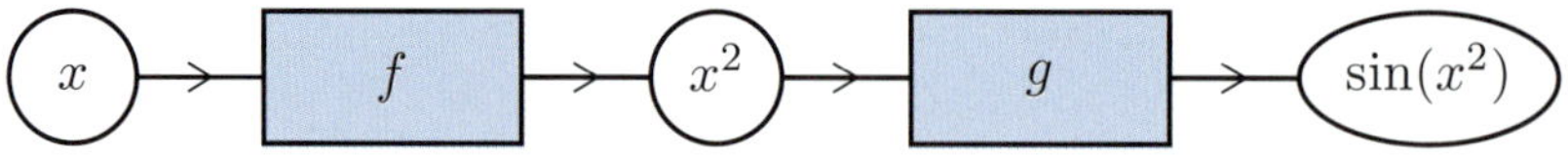

Figure 15 A composite of the functions $f(x) = x^2$ and $g(x) = \sin x$

The functions $f(x) = x^2$ and $g(x) = \sin x$ are functions that you can differentiate.

In this subsection you'll meet another rule for differentiation, called the *chain rule*, which allows you to differentiate functions that are composites of functions that you can already differentiate. You'll see towards the end of the subsection why it's called the chain rule. It's also sometimes called the *composite rule* or the *function of a function rule*.

The first step in learning to use the chain rule is to learn to recognise when a function is a composite of two functions, and how to decompose such a function into its constituent functions. (*Decomposing* is the opposite of *composing*.) To see how to do this, let's consider again the function

$$k(x) = \sin(x^2).$$

While you're learning about decomposing functions, it's easier to work with functions expressed using two variables rather than function notation, so let's first express this function as

$$y = \sin(x^2).$$

To recognise that it's a composite of two functions, you observe that it's of the form

$$y = \sin(\text{something}),$$

where the 'something' is an expression that contains the input variable x.

You can then decompose it into its two constituent functions.
A convenient way to do this is to set the 'something' equal to an extra variable, say u. This gives

$$y = \sin(u) \quad \text{where} \quad u = x^2.$$

These two equations specify the two constituent functions.

In general, suppose that you have an equation that specifies y as a function of x. If the equation is of the form

$$y = \text{a function of 'something'},$$

where the 'something' is a function of the input variable x, then the original function is a composite function. To decompose it into its two constituent functions, you set the 'something' equal to u. In other words, you write y as a function of u, where u is a function of x.

Here are some more examples that illustrate how to do this.

Example 6 *Decomposing composite functions*

For each of the following equations, write y as a function of u, where u is a function of x.

(a) $y = \cos(\ln x)$ (b) $y = e^{2x}$ (c) $y = \sin^2 x$

Solution

(a) $y = \cos(\text{something})$, so set the something $= u$.

$y = \cos u$ where $u = \ln x$.

(b) $y = e^{\text{something}}$, so set the something $= u$.

$y = e^u$ where $u = 2x$.

(c) Remember that $\sin^2 x$ means $(\sin x)^2$. Therefore $y = \text{something}^2$, so set the something $= u$.

$y = u^2$ where $u = \sin x$.

Here are some examples for you to try.

Activity 11 *Decomposing composite functions*

For each of the following equations, write y as a function of u, where u is a function of x.

(a) $y = \ln(x^4)$ (b) $y = \sin(x^2 - 1)$ (c) $y = \cos(e^x)$

(d) $y = \cos^3 x$ (e) $y = e^{\tan x}$ (f) $y = \sqrt{\ln x}$

(g) $y = (x + \sqrt{x})^9$ (h) $y = \sqrt[3]{x^2 - x - 1}$ (i) $y = e^{\sqrt{x}}$

The chain rule, which allows you to differentiate composite functions, is given below. It's given in Leibniz notation, because it's easier to remember in this form, but you'll see it in Lagrange notation later in this subsection.

Chain rule (Leibniz notation)

If y is a function of u, where u is a function of x, then

$$\frac{dy}{dx} = \frac{dy}{du}\frac{du}{dx}$$

for all values of x where y as a function of u, and u as a function of x, are differentiable.

! $\frac{dy}{dx}$ represent derivatives not fractions.

As you can see, in Leibniz notation the chain rule looks like an obvious statement about fractions, which makes it easy to remember. You should keep in mind, though, that this is *not* what it is! The pieces of notation $\mathrm{d}y/\mathrm{d}x$, $\mathrm{d}y/\mathrm{d}u$ and $\mathrm{d}u/\mathrm{d}x$ represent derivatives, not fractions.

too difficult for this level of maths.

As with the other rules for combining derivatives, the chain rule can be proved by using the idea of differentiation from first principles. Unfortunately, this is quite tricky to do, and the proof isn't included in this module.

Each constituent function will have rate of change for every value of x. Then total value is the product of those gradients of constituent functions.

However, here's an informal way to see that the chain rule makes sense. Suppose that y is a function of u, where u is a function of x, as in the statement of the chain rule. Suppose that for a particular value of x, the value of $\mathrm{d}u/\mathrm{d}x$ is 2, and for the value of u that corresponds to this particular value of x, the value of $\mathrm{d}y/\mathrm{d}u$ is 3. Since y is increasing at the rate of 3 units for every unit that u increases, and u is increasing at the rate of 2 units for every unit that x increases, you'd expect y to be increasing at the rate of $3 \times 2 = 6$ units for every unit that x increases. This is what the chain rule tells you.

The next example shows you how to use the chain rule.

Example 7 *Using the chain rule*

Differentiate the function $y = \sin(x^2)$.

Solution

Recognise that the given function is a function of 'something', where the 'something' involves the input variable x. Hence decompose the given function.

Here $y = \sin u$ where $u = x^2$.

Find the derivatives of the constituent functions.

This gives

$$\frac{\mathrm{d}y}{\mathrm{d}u} = \cos u, \qquad \frac{\mathrm{d}u}{\mathrm{d}x} = 2x.$$

Apply the chain rule, $\frac{\mathrm{d}y}{\mathrm{d}x} = \frac{\mathrm{d}y}{\mathrm{d}u}\frac{\mathrm{d}u}{\mathrm{d}x}$.

chain rule is multiplication of the derivatives of two constituent functions, $\frac{dy}{dx} = \frac{dy}{du} \cdot \frac{du}{dx}$

By the chain rule,

$$\frac{\mathrm{d}y}{\mathrm{d}x} = (\cos u) \times 2x$$

Express the answer in terms of x, by substituting for u in terms of x. Here we have $u = x^2$.

$$= (\cos(x^2)) \times 2x$$

Simplify the answer.

$$= 2x\cos(x^2).$$

Notice that when you apply the chain rule as illustrated in Example 7, you first obtain an answer that's expressed in terms of both u and x. You then have to substitute for u, using the equation for u in terms of x that you wrote down when you decomposed the function, to obtain a final answer in terms of x.

Activity 12 *Using the chain rule*

Differentiate the following functions.

(a) $y = e^{5x}$ (b) $y = \sin(2x^2 - 3)$ (c) $y = \sin\left(\frac{x}{4}\right)$

(d) $y = \cos(e^x)$ (e) $y = \ln(\sin x)$ (f) $y = \sin^3 x$

Hint: In part (f), remember that $\sin^3 x$ means $(\sin x)^3$.

You've now seen how to apply the chain rule by first introducing an extra variable, usually u, which you then remove again to obtain the final answer. Once you're familiar with using the chain rule, you can often apply it without introducing this extra variable at all.

To see how to do this, consider again the equation $y = \sin(x^2)$. Think of the expression that you would have written as u, that is, the expression that was referred to as 'something' earlier, as one 'big variable':

$$y = \sin(\,x^2\,)$$

To find $\mathrm{d}y/\mathrm{d}x$, first differentiate with respect to this big variable; that is, with respect to the 'something'. This gives

$$\cos(\,x^2\,)$$

By the chain rule, to obtain the final answer, you now need to multiply by the derivative of the something with respect to x. This gives

$$\frac{\mathrm{d}y}{\mathrm{d}x} = \cos(x^2) \times 2x.$$

This is the final answer, but, as before, you should simplify it slightly to give

$$\frac{\mathrm{d}y}{\mathrm{d}x} = 2x\cos(x^2).$$

no difference in solution if you choose either Leibniz or Lagrange.

Here are two more examples. Notice that in the second of these examples the function is expressed using function notation rather than using two variables. This makes no difference to how you apply the ideas above.

Example 8 *Using the chain rule without introducing an extra variable*

Differentiate the following functions.

(a) $y = \ln(2x - 1)$ (b) $k(x) = \cos^4 x$

Solution

(a) Think of the 'something', $2x - 1$, as a 'big variable'. Differentiate with respect to this variable. Then multiply by the derivative of $2x - 1$ with respect to x.

By the chain rule,

$$\frac{dy}{dx} = \frac{1}{2x-1} \times 2$$

Simplify the answer.

$$= \frac{2}{2x-1}.$$

(b) The function is $k(x) = (\cos x)^4$, so think of the 'something', $\cos x$, as a 'big variable'. Differentiate with respect to this variable. Then multiply by the derivative of $\cos x$ with respect to x.

By the chain rule,

$$k'(x) = (4\cos^3 x)(-\sin x)$$

Simplify the answer.

$$= -4\sin x \cos^3 x.$$

Try this quicker way of using the chain rule in the following activity.

Activity 13 *Using the chain rule without introducing an extra variable*

Differentiate the following functions.

(a) $f(x) = \cos(4x)$ (b) $f(x) = e^{2x+1}$ (c) $r(x) = \cos^5 x$

(d) $g(x) = \ln(x^2 + x + 1)$ (e) $v(t) = \sin(e^t)$ (f) $v = \ln(r^5)$

(g) $r = \tan^2 \theta$ (h) $R = e^{\tan q}$ (i) $A = (1 + p^2)^9$

You should try to get used to applying the chain rule in the quick way demonstrated in Example 8, as you'll find this skill useful in calculus. However, sometimes, if a function that you want to differentiate using the chain rule isn't straightforward, then you might find it easier to revert to introducing an extra variable. It's fine to do that if you find it helpful.

Sometimes you can make it easier to apply the chain rule without introducing an extra variable if you start by rearranging the formula for the function. This applies in particular to composites that involve power functions. Here's an example.

Example 9 *Applying the chain rule to a composite involving a power function*

Differentiate the function

$$f(x) = \frac{1}{\sqrt{\sin x}}.$$

Solution

The function f is of the form $f(x) = 1/\sqrt{\text{something}}$, where the 'something' is $\sin x$. So use the chain rule. Start by rearranging the formula for f into the form 'something^n'.

The function is

$$f(x) = \frac{1}{\sqrt{\sin x}};$$

that is,

$$f(x) = (\sin x)^{-1/2}.$$

Apply the chain rule.

By the chain rule,

$$\begin{aligned} f'(x) &= -\tfrac{1}{2}(\sin x)^{-3/2}(\cos x) \\ &= -\frac{\cos x}{2\sin^{3/2} x}. \end{aligned}$$

With practice, you may be able to do the sort of initial rearranging needed in Example 9 in your head.

Activity 14 *Applying the chain rule to composites involving power functions*

Differentiate the following functions.

(a) $f(x) = \sqrt{\ln x}$ (b) $p(w) = \dfrac{1}{(w^3+7)^4}$ (c) $f(x) = \cos(\sqrt{x})$

(d) $f(x) = \dfrac{1}{\sqrt{e^x}}$ (e) $a(t) = \dfrac{1}{(2-3t-3t^2)^{1/3}}$ (f) $C(p) = e^{1/p^2}$

Gerald was never very good at using the chain rule.

Now here's an explanation of where the name 'chain rule' comes from. Consider a function that's a composite of *three* functions. Suppose that

y is a function of u,
u is a function of v,
v is a function of x.

You can find the derivative of y with respect to x by applying the chain rule twice:

$$\frac{dy}{dx} = \frac{dy}{du}\frac{du}{dx} \quad \text{and} \quad \frac{du}{dx} = \frac{du}{dv}\frac{dv}{dx}$$

so, using the second equation to substitute in the first equation, you obtain

$$\frac{dy}{dx} = \frac{dy}{du}\frac{du}{dv}\frac{dv}{dx}.$$

As you can see, this gives a 'chain' of derivatives. You can keep adding 'links in the chain' indefinitely. One of the examples in Subsection 2.5, Example 12, illustrates how you can apply the chain rule twice in this way.

Finally in this subsection, let's translate the chain rule into Lagrange notation. You can do this by thinking about the way of applying it without introducing an extra variable. Suppose that f and g are functions, and you want to differentiate the expression $g(f(x))$ with respect to x. The chain rule says that you can do this by differentiating it with respect to the 'something' $f(x)$, then multiplying the result by the derivative of $f(x)$ with respect to x. This gives the version of the chain rule below.

Chain rule (Lagrange notation)

If $k(x) = g(f(x))$, where f and g are functions, then

$$k'(x) = g'(f(x))f'(x)$$

for all values of x such that f is differentiable at x and g is differentiable at $f(x)$.

The first use of the chain rule seems to be by Gottfried Wilhelm Leibniz in a memoir of 1676, in which he calculated (with mistakes!) the derivative of $\sqrt{a+bz+cz^2}$. Further examples of the use of the rule appear in the first calculus textbook, written by Guillaume de l'Hôpital and published in 1696, but nowhere is the rule actually stated. (There is more about l'Hôpital's book in Subsection 3.4.) It was not until the 19th century that the chain rule was commonly stated explicitly, and at that time it was generally described as a 'rule for differentiating a function of a function'. The term chain rule seems to have originated as a translation of the German word *Kettenregel*, which was used in German calculus books in the early 20th century.

2.4 Differentiating functions of linear expressions

This subsection is about a type of function that occurs frequently, and which you can differentiate by using the chain rule.

Consider the functions

$$k(x) = \sin(3x) \quad \text{and} \quad h(x) = e^{-2x}.$$

Each of them is a composite function, and in each of them the 'something' is a constant multiple of the input variable x; the 'somethings' are $3x$ and $-2x$, respectively. It's worth noticing, and remembering, what happens when you use the chain rule to differentiate functions like these. For example,

if $k(x) = \sin(3x)$, then $k'(x) = \cos(3x) \times 3 = 3\cos(3x)$,

if $h(x) = e^{-2x}$, then $h'(x) = e^{-2x} \times (-2) = -2e^{-2x}$.

You can see that to differentiate a composite function in which the 'something' is ax, where a is a constant, you differentiate with respect to ax, and then multiply by a.

It's worth remembering this special case of the chain rule, which can be stated as below.

Derivative of a function of the form $k(x) = f(ax)$

If $k(x) = f(ax)$, where a is a non-zero constant, then

$$k'(x) = af'(ax),$$

for all values of x such that f is differentiable at ax.

You can use this rule to quickly differentiate any function of the type discussed above. For example, it tells you that

if $k(x) = \sin(5x)$, then $k'(x) = 5\cos(5x)$.

Try using the rule above in the next activity.

Activity 15 *Differentiating functions of the form $k(x) = f(ax)$*

Differentiate the following functions.

(a) $k(x) = \cos(7x)$ (b) $r(\theta) = \sin\left(\dfrac{\theta}{2}\right)$ (c) $g(u) = e^{3u/2}$

(d) $h(x) = e^{-x}$ (e) $p(\alpha) = \cos(-8\alpha)$ (f) $w(x) = \ln(7x)$

(g) $f(t) = \sin\left(\dfrac{-2t}{3}\right)$ (h) $k(\phi) = \tan(3\phi)$

In fact you can slightly extend the special case of the chain rule above. Notice that, by the chain rule,

if $k(x) = \sin(3x - 2)$, then $k'(x) = \cos(3x - 2) \times 3 = 3\cos(3x - 2)$,
if $h(x) = e^{-2x+5}$, then $h'(x) = e^{-2x+5} \times (-2) = -2e^{-2x+5}$.

You can see that, in general, to differentiate a composite function in which the 'something' is $ax + b$, where a and b are constants, you differentiate with respect to $ax + b$, and then multiply by a. This special case of the chain rule is summarised below.

Derivative of a function of the form $k(x) = f(ax + b)$

If $k(x) = f(ax + b)$, where a and b are constants with a non-zero, then

$$k'(x) = af'(ax + b),$$

for all values of x such that f is differentiable at $ax + b$.

For example, this rule tells you that

if $k(x) = \sin(5x - 8)$, then $k'(x) = 5\cos(5x - 8)$.

Try using it in the next activity.

Activity 16 *Differentiating functions of the form $k(x) = f(ax + b)$*

Differentiate the following functions.

(a) $k(x) = \cos(7x + 4)$ (b) $h(x) = e^{-x-3}$ (c) $r(\theta) = \sin\left(\dfrac{\theta - 1}{2}\right)$

(d) $s(\theta) = \sin\left(\dfrac{2 - \theta}{3}\right)$

ax + b => line graph.

Remember that an expression of the form $ax + b$, where a and b are constants, is called a *linear* expression in x. So in this subsection you've seen how to quickly differentiate functions of linear expressions.

2.5 Using the differentiation rules together

You can differentiate a wide range of functions by combining the various differentiation rules that you've met. For example, you can use more than one differentiation rule, or more than one application of the same rule. This subsection starts with three examples that show you the kinds of things that you can do. important to check next examples!

When you're combining differentiation rules in this way, it can be useful to use the notation $\mathrm{d}/\mathrm{d}x$ to indicate the derivative of a function that appears as part of your working, then actually do the differentiation in the next line. This is illustrated in the examples.

d/dx indicates derivative part of the working,

Example 10 *Combining the differentiation rules*

Find the derivative of

$$f(x) = x^2 \sin(x^3).$$

Solution

The function is of the form something × something. So apply the product rule. The second 'something', $\sin(x^3)$, isn't straightforward to differentiate, so use the notation $\mathrm{d}/\mathrm{d}x$ to indicate its derivative.

$$f'(x) = x^2 \frac{\mathrm{d}}{\mathrm{d}x}(\sin(x^3)) + (\sin(x^3))(2x) \quad \text{(by the product rule)}$$

Now differentiate $\sin(x^3)$. It's of the form sin(something), so apply the chain rule.

$$= x^2(\cos(x^3))(3x^2) + (\sin(x^3))(2x) \quad \text{(by the chain rule)}$$

Simplify the answer.

$$= 3x^4\cos(x^3) + 2x\sin(x^3).$$

do 33, 39 P.

Example 11 *Combining the differentiation rules again*

Find the derivative of

$$y = \left(\frac{x^2-2}{x^2+1}\right)^4.$$

Solution

The function is of the form (something)^4. So apply the chain rule. The 'something' isn't straightforward to differentiate, so use the notation $\mathrm{d}/\mathrm{d}x$ to indicate its derivative.

$$\frac{\mathrm{d}y}{\mathrm{d}x} = 4\left(\frac{x^2-2}{x^2+1}\right)^3 \times \frac{\mathrm{d}}{\mathrm{d}x}\left(\frac{x^2-2}{x^2+1}\right) \quad \text{(by the chain rule)}$$

Now differentiate $(x^2-2)/(x^2+1)$. It's of the form something/something, so apply the quotient rule.

$$= 4\left(\frac{x^2-2}{x^2+1}\right)^3 \times \left(\frac{(x^2+1)(2x)-(x^2-2)(2x)}{(x^2+1)^2}\right)$$

(by the quotient rule)

Simplify the answer.

$$\begin{aligned} &= 4\left(\frac{x^2-2}{x^2+1}\right)^3 \times \frac{(2x)(x^2+1-x^2+2)}{(x^2+1)^2} \\ &= 4\left(\frac{x^2-2}{x^2+1}\right)^3 \times \frac{6x}{(x^2+1)^2} \\ &= \frac{24x(x^2-2)^3}{(x^2+1)^5}. \end{aligned}$$

Example 12 *Combining the differentiation rules yet again*

Find the derivative of

$$g(t) = \cos(e^{2t+1}).$$

Solution

The function is of the form cos(something). So apply the chain rule. The 'something' isn't straightforward to differentiate, so use the notation $\mathrm{d}/\mathrm{d}t$ to indicate its derivative.

$$g'(t) = -\sin(e^{2t+1}) \times \frac{\mathrm{d}}{\mathrm{d}t}(e^{2t+1}) \quad \text{(by the chain rule)}$$

Now differentiate e^{2t+1}. It's of the form $e^{\text{something}}$. So apply the chain rule again.

$$= -\sin(e^{2t+1}) \times e^{2t+1} \times 2 \quad \text{(by the chain rule again)}$$

Simplify the answer.

$$= -2e^{2t+1}\sin(e^{2t+1}).$$

You can practise combining the differentiation rules in the next activity. If you find it tricky to work out which rules you need to apply, then the checklist below might help.

Checklist for differentiating a function

1. Is it a standard function (is its derivative given in the *Handbook*)?
2. Can you use the constant multiple rule and/or sum rule?
3. Can you rewrite it to make it easier to differentiate? For example, multiplying out brackets may help.
4. Is it of the form $f(ax)$ or $f(ax+b)$, where a and b are constants and f is a function? If so, use the rule for differentiating a function of a linear expression.
5. Can you use the product rule (is it of the form $f(x) = \text{something} \times \text{something}$)?
6. Can you use the quotient rule (is it of the form $f(x) = \text{something/something}$)?
7. Can you use the chain rule (is it of the form $f(x) = \text{a function of something}$)?

When you use a differentiation rule, you usually have to find the derivatives of simpler functions. You can apply the checklist above to each of these simpler functions in turn.

Note also that when you need to apply more than one rule to differentiate a function, the first rule that you should apply is the one that applies to the overall structure of the rule of the function. For example, as you saw in Example 10, the overall structure of the function $f(x) = x^2\sin(x^3)$ is that it's a product, since $x^2\sin(x^3)$ is the product of x^2 and $\sin(x^3)$. So, to differentiate this function, you start by applying the product rule, ignoring for the moment the fact that one of the two functions in the product is itself a composite. After you've done that, you can deal with the composite.

Activity 17 *Combining the differentiation rules*

Differentiate the following functions.

(a) $g(x) = (x^2 + 1)(x^3 + 1)$ (b) $f(x) = \dfrac{x}{(x-2)^3}$

(c) $g(x) = \cos(x \ln x)$ (d) $h(x) = e^{x/2} \sin(3x)$ (e) $q(u) = \dfrac{\cos(4u)}{e^{3u}}$

(f) $f(x) = e^{\sin(5x)}$ (g) $h(x) = \cos^3(x^2)$ (h) $z(x) = \dfrac{e^{-2x}}{5}$

(i) $G(x) = 3 \sin x \cos x + 2 \sin(3x)$ (j) $p(t) = te^t \sin t$

(k) $r = \sin(2\theta) \cos \theta$ (l) $h = ze^{3z}$ (m) $v = \dfrac{\sin \sqrt{u}}{\sqrt{u}}$

Sometimes you have to differentiate functions that involve constants represented by letters. You can do this in the usual way, treating the constants like any other numbers.

Activity 18 *Differentiating functions involving constants*

Differentiate the following functions.

(a) $g(x) = e^{-kx}$, where k is a constant.

(b) $f(x) = x \cos(\pi x)$.

(c) $h = \sin^2(a\theta)$, where a is a constant.

(d) $y = \cos(3x) + c$, where c is a constant.

Here's another technique that's occasionally useful when you want to differentiate a complicated function. Whenever you have a function that you can differentiate using the quotient rule, an alternative way to differentiate it is to rearrange its rule to express it as a *product* of two functions, and then apply the product rule.

In the next activity you're asked to differentiate a function in this way. (You were asked to differentiate the same function using the quotient rule in Activity 9(e).)

Activity 19 *Differentiating a quotient by using the product rule*

Differentiate the function $p(t) = \dfrac{\ln t}{t^2}$ by writing it as $p(t) = t^{-2} \ln t$ and using the product rule.

In fact, the quotient rule is really just a combination of the product rule and a special case of the chain rule. Here's a proof of it, as promised in Subsection 2.2.

A proof of the quotient rule

Suppose that the function k has rule $k(x) = f(x)/g(x)$. This rule can be written as

$$k(x) = \frac{1}{g(x)} \times f(x);$$

that is,

$$k(x) = (g(x))^{-1} f(x).$$

So, for each value of x at which both f and g are differentiable and $g(x) \neq 0$,

$$\begin{aligned} k'(x) &= (g(x))^{-1} f'(x) + f(x) \frac{\mathrm{d}}{\mathrm{d}x}\left((g(x))^{-1}\right) \quad \text{(by the product rule)} \\ &= (g(x))^{-1} f'(x) + f(x)\left(-(g(x))^{-2}\right) g'(x) \quad \text{(by the chain rule)} \\ &= \frac{f'(x)}{g(x)} - \frac{f(x)g'(x)}{(g(x))^2} \\ &= \frac{g(x)f'(x) - f(x)g'(x)}{(g(x))^2}, \end{aligned}$$

which is the quotient rule.

A proof of the formula for the derivative of a power function

Finally in this subsection, here's a proof of the general formula for the derivative of a power function, as promised in Unit 6. The formula states that, for any constant n,

$$\frac{\mathrm{d}}{\mathrm{d}x}(x^n) = nx^{n-1}.$$

The proof given here is valid only for $x > 0$, as you'll see. However, for a power function $y = x^n$ that's defined when $x < 0$, you can use the fact that the formula holds for $x > 0$, together with the symmetry of the graph of $y = x^n$, to deduce that the formula also holds for $x < 0$; the details are not given here. To see that the formula also holds when $x = 0$, you can use the fact that at $x = 0$ the gradient of the graph of $y = x^n$ is 0.

To prove the formula for $x > 0$, you can use the fact that $x = e^{\ln x}$ for such values of x to write the expression x^n in the following form:

$$x^n = (e^{\ln x})^n = e^{n \ln x}.$$

This gives

$$\begin{aligned}\frac{\mathrm{d}}{\mathrm{d}x}(x^n) &= \frac{\mathrm{d}}{\mathrm{d}x}(e^{n\ln x}) \\ &= (e^{n\ln x})\frac{\mathrm{d}}{\mathrm{d}x}(n\ln x) \quad \text{(by the chain rule)} \\ &= (e^{n\ln x}) \times n \times \frac{\mathrm{d}}{\mathrm{d}x}(\ln x) \quad \text{(by the constant multiple rule)} \\ &= (e^{n\ln x}) \times n \times \frac{1}{x} \\ &= x^n \times n \times \frac{1}{x} \\ &= nx^{n-1},\end{aligned}$$

which is the formula stated above.

Note that the two proofs that you've just seen, namely the proof of the quotient rule and the proof of the formula for the derivative of a power function, both depend on the chain rule. Since this module doesn't include a proof of the chain rule, strictly the proofs of these two results are incomplete.

3 More differentiation

In this section you'll see how to use differentiation to solve a type of problem that occurs frequently in applications of mathematics, how to use differentiation on a computer, and how to find the derivative of the inverse of a function if you know the derivative of the original function.

3.1 Optimisation problems

In mathematics, and its applications, it's often helpful to solve problems that involve identifying the best possible option from a choice of suitable possibilities. Problems of this type are known as **optimisation** problems.

For example, the manufacturer of an electronic device might want to identify the best price that it should charge for its device, if its goal is to achieve the maximum profit. Increasing the price might increase the profit, but increasing it further might decrease the profit, by causing fewer of the devices to be sold. This problem is an example of a **maximisation** problem, because it involves identifying the circumstances under which the *maximum* value of a quantity is obtained. Similarly, some optimisation problems are **minimisation** problems, which involve identifying the circumstances under which the *minimum* value of a quantity is obtained.

You can often solve optimisation problems by modelling them mathematically and then applying calculus. In particular, you can often use the strategy below, which you met in Unit 6, or an adaptation of it.

Strategy:
To find the greatest or least value of a function on an interval of the form $[a, b]$

(This strategy is valid when the function is continuous on the interval, and differentiable at all values in the interval except possibly the endpoints.)

1. Find the stationary points of the function.
2. Find the values of the function at any stationary points inside the interval, and at the endpoints of the interval.
3. Find the greatest or least of the function values found.

The next example illustrates how you can use this strategy to solve an optimisation problem. You might like to try guessing the answer to the problem posed in the example, and then read on to see whether you're right.

Example 13 *Solving an optimisation problem*

A rectangular sheet of metal, of width 1 m, is to be bent along two straight lines parallel to the sides of the sheet, to form an open channel with a rectangular cross-section, as shown below. How far from the sides of the sheet should the metal be bent to achieve a channel of maximum volume?

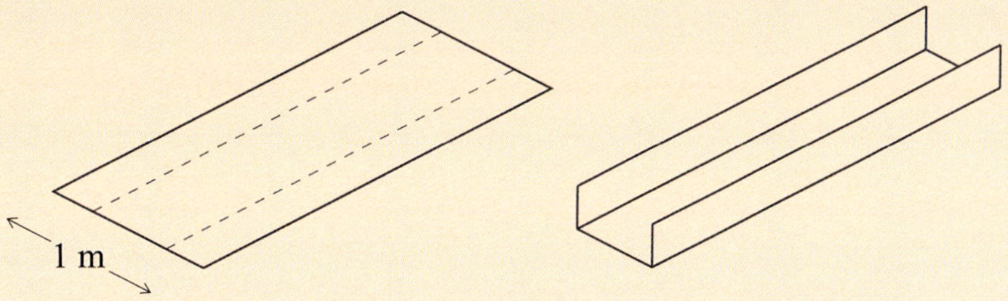

Solution

Simplify the problem if possible. Draw a diagram.

The volume of the channel is the area of its cross-section times its (fixed) length. So to achieve a channel of maximum volume, we have to achieve the maximum area of the cross-section.

The diagrams below show the cross-sections of the metal sheet and the channel. We have to determine how far from the side the metal sheet should be bent to achieve the maximum area of the cross-section.

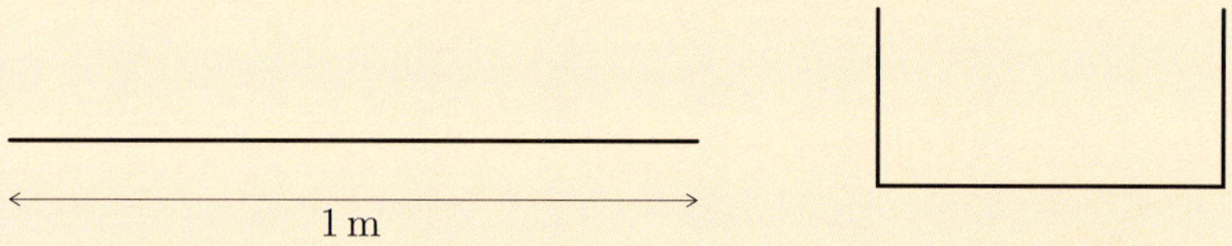

Identify the quantity that you can change, and represent it by a variable, noting the possible values that it can take. Identify the quantity to be maximised, and represent it by a variable too.

Let the distance from the side of the metal to where it is bent be x (in m). The value of x must be between 0 and $\frac{1}{2}$.

Let the area of the cross-section be A (in m^2).

Find a formula for the variable to be maximised (the dependent variable) in terms of the variable that you can change (the independent variable). Start by annotating the diagram.

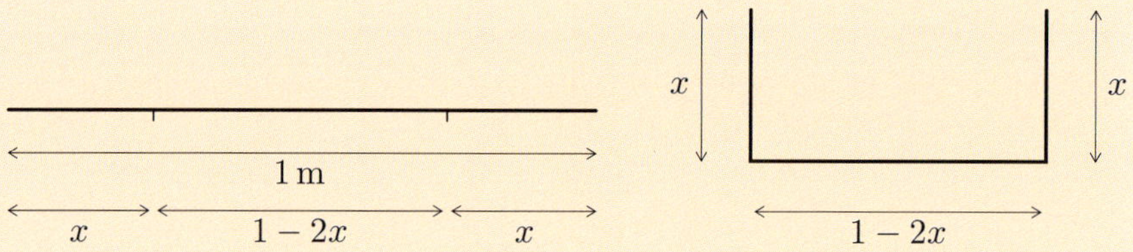

The area of the cross-section is given by

$$A = x(1 - 2x).$$

Formulate the problem in terms of the variables.

We have to find the value of x, between 0 and $\frac{1}{2}$, that gives the maximum value of A.

Use calculus to solve the problem. You can use the strategy in the box given before this example.

The variable A is a polynomial function of x, so it is continuous on its whole domain and differentiable at every value of x. So the maximum value of A must occur either at an endpoint of the interval $[0, \frac{1}{2}]$, or at a stationary point.

Find the stationary points.

The formula for A is

$$\begin{aligned} A &= x(1-2x) \\ &= x - 2x^2, \end{aligned}$$

so

$$\frac{dA}{dx} = 1 - 4x.$$

At a stationary point, $dA/dx = 0$, which gives

$$1 - 4x = 0;$$

that is,

$$x = \tfrac{1}{4}.$$

So the only stationary point occurs when $x = \frac{1}{4}$.

Consider the values of the function at any stationary points inside the interval, and at the endpoints of the interval.

The endpoints of the interval are 0 and $\frac{1}{2}$, so the maximum value of A occurs when $x = 0$, $x = \frac{1}{4}$ or $x = \frac{1}{2}$.

When $x = 0$, $A = 0$, and similarly when $x = \frac{1}{2}$, $A = 0$.

When $x = \frac{1}{4}$,

$$A = \tfrac{1}{4} \times (1 - 2 \times \tfrac{1}{4}) = \tfrac{1}{4} \times \tfrac{1}{2} = \tfrac{1}{8} = 0.125.$$

Find the greatest of the function values found.

The greatest of the three values of A found is 0.125, which is achieved when $x = \frac{1}{4}$.

So the maximum value of A is achieved when $x = \frac{1}{4}$.

State a conclusion that answers the original problem.

To achieve a channel of maximum volume, the metal should be bent at a distance of 25 cm from each side.

In fact, in Example 13 it isn't necessary to actually work out the values of A when $x = 0$, $x = \frac{1}{4}$ and $x = \frac{1}{2}$. When $x = 0$ the height of the channel is zero, and when $x = \frac{1}{2}$ its width is zero, so in both these cases the area A will be zero. When $x = \frac{1}{4}$, the height and the width of the channel are both positive, so the area A will also be positive. So you can see immediately that, of these three values of x, the value $x = \frac{1}{4}$ will give the greatest value of A.

The solution to Example 13 shows that the channel has maximum volume when it is twice as wide as it is high.

When you're solving a maximisation or minimisation problem, it can be helpful to sketch a graph of the function involved, or plot it using a computer. This will give you a rough idea of how the quantity that you want to maximise or minimise changes as the quantity on which it depends changes. For example, Figure 16 shows a graph of the equation $A = x(1 - 2x)$ from Example 13. You can see that, as x increases, the value of A first increases and then decreases, as you'd expect, and the maximum value of A seems to occur when x is about 0.25, which accords with the answer found in Example 13.

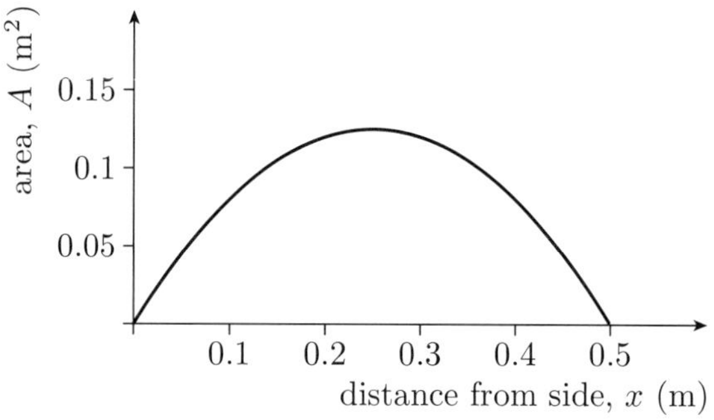

Figure 16 The graph of $A = x(1 - 2x)$ for $0 \leq x \leq \frac{1}{2}$

Notice that in fact you could have solved the problem in Example 13 without using calculus, because the function involved turned out to be a quadratic function. You could have found the value of x that gives the maximum value of A by using a method other than calculus to find the vertex of the parabola that's the graph of the function. You saw some methods for doing this in Unit 2. However, once you're familiar with calculus, it's usually easier to use it.

Each of the next two activities poses an optimisation problem for you to solve by using calculus. The first activity involves maximising the volume of a box, and the second one involves maximising the area of the cross-section of an open channel, in a similar way to Example 13. In both activities, you might like to try guessing the answers to the problems posed before you start work on the solutions.

Activity 20 *Solving an optimisation problem*

Suppose that you have a sheet of cardboard that measures 1 m by 1 m. You intend to cut a square from each of the four corners and then fold the resulting shape to form a box (without a top), as shown. What should the side length of the squares be, to give a box of maximum volume?

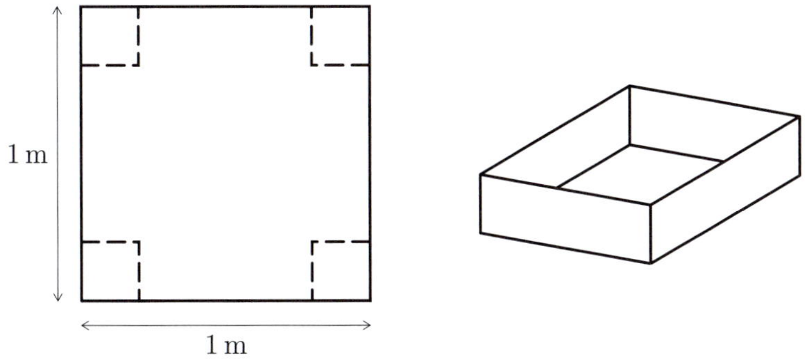

In the next activity you should use the fact that the area of a triangle with sides of lengths a and b and included angle θ (as illustrated in Figure 17) is given by the formula $A = \frac{1}{2}ab\sin\theta$. You met this formula in Unit 4.

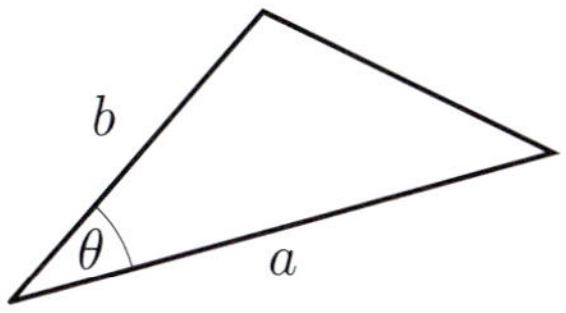

Figure 17 A triangle with sides of lengths a and b and included angle θ

Activity 21 *Solving another optimisation problem*

Consider again the problem of the open channel in Example 13. Suppose that instead of bending the metal along two straight lines, you intend to bend it along one straight line in the middle, to form an open channel with a v-shaped cross-section, as shown below. What should be the angle between the two sides of the v-shape, to give a channel with maximum volume?

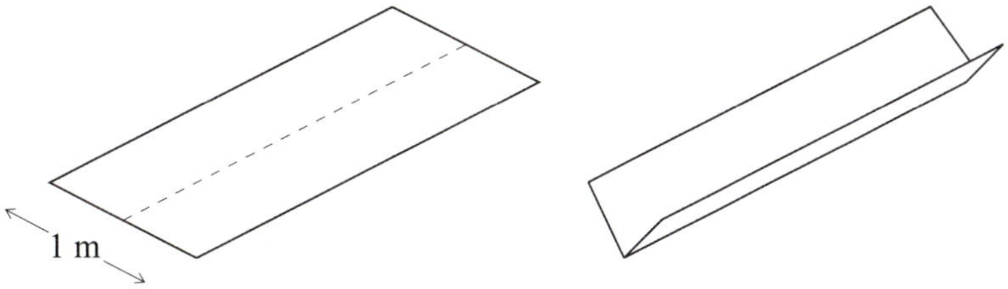

Here's a summary of the main steps that you can use to solve an optimisation problem. Of course, you should combine these steps with other problem-solving techniques as appropriate, such as simplifying the problem if possible, drawing a diagram and drawing a graph.

Strategy:
To solve an optimisation problem

1. Identify the quantity that you can change, and represent it by a variable, noting the possible values that it can take. Identify the quantity to be maximised or minimised, and represent it by a variable. These variables are the independent and dependent variables, respectively.
2. Find a formula for the dependent variable in terms of the independent variable.
3. Use the techniques of differential calculus to find the value of the independent variable that gives the maximum/minimum value of the dependent variable.
4. Interpret your answer in the context of the problem.

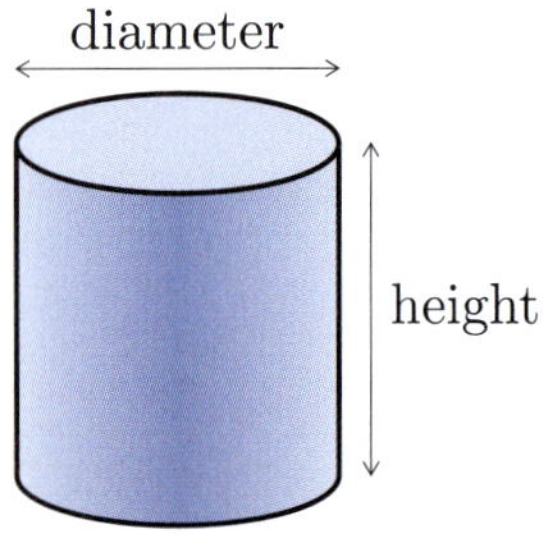

Figure 18 A cylinder

The optimisation techniques that you've met in this subsection are useful in the design of everyday objects. For example, suppose that you want to design the shape of a food tin so that it has a certain volume but uses the minimum amount of metal. You can use calculus to determine the optimal shape of the tin. If you simplify the problem by assuming that the tin is simply a hollow cylinder, as shown in Figure 18, then it turns out that the optimal shape is for the diameter of the cylinder to be equal to its height. (You might like to try to prove this.) Food tins tend to have proportions close to this, though to optimise accurately the shape of a food tin you would need to take into account the extra metal needed for the joins, and any differences in the thicknesses of metal needed for the sides, base and top.

3.2 Differentiation using a computer

Sometimes it's convenient to use a computer to find the derivatives of functions, and to work with these derivatives. The next activity shows you how to use the module computer algebra system to do this.

Activity 22 *Using a computer for differentiation*

Work through Section 8 of the *Computer algebra guide*.

To illustrate how computers can be used to solve calculus problems, in the rest of this subsection we'll use a computer to solve the following minimisation problem.

> A man is in a forest, 1 km away from the nearest point on a straight road that passes through the forest, as shown in Figure 19. He wants to get to a point 2 km further along the road, as quickly as possible, by first running in a straight line through the forest and then running along the road. He can run at 8 kilometres per hour through the forest, and at 16 kilometres per hour along the road. At what point along the road should he aim to join it?

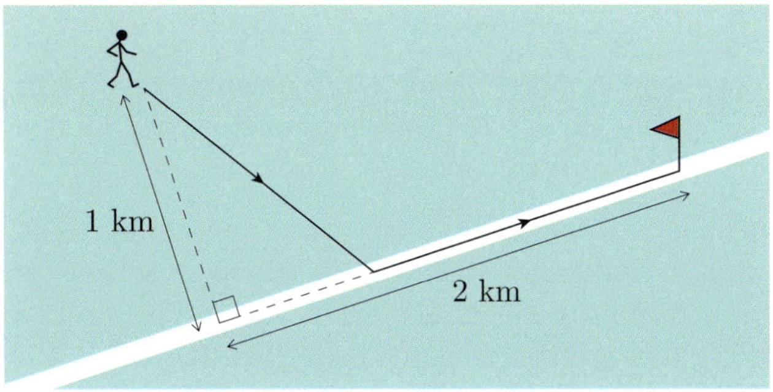

Figure 19 The position of a man in a forest and his path to his destination

Before we can use a computer, we need to carry out the first two steps of the strategy given at the end of the last subsection, to convert the problem from a word problem into a problem expressed in terms of variables and a function.

To do this, let x be the distance in kilometres between the point where the man joins the road and the point on the road that is closest to his initial position, as shown in Figure 20. Then x can take any value in the interval $[0, 2]$. Let T be the total time in hours taken by the man to reach his destination.

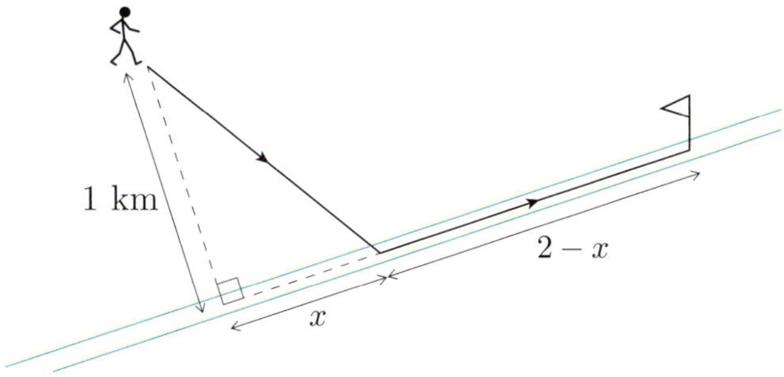

Figure 20 An annotated diagram of the man's position and path

We now have to express T in terms of x. By Pythagoras' theorem, the distance in kilometres that the man runs through the forest is

$$\sqrt{1^2 + x^2}.$$

The distance in kilometres that he runs along the road is simply

$$2 - x.$$

For each of the two parts of the man's run, the time that he takes is equal to the distance that he runs divided by the speed at which he runs. Hence the total time T that he takes to reach his destination is given by

$$T = \frac{\sqrt{1 + x^2}}{8} + \frac{2 - x}{16}.$$

To solve the problem, we now need to use calculus to find the value of x in the interval $[0, 2]$ that gives the minimum value of T. That is, we need to carry out the final step of the strategy.

You could do that without using a computer, and you might like to try this, if you want a challenge. However, in the next activity you're asked to do it by using a computer. This is quicker and easier, and it will give you practice in using a computer to solve problems like this. A by-hand solution is provided at the end of the solution to the activity.

Activity 23 *Solving an optimisation problem using a computer*

Solve the problem discussed above, by following the steps below. Carry out parts (a), (b), (d) and (e) using the module computer algebra system.

(a) Define a function f as follows:

$$f(x) = \frac{\sqrt{1 + x^2}}{8} + \frac{2 - x}{16}.$$

(b) Plot the graph of f, for values of x in the interval $[0, 2]$ at least.

(c) Use your plot to estimate roughly the value of x in $[0, 2]$ that gives the minimum value of $f(x)$.

(d) Find the derivative f' of f.

(e) Solve the equation $f'(x) = 0$.

(f) Hence state, to the nearest 10 metres, how far along the road the man should join it.

3.3 Derivatives of inverse functions

In this subsection you'll meet a rule, known as the *inverse function rule*, which you can use to work out a formula for the derivative of an inverse function when you know a formula for the derivative of the original function.

The inverse function rule can be derived from the chain rule, as follows. We'll use Leibniz notation, since the inverse function rule is usually easier to work with in Leibniz notation.

Suppose that y is an invertible function of x. Then x is also a function of y. As you know, the derivative of x with respect to x is 1, and we can write this fact as

$$\frac{\mathrm{d}x}{\mathrm{d}x} = 1.$$

However, by the chain rule,

$$\frac{\mathrm{d}x}{\mathrm{d}x} = \frac{\mathrm{d}x}{\mathrm{d}y}\frac{\mathrm{d}y}{\mathrm{d}x}.$$

Therefore

$$\frac{\mathrm{d}x}{\mathrm{d}y}\frac{\mathrm{d}y}{\mathrm{d}x} = 1.$$

Hence, for values of x for which $\mathrm{d}x/\mathrm{d}y$ exists and is non-zero,

$$\frac{\mathrm{d}y}{\mathrm{d}x} = \frac{1}{\mathrm{d}x/\mathrm{d}y}.$$

So we have the following rule.

Inverse function rule (Leibniz notation)

If y is an invertible function of x, then

$$\frac{\mathrm{d}y}{\mathrm{d}x} = \frac{1}{\mathrm{d}x/\mathrm{d}y},$$

for all values of x such that $\dfrac{\mathrm{d}x}{\mathrm{d}y}$ exists and is non-zero.

You can see that, as with the chain rule, the Leibniz form of the inverse function rule looks like an obvious fact about fractions. As with the chain rule, you should keep in mind that that's not what it is.

Here's an informal way to see that the inverse function rule makes sense. Suppose that y is an invertible function of x, which means that x is also a function of y. Suppose that, for a particular value of x and its corresponding value of y, the value of x is increasing at the rate of 2 units for every unit that y increases, as illustrated in Figure 21. Then, at these values of x and y, you'd expect the value of y to be increasing at the rate of $\frac{1}{2}$ unit for every unit that x increases. This is what the inverse function rule tells you.

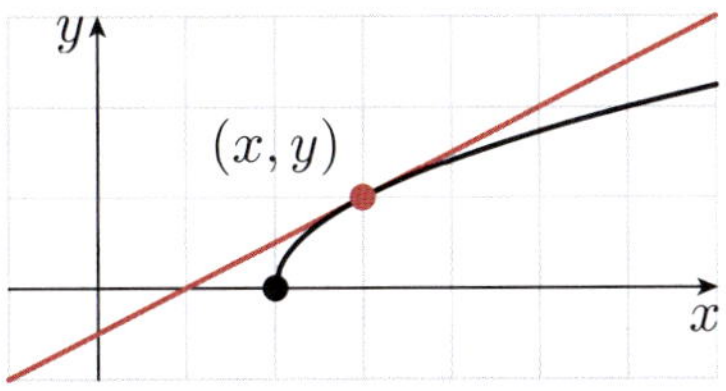

Figure 21 A point (x, y) on the graph of a function

You'll see the inverse function rule in Lagrange notation later in this subsection. First let's look at some examples of how you can use it.

Using the inverse function rule

In the next example, the inverse function rule in Leibniz notation is used to work out that the derivative of $\ln x$ is $1/x$. This is a fact that you met in Subsection 1.2.

Example 14 *Using the inverse function rule*

Use the inverse function rule, and the derivative of the exponential function, to differentiate $y = \ln x$.

Solution

Express x in terms of y.

We have $y = \ln x$, so

$$x = e^y.$$

Differentiate with respect to y.

Therefore

$$\frac{dx}{dy} = e^y.$$

Use the inverse function rule.

By the inverse function rule,

$$\frac{dy}{dx} = \frac{1}{e^y}.$$

Use the relationship between x and y to express the derivative in terms of x.

Hence, since $x = e^y$,

$$\frac{dy}{dx} = \frac{1}{x}.$$

As illustrated in Example 14, when you use the inverse function rule, you obtain an expression for the derivative that's expressed in terms of y rather than x. You then have to use the relationship between x and y to express the derivative in terms of x.

In the next example, the inverse function rule is used to find the derivative of the function $f(x) = \sin^{-1}(x)$. You saw in Unit 4 that this function has domain $[-1, 1]$. Its graph is shown in Figure 22.

Example 15 *Using the inverse function rule again*

Use the inverse function rule, and the derivative of the sine function, to differentiate $y = \sin^{-1} x$.

Solution

Express x in terms of y.

We have $y = \sin^{-1} x$, so

$$x = \sin y.$$

Differentiate x with respect to y.

Therefore

$$\frac{\mathrm{d}x}{\mathrm{d}y} = \cos y.$$

Use the inverse function rule.

By the inverse function rule,

$$\frac{\mathrm{d}y}{\mathrm{d}x} = \frac{1}{\cos y},$$

provided that $\cos y \neq 0$.

Use the relationship between x and y to express the derivative in terms of x. In this case the relationship is given by $x = \sin y$ (or $y = \sin^{-1} x$), and you need to express $\cos y$ in terms of x. You could write $\cos y = \cos(\sin^{-1} x)$, but you can obtain a simpler expression by using the identity $\sin^2 y + \cos^2 y = 1$, as follows.

The identity $\cos^2 y + \sin^2 y = 1$ gives

$$\cos y = \pm\sqrt{1 - \sin^2 y} = \pm\sqrt{1 - x^2}.$$

The + sign applies here, because y takes values only in the interval $[-\pi/2, \pi/2]$ (since $y = \sin^{-1}(x)$) and so $\cos y$ is always non-negative. Hence

$$\cos y = \sqrt{1 - x^2}.$$

Therefore

$$\frac{\mathrm{d}y}{\mathrm{d}x} = \frac{1}{\sqrt{1 - x^2}},$$

provided that $x \neq \pm 1$.

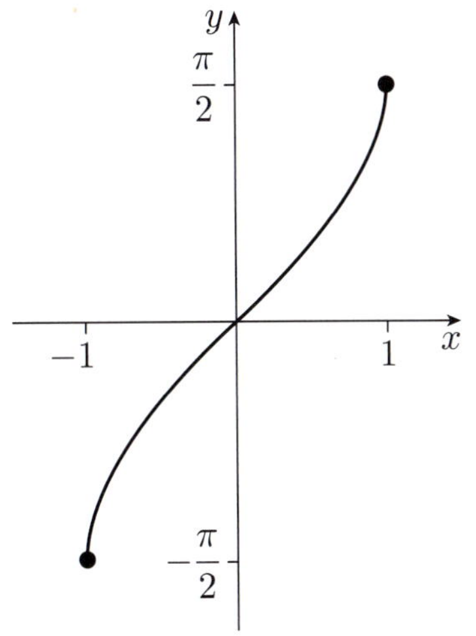

Figure 22 The graph of $y = \sin^{-1} x$

Notice that the formula for the derivative of $\sin^{-1} x$ found in Example 15 doesn't apply at the endpoints of the domain of this function, that is, when $x = -1$ and $x = 1$. This is because the function $f(x) = \sin^{-1} x$ isn't right-differentiable at $x = -1$ and isn't left-differentiable at $x = 1$, as you can see, roughly, from the shape of its graph, in Figure 22. The 'one-sided tangents' at these values of x are vertical.

Here's another point to note about Example 15. Strictly, it involves the sine function with its domain restricted to the interval $[-\pi/2, \pi/2]$, rather than the sine function with its usual domain, $\mathbb{R}$. As you saw in Unit 4, the sine function with its usual domain doesn't have an inverse function.

In the next activity you're asked to find the derivatives of the functions $f(x) = \cos^{-1} x$ and $f(x) = \tan^{-1} x$. You saw in Unit 4 that these functions have domains $[-1, 1]$ and $\mathbb{R}$, respectively. Their graphs are shown in Figures 23 and 24, respectively.

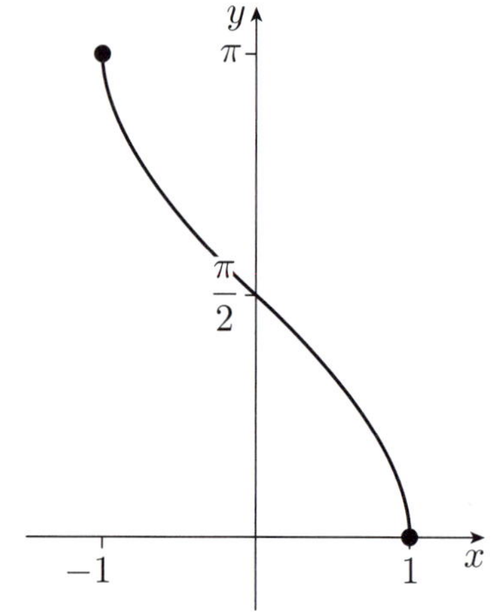

Figure 23 The graph of $y = \cos^{-1} x$

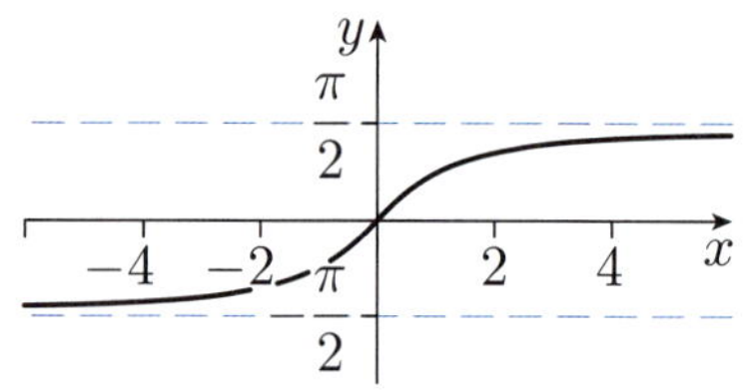

Figure 24 The graph of $y = \tan^{-1} x$

Activity 24 *Using the inverse function rule*

Use the inverse function rule, and the derivatives of the cosine and tangent functions, to differentiate the following functions.

(a) $y = \cos^{-1} x$ (b) $y = \tan^{-1} x$

Hint: To express the derivatives in terms of x, use the identity $\sin^2 y + \cos^2 y = 1$ in part (a), and the identity $\tan^2 y + 1 = \sec^2 y$ in part (b).

The derivatives found in Example 15 and Activity 24 are repeated below, and can now be included in the list of derivatives of standard functions that you have available to work with.

$$\frac{\mathrm{d}}{\mathrm{d}x}(\sin^{-1} x) = \frac{1}{\sqrt{1-x^2}}$$

$$\frac{\mathrm{d}}{\mathrm{d}x}(\cos^{-1} x) = -\frac{1}{\sqrt{1-x^2}}$$

$$\frac{\mathrm{d}}{\mathrm{d}x}(\tan^{-1} x) = \frac{1}{1+x^2}$$

In the next activity you're asked to combine these standard derivatives with some of the rules for differentiation that you've seen, to find the derivatives of some more functions.

Activity 25 *Finding derivatives of functions involving inverse trigonometric functions*

Differentiate the following functions.

(a) $f(x) = \sin^{-1}(3x)$ (b) $f(x) = e^x \cos^{-1} x$

The inverse function rule in Lagrange notation

Let's now translate the inverse function rule into Lagrange notation. To do this, we set $y = f^{-1}(x)$. Then

$$\frac{dy}{dx} = (f^{-1})'(x),$$

where $(f^{-1})'$ denotes the derivative of f^{-1}, as you'd expect.

Also, the equation $y = f^{-1}(x)$ is equivalent to the equation $x = f(y)$, so

$$\frac{dx}{dy} = f'(y) = f'(f^{-1}(x)).$$

This gives the following form of the inverse function rule.

Inverse function rule (Lagrange notation)

If f is a function with inverse function f^{-1}, then

$$(f^{-1})'(x) = \frac{1}{f'(f^{-1}(x))},$$

for all values of x such that $f'(f^{-1}(x))$ exists and is non-zero.

This is a formula for the derivative of f^{-1} in terms of the derivative of f, but it may look rather complicated and difficult to understand. Here's a helpful way to understand what it tells you, and why it holds.

First, it's important to realise the following fact. If two straight lines (drawn on axes with equal scales) are reflections of each other in the line $y = x$, then their gradients are reciprocals of each other. This is because if you consider any pair of points on one line, and their reflections on the other line, then the values of the run and the rise are interchanged for the two pairs of points, as illustrated in Figure 25(a). The lines in this diagram have gradients 2 and $\frac{1}{2}$.

Another fact, which you met in Unit 3 and will be useful here, is that if two points (plotted on axes with equal scales) are reflections of each other in the line $y = x$, then they're obtained from each other by interchanging x- and y-coordinates. For example, Figure 25(b) shows the point $(3, 1)$ and its reflection $(1, 3)$ in the line $y = x$.

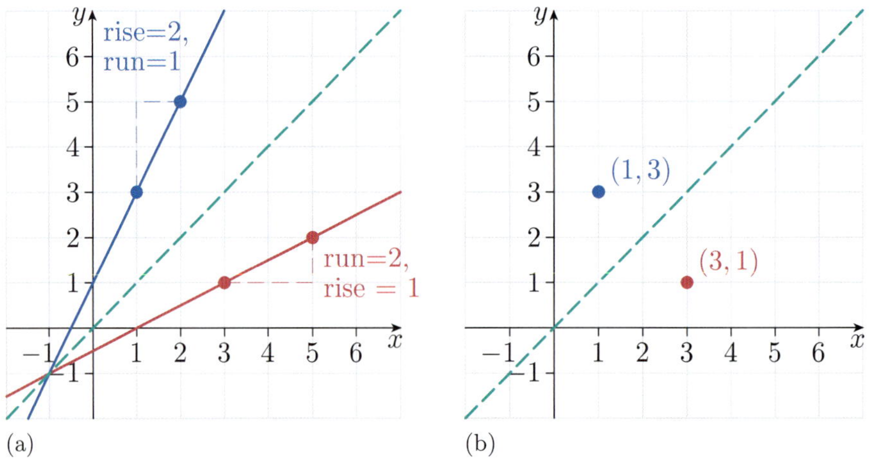

Figure 25 (a) Two straight lines that are reflections of each other in the line $y = x$ (b) Two points that are reflections of each other in the line $y = x$

Now remember that the graphs of a function and its inverse function are reflections of each other in the line $y = x$, when they're drawn on axes with equal scales. For example, Figure 26 shows the graphs of a function f and its inverse function f^{-1}. (The function is in fact $f(x) = x^2 + 2$ $(x \geq 0)$, with inverse function $f^{-1}(x) = \sqrt{x-2}$.) The two tangents shown are reflections of each other in the line $y = x$, so their gradients are reciprocals of each other. The gradient of the red tangent is the gradient of the graph of f^{-1} at the point $(3, 1)$, which is $(f^{-1})'(3)$. The gradient of the blue tangent is the gradient of the graph of f at the point $(1, 3)$, which is $f'(1)$. Hence

$$(f^{-1})'(3) = \frac{1}{f'(1)}.$$

This is a particular instance of the inverse function rule, with $x = 3$ and $f^{-1}(x) = f^{-1}(3) = 1$.

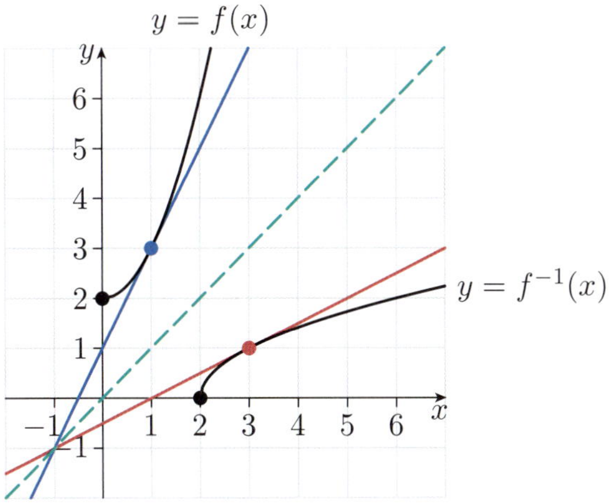

Figure 26 The graphs of a function f and its inverse function f^{-1}, and their tangents at $(1, 3)$ and $(3, 1)$, respectively

In general, consider the graphs of any invertible function f and its inverse function f^{-1}, as illustrated in Figure 27. Consider two (non-vertical) tangents that are reflections of each other in the line $y = x$, as illustrated. Their gradients are reciprocals of each other. The gradient of the red tangent is the gradient of the graph of f^{-1} at the point $(x, f^{-1}(x))$, which is $(f^{-1})'(x)$. The gradient of the blue tangent is the gradient of the graph of f at the point $(f^{-1}(x), x)$, which is $f'(f^{-1}(x))$. Hence

$$(f^{-1})'(x) = \frac{1}{f'(f^{-1}(x))}.$$

This is the inverse function rule.

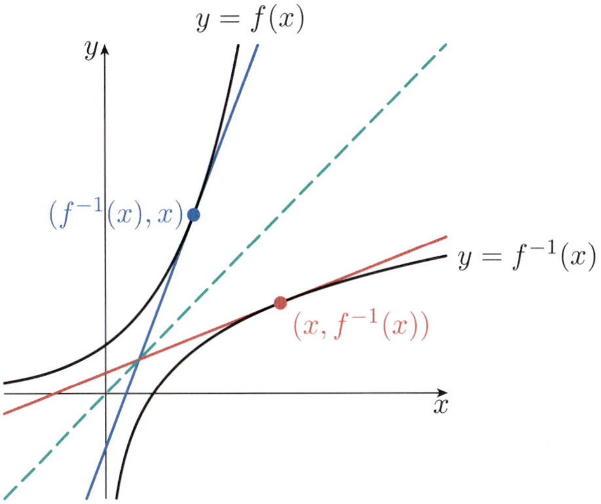

Figure 27 The graphs of a function f and its inverse function f^{-1}, and their tangents at $(f^{-1}(x), x)$ and $(x, f^{-1}(x))$, respectively

3.4 Table of standard derivatives

You've now reached the end of the material that introduces differential calculus in this module, though you'll use the ideas and methods again in Unit 11, *Taylor polynomials*. In the next section, you'll make a start on learning about integral calculus. To round off your introduction to differential calculus, here's a table of the standard derivatives that you've met. You should try to memorise at least the first six (those above the line in the middle of the table), as they're used frequently in mathematics. This table of standard derivatives is also included in the *Handbook*.

Standard derivatives

Function $f(x)$	Derivative $f'(x)$
a (constant)	0
x^n	nx^{n-1}
e^x	e^x
$\ln x$	$\dfrac{1}{x}$
$\sin x$	$\cos x$
$\cos x$	$-\sin x$
$\tan x$	$\sec^2 x$
$\operatorname{cosec} x$	$-\operatorname{cosec} x \cot x$
$\sec x$	$\sec x \tan x$
$\cot x$	$-\operatorname{cosec}^2 x$
$\sin^{-1} x$	$\dfrac{1}{\sqrt{1-x^2}}$
$\cos^{-1} x$	$-\dfrac{1}{\sqrt{1-x^2}}$
$\tan^{-1} x$	$\dfrac{1}{1+x^2}$

The first calculus textbook

In about 1690 the wealthy Marquis de l'Hôpital (1661–1704), a minor French nobleman, became interested in calculus as a result of journal articles by Gottfried Wilhelm Leibniz and the Swiss mathematicians Jacob and Johann Bernoulli. However, these articles contained little by way of explanation and so l'Hôpital paid Johann Bernoulli, then living in Paris, to tutor him in the new subject. When Bernoulli moved from Paris to Groningen to take up a professorship, they came to an arrangement whereby for a considerable monthly sum Bernoulli would send l'Hôpital material on calculus, including his new results, and would deny access to the material to anyone else. In effect, l'Hôpital bought the rights to Bernoulli's discoveries.

By 1696 l'Hôpital felt he understood enough of the mathematics to publish a book on it, *Analyse des Infiniment Petits* (*Analysis of Infinitely Small Quantities*). Although in the book l'Hôpital acknowledges his debt to Bernoulli, after l'Hôpital's death Bernoulli complained that l'Hôpital had won undeserved praise for the text. But because l'Hôpital was known to have been a competent mathematician and because Bernoulli was known to be a contentious egotist, the latter's complaints were generally dismissed. However, documents discovered in the 20th century have revealed that Bernoulli was in fact right to claim much of the book for himself.

Guillaume De l'Hôpital (1661–1704)

Johann Bernoulli (1667–1748)

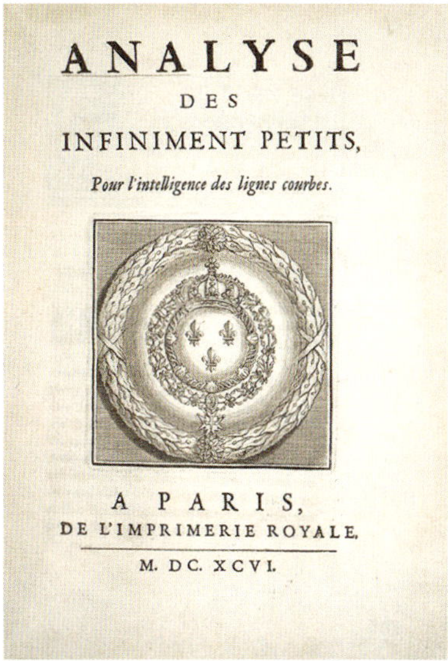

ANALYSE
DES
INFINIMENT PETITS,
Pour l'intelligence des lignes courbes.

A PARIS,
DE L'IMPRIMERIE ROYALE.
M. DC. XCVI.

The title page of *Analyse des Infiniment Petits* (1696)

4 Reversing differentiation

In Unit 6 you learned that if you know the values taken by a continuously changing quantity throughout a period of change, then you can use differentiation to find the values taken by its rate of change throughout the same period. For example, you saw that if you have a formula for the displacement of a moving object in terms of time, then you can use differentiation to find a formula for its velocity in terms of time. You saw that the area of calculus that deals with this type of process is called *differential calculus*.

In the rest of this unit, and in Unit 8, we'll consider the opposite process. We'll consider situations where you know the values taken by the rate of change of a continuously changing quantity throughout a period of change, and you want to find the values taken by the quantity throughout the same period. For example, you might have a formula or a graph that tells you the values taken by the velocity of an object (such as the walking man discussed in Unit 6) over some time period, and you want to use this information to work out the values taken by the displacement of the object over the same time period. As mentioned earlier, the area of calculus that deals with this type of process is called *integral calculus*.

reverse process → you know value taken by rate of change → but want to know value taken by quantity through same period.

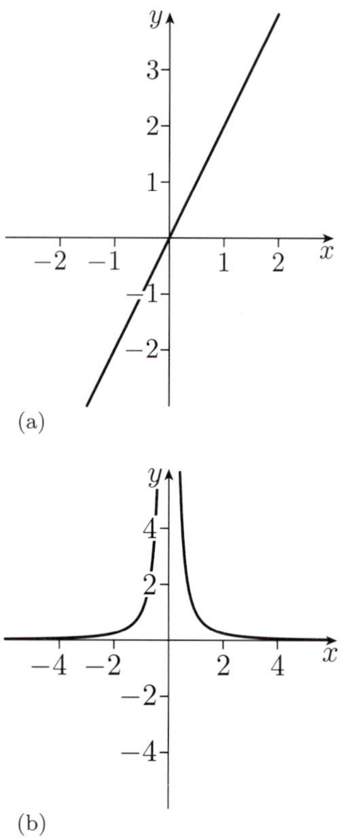

Figure 28 The graphs of
(a) $f(x) = 2x$
(b) $f(x) = 1/x^2$

As you'll see, when you're working with integral calculus, it can be useful to consider whether a function is *continuous*. A **continuous** function is one whose graph has no discontinuities (that is, 'breaks') in it. In other words, informally, it's a function whose graph you can draw without taking your pen off the paper (though you might need an infinitely large piece of paper, as the graph might be infinitely long!). For example, the function $f(x) = 2x$ is a continuous function, as shown in Figure 28(a), but the function $f(x) = 1/x^2$ isn't, as its graph has a break at $x = 0$, as shown in Figure 28(b).

continuous – no breaks as h gets close to zero.

4.1 Antiderivatives and indefinite integrals

In integral calculus you usually start with a function f, and you want to find another function, say F, whose derivative is f. Such a function F is called an **antiderivative** of the original function f. For example, an antiderivative of the function $f(x) = 2x$ is the function $F(x) = x^2$, because the derivative of x^2 is $2x$.

The process of finding an antiderivative of a function is called **antidifferentiation**, or, more commonly, **integration**. So integration is the reverse of differentiation. ✓ reverse of differentiation.

The first thing to realise is that a function can have more than one antiderivative. To see this, try the following activity.

Activity 26 *Differentiating functions whose formulas are the same apart from a constant term*

Differentiate the following functions.

(a) $F(x) = x^2$ (b) $F(x) = x^2 + 3$ (c) $F(x) = x^2 - \frac{5}{7}$

You should have found that all three functions in Activity 26 have derivative $F'(x) = 2x$. In fact, you can see that any function of the form

$$F(x) = x^2 + c,$$

where c is a constant, has derivative $F'(x) = 2x$. This is because the derivative of a constant term is zero. So each function of the form $F(x) = x^2 + c$ is an antiderivative of the function $f(x) = 2x$.

Another way to think about the fact that all the functions of the form $F(x) = x^2 + c$ have the same derivative is to remember that adding a constant to the formula of a function has the effect of translating its graph vertically. So the graphs of all the functions of the form $F(x) = x^2 + c$ are vertical translations of each other, as illustrated in Figure 29(b). Translating a graph vertically doesn't change its gradient at each x-value, of course. In other words, it doesn't change the derivative of the function that the graph represents. So each of the functions in Figure 29(b) has the same derivative, namely $f(x) = 2x$, whose graph is shown in Figure 29(a).

Derivative of an antiderivative is F'(x) not f(x)!

Fact is $x^2 + c$ antiderivative is 2x as constant is zero

Vertical scaling by adding constant will ± y-axis value for each x. But the gradient will be the same. Thus does not change derivative of f(x).

Another thing to appreciate about all the functions of the form $F(x) = x^2 + c$, where c is a constant, is that these functions are the *only* antiderivatives of the function $f(x) = 2x$. That's because any antiderivative of $f(x) = 2x$ has the same gradient at each x-value as the function $F(x) = x^2$, and the functions of the form $F(x) = x^2 + c$ are the only functions with this property.

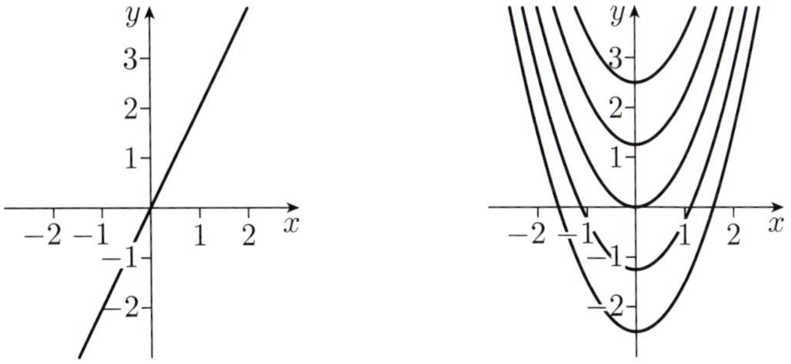

(a) $f(x) = 2x$ (b) $F(x) = x^2 + c$

Figure 29 The graphs of (a) a function (b) some of its antiderivatives

So the formula $F(x) = x^2 + c$ describes the *complete family* of antiderivatives of the function $f(x) = 2x$. We call the general function $F(x) = x^2 + c$ the **indefinite integral** of the function $f(x) = 2x$. The word 'indefinite' refers to the fact that the constant c can take any value.

In general, consider any function f that has an antiderivative. You can see that you can add any constant that you like to the formula for the antiderivative, and you'll get the formula for another antiderivative of f. Equivalently, you can translate the graph of the antiderivative vertically by any amount, and you'll get the graph of another antiderivative of f.

These antiderivatives are the *only* antiderivatives of the function f, provided that f is a *continuous* function.

So, if f is a function that has an antiderivative, and f is continuous, then the general function

$$F(x) = (\text{formula for any particular antiderivative of } f) + c,$$

where c represents any constant, and whose domain is the same as the domain of f, describes the complete family of antiderivatives of f. We call this general function F the **indefinite integral** of the function f.

The constant c in an indefinite integral is called an **arbitrary constant**, or the **constant of integration**. It's usually denoted by c, but you can use any letter. Like the word 'indefinite' in 'indefinite integral', the word 'arbitrary' in 'arbitrary constant' refers to the fact that the constant c can take any value.

You don't need to be concerned about what happens when the function f *isn't* continuous, because normally when we're solving problems in integral calculus we work only with continuous functions. These are usually the only functions that we need for the sorts of calculations that we want to perform. However, if you'd like to know what the problem is with functions that aren't continuous, then read the explanation at the end of this subsection, when you reach it.

Before you go on, it's important to make sure that you fully understand the difference between an antiderivative and an indefinite integral of a function. The definitions that you've seen are summarised below.

Antiderivatives and indefinite integrals

Suppose that f is a function.

An **antiderivative** of f is any specific function whose derivative is f.

If f has an antiderivative, and f is continuous, then the **indefinite integral** of f is the *general* function obtained by adding an arbitrary constant c to the formula for an antiderivative of f. It describes the complete family of antiderivatives of f.

Example 16 *Understanding antiderivatives and indefinite integrals*

(a) Show that the function $F(x) = x^3$ is an antiderivative of the function $f(x) = 3x^2$.

(b) What is the indefinite integral of the function $f(x) = 3x^2$?

Solution

(a) Since

$$\frac{d}{dx}(x^3) = 3x^2,$$

it follows that $F(x) = x^3$ is an antiderivative of $f(x) = 3x^2$.

(b) The indefinite integral of $f(x) = 3x^2$ is $F(x) = x^3 + c$.

Activity 27 *Understanding antiderivatives and indefinite integrals*

(a) Show that the function $F(x) = \frac{1}{2}\sin(2x)$ is an antiderivative of the function $f(x) = \cos(2x)$.

(b) What is the indefinite integral of the function $f(x) = \cos(2x)$?

(c) Write down an antiderivative of the function $f(x) = \cos(2x)$ other than the antiderivative in part (a).

Even though the idea of an indefinite integral really applies only to continuous functions, it's convenient in practice to state indefinite integrals of other functions. For example, consider again the function $f(x) = 1/x^2$, which isn't continuous. Its graph is repeated in Figure 30. An antiderivative of this function is the function $F(x) = -1/x$, as you can check by differentiating this function F. Even though f isn't continuous, we do still say

the function $f(x) = 1/x^2$ has indefinite integral $F(x) = -1/x + c$.

This is shorthand for

any continuous function f with rule $f(x) = 1/x^2$ has indefinite integral F with rule $F(x) = -1/x + c$.

For example, the continuous function $f(x) = 1/x^2$ $(x > 0)$, whose graph is shown in Figure 31, has indefinite integral $F(x) = -1/x + c$ $(x > 0)$.

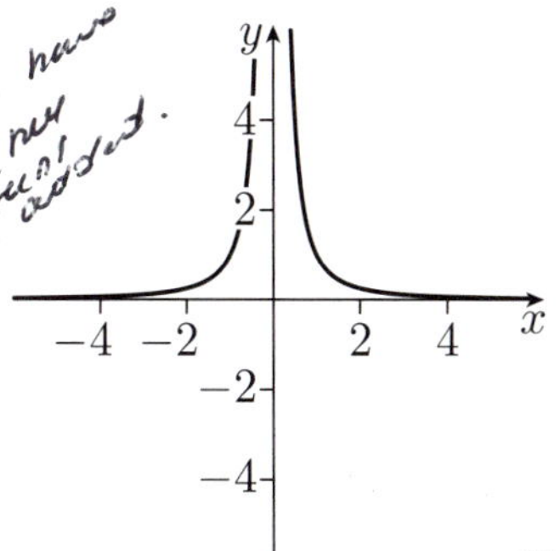

Figure 30 The graph of the function $f(x) = 1/x^2$

We use this shorthand for any function that isn't continuous.

As you've seen, to find an indefinite integral of a function, you just find a formula for any antiderivative of the function, and add an arbitrary constant, usually c.

You can find antiderivatives of many functions by using what you already know about derivatives. In the next two subsections you'll practise finding antiderivatives, and hence indefinite integrals, of some simple functions in this way. Before you make a start on that, here are a few things that you need to know.

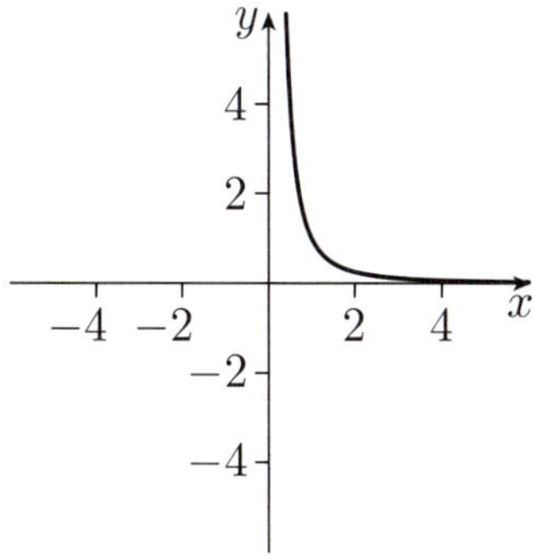

Figure 31 The graph of the function $f(x) = 1/x^2$ $(x > 0)$

First, note that when we're discussing antiderivatives and indefinite integrals, we use conventions similar to those for derivatives. For example, the terms 'antiderivative' and 'indefinite integral' can be applied directly to expressions that could appear on the right-hand side of the rule of a function, as well as to the function itself. For instance, rather than saying that

the indefinite integral of $f(x) = 2x$ is $F(x) = x^2 + c$,

we can simply say that

the indefinite integral of $2x$ is $x^2 + c$.

Also, as you'd expect, you can use any letters for functions and variables – you're not restricted to the standard ones, f and x.

When we're working with antiderivatives and indefinite integrals in the rest of this unit, we'll usually use a lower-case letter, such as f, to denote the function that we start with, and the corresponding upper-case letter, such as F, to denote any of its antiderivatives, or its indefinite integral. You'll meet a more adaptable, though more complicated, notation for indefinite integrals in Unit 8. It involves the *integral sign*, $\int$, which you may have seen elsewhere.

Functions that aren't continuous

If you'd like to know what the problem is with functions that aren't continuous, then read the discussion below. Otherwise just go on to Subsection 4.2.

A function f that has an antiderivative but which isn't continuous may have other antiderivatives as well as those obtained by vertically translating the graph of the original antiderivative. For example, consider the function $f(x) = 1/x^2$, whose graph is shown in Figure 32(a). One antiderivative of this function f is the function $F(x) = -1/x$, as mentioned above. The graph of this function F is shown in Figure 32(b). Every function obtained by vertically translating this graph is an antiderivative of f, but so is every function obtained by vertically translating each of the *two pieces* of the graph *individually*, since that doesn't change the gradient at each x-value. For example, Figure 32(c) shows an antiderivative of f obtained by translating the left-hand piece of the graph of $F(x) = -1/x$ down by 2 units, and the right-hand piece up by 1 unit.

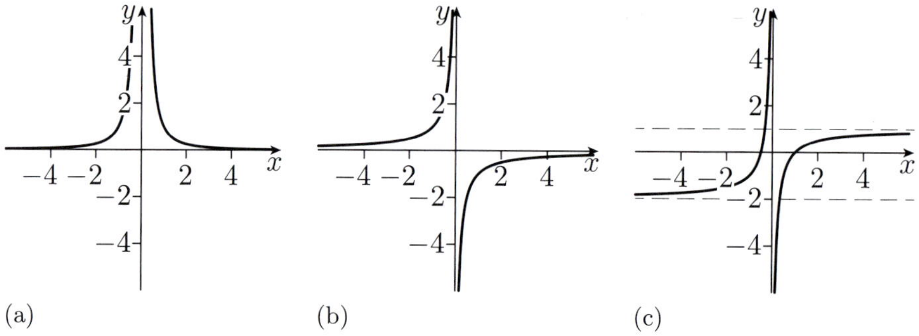

Figure 32 The graphs of (a) $f(x) = 1/x^2$ (b) the antiderivative $F(x) = -1/x$ of $f(x) = 1/x^2$ (c) another antiderivative of $f(x) = 1/x^2$

So if f is a function that has an antiderivative, but which isn't continuous, then the general function

$$F(x) = (\text{formula for any particular antiderivative of } f) + c,$$

where c represents any constant, and whose domain is the same as the domain of f, usually *doesn't* describe the complete family of antiderivatives of f, as stated earlier. As mentioned earlier, this isn't a problem, because normally when we're solving problems in integral calculus we work only with continuous functions.

4.2 Antiderivatives of power functions

In this section you'll see how to find antiderivatives, and hence indefinite integrals, of power functions.

You can find antiderivatives of power functions by using what you already know about derivatives of power functions. For example, suppose that you want to find an antiderivative of the power function $f(x) = x^6$. Consider what happens when you differentiate the power function whose exponent is 1 more than 6, namely $g(x) = x^7$:

$$\frac{\mathrm{d}}{\mathrm{d}x}\left(x^7\right) = 7x^6.$$

This equation tells you that x^7 is 'nearly' an antiderivative of x^6. You can deduce from it that if you multiply x^7 by the constant $\frac{1}{7}$, then differentiating will give the answer that you want, x^6:

$$\frac{\mathrm{d}}{\mathrm{d}x}\left(\frac{1}{7}x^7\right) = \frac{1}{7} \times 7x^6 = x^6.$$

So an antiderivative of $f(x) = x^6$ is $F(x) = \frac{1}{7}x^7$. Hence the indefinite integral of $f(x) = x^6$ is

$$F(x) = \tfrac{1}{7}x^7 + c.$$

You can find a general formula for an antiderivative of a power function by applying the thinking above to the general power function $f(x) = x^n$. Differentiating the power function $g(x) = x^{n+1}$ gives

$$\frac{\mathrm{d}}{\mathrm{d}x}\left(x^{n+1}\right) = (n+1)x^n.$$

So if you multiply x^{n+1} by $1/(n+1)$, then differentiating will give the answer that you want, x^n:

$$\frac{\mathrm{d}}{\mathrm{d}x}\left(\frac{1}{n+1}x^{n+1}\right) = \frac{1}{n+1} \times (n+1)x^n = x^n.$$

This equation gives a general formula for an antiderivative, and hence the indefinite integral, of a power function, which is stated below. The formula applies for every value of the exponent n, except $n = -1$. It doesn't apply for $n = -1$ because division by zero is undefined. (The case $n = -1$ is more complicated, and is dealt with in Subsection 6.1.)

Indefinite integral of a power function

For any number n except $n = -1$,

the indefinite integral of x^n is $\dfrac{1}{n+1}x^{n+1} + c$.

Here's a helpful informal way to remember this formula.

Indefinite integral of a power function (informal)

To find an antiderivative of a power function (except the power function $f(x) = x^{-1}$), first increase the power by 1, then divide by the new power.

As usual, to obtain the indefinite integral, add the arbitrary constant c.

Example 17 *Finding indefinite integrals of power functions*

Find the indefinite integrals of the following functions.

(a) $f(x) = x^{10}$ (b) $g(x) = \dfrac{1}{x^5}$ (c) $h(x) = \sqrt{x}$

Solution

(a) To find an antiderivative, increase the power by 1, then divide by the new power. To obtain the indefinite integral, add the arbitrary constant c.

The indefinite integral is

$$F(x) = \frac{1}{11}x^{11} + c.$$

(b) First write the function in the form $g(x) = x^n$.

The function is $g(x) = x^{-5}$.

Proceed as in part (a).

The indefinite integral is

$$G(x) = \frac{1}{-4}x^{-4} + c$$

Simplify the answer.

$$= -\frac{1}{4x^4} + c.$$

(c) First write the function in the form $y = x^n$.

The function is $h(x) = x^{1/2}$.

Proceed as in part (a).

The indefinite integral is

$$H(x) = \frac{1}{3/2}x^{3/2} + c$$

Simplify the answer.

$$= \tfrac{2}{3}x^{3/2} + c.$$

Activity 28 *Finding indefinite integrals of power functions*

Find the indefinite integrals of the following functions.

(a) $f(x) = x^9$ (b) $p(u) = \dfrac{1}{u^3}$ (c) $f(x) = x^{2/3}$

(d) $f(x) = \sqrt[4]{x}$ (e) $g(t) = t^{-2/3}$ (f) $b(v) = \dfrac{1}{\sqrt{v}}$

Activity 29 *Finding different antiderivatives of a function*

Let f be the function with rule $f(x) = x^5$.

(a) Find the indefinite integral of f.

(b) Hence write down three different antiderivatives of f.

Notice that the formula for the indefinite integral of a power function tells you that the indefinite integral of the function

$$f(x) = 1 \quad \text{(which is the same as } f(x) = x^0\text{)}$$

is

$$F(x) = x + c.$$

(As in earlier units, we assume that $0^0 = 1$, so the formula works when $x = 0$.) Another way to obtain the indefinite integral above is, of course, to use the fact that the derivative of x is 1.

In Subsection 6.1 later in the unit, you'll see a formula for the indefinite integral of the only power function that isn't covered by the general formula that you've met in this subsection. This is the power function $f(x) = x^{-1}$, that is, the reciprocal function $f(x) = 1/x$.

4.3 Constant multiple rule and sum rule for antiderivatives

Just as with derivatives, once you have formulas for antiderivatives of some functions, you can combine them to obtain formulas for antiderivatives of further functions.

Unfortunately, though, it's generally more tricky to combine antiderivatives than it is to combine derivatives. For example, there's no straightforward rule that you can use to find an antiderivative of a product of two functions if you already know an antiderivative of each of the two individual functions. In other words, there's no 'product rule for antiderivatives'. Similarly, there's no 'quotient rule for antiderivatives' and no 'chain rule for antiderivatives'.

However, there is a constant multiple rule and a sum rule for antiderivatives, and they work in the same simple ways as the similar rules for derivatives, as follows.

Constant multiple rule for antiderivatives

If $F(x)$ is an antiderivative of $f(x)$, and k is a constant, then $kF(x)$ is an antiderivative of $kf(x)$.

Sum rule for antiderivatives

If $F(x)$ and $G(x)$ are antiderivatives of $f(x)$ and $g(x)$, respectively, then $F(x) + G(x)$ is an antiderivative of $f(x) + g(x)$.

(You might have noticed that the letter used for the constant in the statement above of the constant multiple rule for antiderivatives, namely k, is different from the letter used in Unit 6 for the constant in the constant multiple rule for derivatives, which was a. It doesn't matter which letter is used, of course, but in integral calculus it's sometimes convenient to reserve the letter a to denote something else, as you'll see in Unit 8.)

These rules follow directly from the similar rules for derivatives. To see why, first suppose that $F(x)$ is an antiderivative of $f(x)$. Then

$$\frac{\mathrm{d}}{\mathrm{d}x}(F(x)) = f(x).$$

It follows from the constant multiple rule for derivatives that, for any constant k,

$$\frac{\mathrm{d}}{\mathrm{d}x}(kF(x)) = kf(x).$$

This equation tells you that $kF(x)$ is an antiderivative of $kf(x)$, which is what the constant multiple rule for antiderivatives says.

Now suppose that $F(x)$ and $G(x)$ are antiderivatives of $f(x)$ and $g(x)$, respectively. Then

$$\frac{\mathrm{d}}{\mathrm{d}x}(F(x)) = f(x) \quad \text{and} \quad \frac{\mathrm{d}}{\mathrm{d}x}(G(x)) = g(x).$$

It follows from the sum rule for derivatives that

$$\frac{\mathrm{d}}{\mathrm{d}x}(F(x) + G(x)) = f(x) + g(x).$$

This equation tells you that $F(x) + G(x)$ is an antiderivative of $f(x) + g(x)$, which is what the sum rule for antiderivatives says.

Remember that, as always when a constant multiple rule and a sum rule hold, the sum rule also holds if you replace the plus signs by minus signs. That is, if $F(x)$ and $G(x)$ are antiderivatives of $f(x)$ and $g(x)$, respectively, then $F(x) - G(x)$ is an antiderivative of $f(x) - g(x)$.

Here's an example of how you can use the two rules above.

Example 18 *Using the constant multiple rule and the sum rule for antiderivatives*

Find the indefinite integrals of the following functions.

(a) $f(x) = 10x^4$ (b) $g(t) = \dfrac{6}{t^3} - 5\sqrt{t} + 1$

Solution

(a) Find an antiderivative by using the constant multiple rule, together with the rule for finding an antiderivative of a power function. To obtain the indefinite integral, add the arbitrary constant c.

The indefinite integral is

$$\begin{aligned} F(x) &= 10 \times \tfrac{1}{5}x^5 + c \\ &= 2x^5 + c. \end{aligned}$$

(b) First express the function as a sum of constant multiples of power functions.

The function is

$$g(t) = 6t^{-3} - 5t^{1/2} + 1.$$

By the sum rule, you can find an antiderivative by working term by term. Use the method of part (a) to find an antiderivative of each term. To obtain the indefinite integral, add the arbitrary constant c.

The indefinite integral is

$$\begin{aligned} G(t) &= 6 \times \frac{1}{-2}t^{-2} - 5 \times \frac{1}{3/2}t^{3/2} + t + c \\ &= -3t^{-2} - 5 \times \tfrac{2}{3}t^{3/2} + t + c \\ &= -\frac{3}{t^2} - \frac{10t^{3/2}}{3} + t + c. \end{aligned}$$

Notice in particular that it follows from the constant multiple rule for antiderivatives that, for any constant a, the indefinite integral of the function

$$f(x) = a \quad \text{(which is the same as } f(x) = a \times 1\text{)}$$

is

$$F(x) = ax + c.$$

Another way to see this is, of course, to use the fact that the derivative of ax is a.

This useful fact is summarised below.

Indefinite integral of a constant function

The indefinite integral of the constant a is $ax + c$.

You can practise using the constant multiple rule and the sum rule for antiderivatives in the activity below. In parts (d)–(g), you have to begin by using an algebraic technique, such as multiplying out brackets or expanding a fraction, to express the function as a sum of constant multiples of power functions.

Remember that you must include an arbitrary constant c in every indefinite integral that you find.

Activity 30 *Using the constant multiple rule and the sum rule for antiderivatives*

Find the indefinite integrals of the following functions.

(a) $f(x) = 8x^3$ (b) $f(x) = 7$ (c) $f(x) = \sqrt{x} - x^2$

(d) $f(x) = (x+2)(x-5)$ (e) $g(t) = (t+1)^2$

(f) $g(x) = \dfrac{x^2+1}{x^4}$ (g) $f(r) = \sqrt[3]{r}\left(10r^2 - \dfrac{1}{r^2}\right)$

You'll meet some more ways to combine antiderivatives in Unit 8.

5 Particular antiderivatives

Sometimes when you're working with a function, you need to find a *particular* antiderivative that it has, rather than *any* antiderivative, or its indefinite integral. For example, this is often what you need to do when the function is a mathematical model of a real-life situation. In this section, you'll learn how to find such particular antiderivatives, and see some examples of how you can use them.

5.1 Finding particular antiderivatives

You can work out which of the infinitely many antiderivatives of a function is the right one if you have some appropriate extra information. Usually the extra information that you have is the value taken by the antiderivative at some value of the input variable. Here's an example.

Example 19 *Finding a particular antiderivative*

Find the antiderivative F of the function $f(x) = x^2 + 5$ such that $F(3) = 20$.

Solution

Find the indefinite integral.

The indefinite integral is

$$F(x) = \tfrac{1}{3}x^3 + 5x + c.$$

Use the extra information to find the required value of the constant c.

Using the fact that $F(3) = 20$ gives

$$\tfrac{1}{3} \times 3^3 + 5 \times 3 + c = 20$$
$$9 + 15 + c = 20$$
$$c = -4.$$

Hence the required antiderivative is

$$F(x) = \tfrac{1}{3}x^3 + 5x - 4.$$

Here are some similar examples for you to try.

Activity 31 *Finding particular antiderivatives*

(a) Find the antiderivative F of the function $f(x) = x + 2$ such that $F(1) = \frac{9}{2}$.

(b) Find the antiderivative F of the function

$$f(x) = 1/x^3 \quad (x \in (0, \infty))$$

such that $F(3) = -\frac{1}{9}$.

5.2 Using integration to work with rates of change

As mentioned at the start of this section, it's often useful to find a particular antiderivative when you're working with a function that models a real-life situation. This is illustrated in the example below.

In this example, we'll consider again a situation that you met in Unit 6, namely the one where a man walks along a straight path, gradually slowing down as he walks. In Section 3.1 of Unit 6, we started with the equation for the man's displacement in terms of time, and we used differentiation to find an equation for his velocity (his rate of change of displacement) in terms of time. Here we'll do the opposite: we'll start with the equation for his velocity in terms of time, and we'll use integration to find an equation for his displacement in terms of time. Remember that integration is the opposite process to differentiation – it's the process of finding an antiderivative of a function.

Example 20 *Using integration to deduce displacement from velocity*

Suppose that a man walks along a straight path, and his velocity v (in kilometres per hour) at time t (in hours) after he begins his walk is given by the equation

$$v = 6 - \tfrac{5}{4}t.$$

Let s be his displacement in kilometres from his starting point.

(a) Find an equation that expresses s in terms of t.

(b) Hence find the man's displacement two hours after he began his walk.

Solution

(a) Write down the given equation for the man's velocity.

The man's velocity v at time t is given by

$$v = 6 - \tfrac{5}{4}t.$$

Use integration to find an equation for his displacement. It will contain an arbitrary constant, because at this stage we don't know which antiderivative is the right one.

Hence his displacement s at time t is given by

$$s = 6t - \tfrac{5}{4} \times \tfrac{1}{2}t^2 + c,$$

that is,

$$s = 6t - \tfrac{5}{8}t^2 + c,$$

where c is a constant.

Use extra information to find the value of the constant c.

When the man begins his walk, his displacement from his starting point is 0 km. That is, when $t = 0$, $s = 0$. Substituting these values into the equation above gives

$$0 = 6 \times 0 - \tfrac{5}{8} \times 0^2 + c,$$

that is,

$$c = 0.$$

Hence the required equation for s in terms of t is

$$s = 6t - \tfrac{5}{8}t^2.$$

(b) When $t = 2$,

$$s = 6 \times 2 - \tfrac{5}{8} \times 2^2 = 12 - \tfrac{5}{2} = \tfrac{19}{2} = 9.5.$$

So the man's displacement two hours after he began his walk is 9.5 km.

The calculation in Example 20 is the sort of calculation that integral calculus allows you to carry out: if you know the values taken by the rate of change of a quantity throughout a period of change, then you can use integration to find the values taken by the quantity throughout the same period.

Here's something else that's illustrated by Example 20. If the only information that you have about the motion of an object is an equation for its velocity in terms of time, then you *don't have enough information* to find an equation for its displacement in terms of time. The best that you can do is to find an equation for its displacement in terms of time that contains an arbitrary constant. To find a definite equation (that is, to choose the correct function from a family of functions whose graphs are vertical translations of each other), you need to use further information, which is usually a fact about the displacement of the object at some moment in time.

This is just what you'd expect: if all you know is an equation for the velocity of the object in terms of time, then you don't have enough information to determine its displacement in terms of time, because it could have started anywhere!

In this sort of situation you do, however, have enough information to find the *change in the displacement* of the object between any two points in time. You'll see some examples of this later in this subsection.

Here's an example similar to Example 20 for you to try.

Activity 32 *Using integration to deduce displacement from velocity*

Suppose that a marble is rolled in a straight line down a long slope, and its velocity v (in metres per second) at time t (in seconds) after it begins rolling is given by the equation

$$v = \tfrac{1}{10}t.$$

The marble starts rolling from a position that is 0.3 metres down the slope.

Let s be its displacement in metres from the top of the slope.

(a) Find an equation that expresses s in terms of t.

(Hint: First find an equation for s in terms of t that contains an arbitrary constant, then use the value of s when $t = 0$.)

(b) Hence find the displacement of the marble from the top of the slope 4 seconds after it begins rolling.

Let's now consider another example of motion in a straight line from Unit 6. You saw that if an object falls from rest under the influence of gravity, and the effects of air resistance are negligible, then its displacement s (in metres) at time t (in seconds) after it began falling is modelled by the equation

$$s = -4.9t^2.$$

Here the displacement of the object is measured from the point from which it starts to fall, and the positive direction along the line of motion is upwards.

You saw that if you differentiate this equation, to obtain an equation for the object's velocity in terms of time, and then differentiate again to obtain an equation for its acceleration in terms of time, then you obtain the fact that the object is moving with a constant acceleration of $-9.8\,\mathrm{m\,s^{-2}}$. The magnitude of this acceleration, $9.8\,\mathrm{m\,s^{-2}}$, is known as the *acceleration due to gravity*.

In the next example, we'll carry out this process in reverse. We'll start by assuming that the effect of gravity on any object falling from rest is to produce a constant acceleration of $-9.8\,\mathrm{m\,s^{-2}}$. We'll use this fact to write down an equation for the object's acceleration in terms of time, integrate it to obtain an equation for the object's velocity in terms of time, and then integrate again to obtain an equation for the object's displacement in terms of time. You'll see that the result is the equation stated above.

Example 21 *Using integration to deduce displacement from acceleration*

Assume that an object falling from rest has a constant acceleration of $-9.8\,\mathrm{m\,s^{-2}}$, where the positive direction is upwards.

Let s be the displacement in metres of the object from its initial position and let t be the time in seconds since it began falling.

Find an equation that expresses s in terms of t.

Solution

Define the variables that you intend to use.

Let v be the velocity of the object in $\mathrm{m\,s^{-1}}$ and let a be its acceleration in $\mathrm{m\,s^{-2}}$.

Write down the equation for the acceleration of the object.

The equation for a in terms of t is

$$a = -9.8.$$

The variable t doesn't appear on the right-hand side of this equation because the acceleration is constant.

Use integration to find an equation for v in terms of t, containing an arbitrary constant.

Hence the equation for v in terms of t is

$$v = -9.8t + c,$$

where c is a constant.

Use extra information to find the value of the constant c.

The object falls from rest, so when it begins falling its velocity is $0\,\mathrm{m\,s^{-1}}$. That is, when $t = 0$, $v = 0$. Substituting these values into the equation above gives

$$0 = -9.8 \times 0 + c, \quad \text{that is,} \quad c = 0.$$

So the equation for v in terms of t is

$$v = -9.8t.$$

Use integration again to find an equation for s in terms of t, containing another arbitrary constant. Don't use the letter c for the arbitrary constant, as it's been used already. Choose a different letter, such as b.

It follows that the equation for s in terms of t is

$$s = -9.8 \times \tfrac{1}{2}t^2 + b,$$

that is,

$$s = -4.9t^2 + b,$$

where b is a constant.

Use extra information to find the value of the constant b.

When the object begins falling, its displacement is 0 m. That is, when $t = 0$, $s = 0$. Substituting these values into the equation gives

$$0 = -4.9 \times 0^2 + b, \quad \text{that is,} \quad b = 0.$$

So the equation for s in terms of t is

$$s = -4.9t^2.$$

This is the required equation for the displacement of the object in terms of time.

The next activity is about the motion of a ball thrown vertically into the air. The ball is an example of a *vertical projectile*. A **projectile** is an object that's launched into the air by a force that ceases after launch, and a **vertical projectile** is a projectile that's launched vertically upwards.

Provided that the effects of air resistance are negligible, a vertical projectile has a constant downwards acceleration due to gravity of about $9.8\,\mathrm{m\,s^{-2}}$ at all times after its launch, just as an object falling from rest has. The only thing that makes its motion different from that of an object falling from rest is that at the start of its motion its velocity is not zero. In the next activity you're asked to use a process similar to that in Example 21 to find a formula for the displacement of a ball thrown vertically into the air, in terms of time.

Activity 33 *Using integration to deduce displacement from acceleration*

Suppose that a ball is thrown vertically upwards with initial speed $12\,\mathrm{m\,s^{-1}}$. Assume that its subsequent motion is modelled as having a constant acceleration of $-9.8\,\mathrm{m\,s^{-2}}$, where the positive direction along the line of motion is upwards.

Let the acceleration, velocity and displacement of the ball at time t (in seconds) after it was thrown be a (in $\mathrm{m\,s^{-2}}$), v (in $\mathrm{m\,s^{-1}}$) and s (in m), respectively, where displacement is measured from the point from which the ball was thrown.

(a) Find an equation that expresses v in terms of t.

(b) Hence find an equation that expresses s in terms of t.

(c) Use the formulas that you found in parts (a) and (b) to find the velocity and the displacement of the ball 1 second after it was thrown.

(d) Use the formula that you found in part (b) to determine how long it takes for the ball to fall back to the point from which it was thrown.

Hint: At the time when the ball has fallen back to this point, what is the value of s?

The next activity involves an economic model of a type that you met in Unit 6. You saw there that marginal cost is the derivative of total cost.

Activity 34 *Using integration to work with marginal cost and total cost*

A company that produces agricultural fertiliser has developed a model for its production costs. According to the model, if the amount of fertiliser that the company is currently producing each week is q (in tonnes), then the marginal weekly cost m (in £ per tonne) of producing extra fertiliser is given by the equation

$$m = 250 - 0.5q.$$

The model applies for values of q between 50 and 200.

Currently the company produces 80 tonnes of fertiliser each week, at a total weekly cost of £30 400.

Let the total weekly cost of producing q tonnes of fertiliser be t (in £).

(a) By using the fact that marginal cost is the derivative of total cost, find a formula for t in terms of q.

(b) Use the formula that you found in part (a) to find the total weekly cost of producing 160 tonnes of fertiliser.

(c) What is the cost per tonne (in other words, the unit cost, not the marginal cost) of producing the fertiliser if the company produces 80 tonnes of fertiliser each week? What is it if it produces 160 tonnes each week?

5.3 Changes in the values of antiderivatives

There's a fact about the antiderivatives of a continuous function that it's important to appreciate.

To understand this fact, consider again the function $f(x) = 2x$, and its antiderivatives, the functions of the form $F(x) = x^2 + c$, whose graphs are shown in Figure 33.

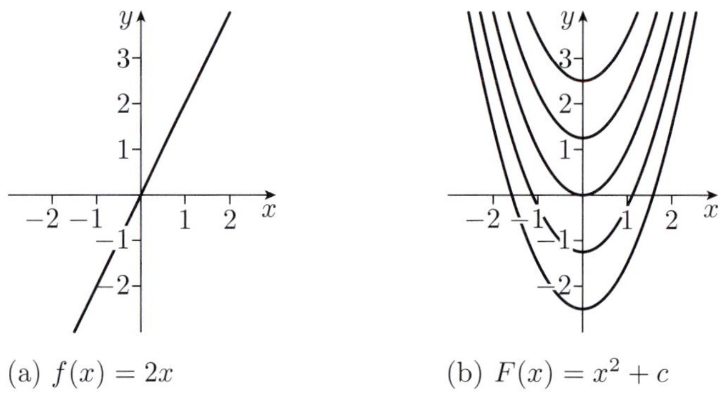

Figure 33 The graphs of a particular continuous function and some of its antiderivatives

Let's choose any two numbers in the domain of the function f, say 1 and 2, as shown in Figure 34(a), and consider the amounts by which the *antiderivatives* of f change as x changes from the first number to the second number. Figure 34(b) illustrates the amount of the change for each of two different antiderivatives of f. You can see that the amount of the change appears to be 3 units in each case.

The important fact to appreciate is that *the amount of the change is the same for all the antiderivatives of* f. That's because the graphs of all the antiderivatives are vertical translations of each other, and translating a graph vertically doesn't affect how much it changes from one value of x to another.

Another way to see that all the antiderivatives of the function $f(x) = 2x$ change by the same amount as x changes from $x = 1$ to $x = 2$ is to use algebra. Consider any antiderivative F of f. Then F is of the form $F(x) = x^2 + c$, where c is some constant. The value of F when $x = 1$ is

$$F(1) = 1^2 + c,$$

and the value of F when $x = 2$ is

$$F(2) = 2^2 + c.$$

So the change in the value of F as x changes from $x = 1$ to $x = 2$ is

$$\begin{aligned} F(2) - F(1) &= (2^2 + c) - (1^2 + c) \\ &= 4 + c - 1 - c \\ &= 3. \end{aligned}$$

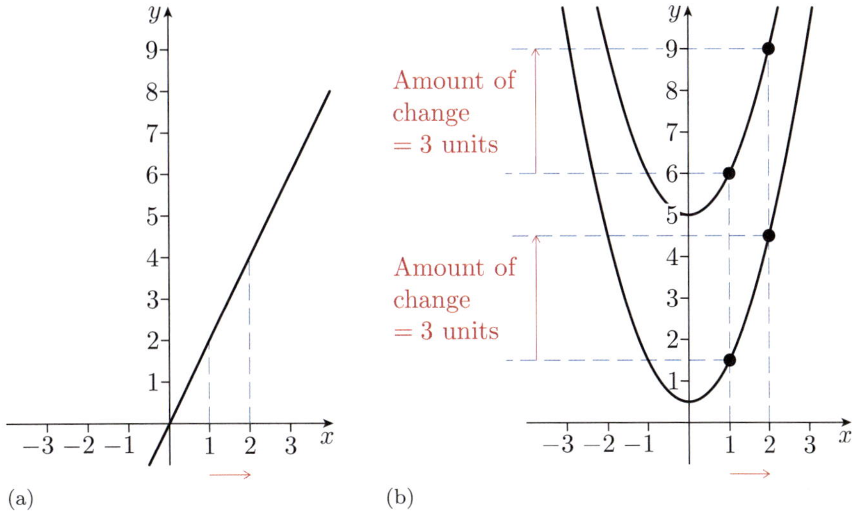

Figure 34 (a) The numbers 1 and 2 in the domain of $f(x) = 2x$ (b) The amounts that two different antiderivatives of $f(x) = 2x$ change as x changes from 1 to 2

You can see that no matter what the value of c is, the two occurrences of it will cancel out when you calculate $F(2) - F(1)$. Hence you'll always get the same answer for $F(2) - F(1)$, namely 3.

In general, we have the following important fact.

Suppose that f is a function that has an antiderivative, and f is continuous. If a and b are numbers in the domain of f, then all the antiderivatives of f change by the same amount as x changes from $x = a$ to $x = b$. In other words, the value of $F(b) - F(a)$ is the same for every antiderivative F of f.

So if you want to work out the amount by which an antiderivative of a continuous function f changes from $x = a$ to $x = b$, then you just need to find *any* antiderivative F of f, and calculate $F(b) - F(a)$.

You can use this method to solve problems in which you know the values taken by the rate of change of a quantity over some period of change, and you want to find the amount by which the quantity changes from one point in this period to another. Here's an example.

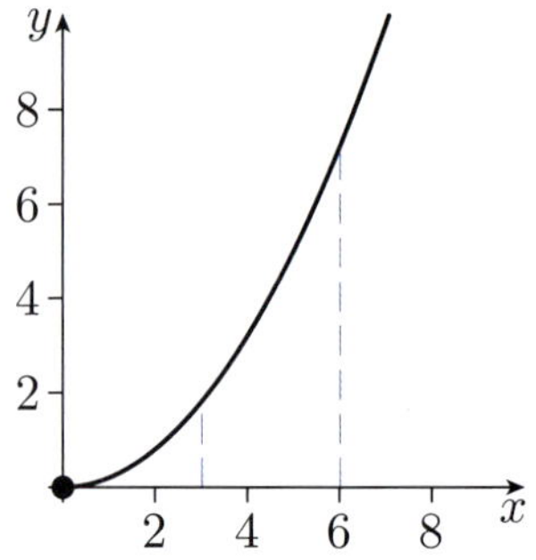

Figure 35 The graph of $y = \frac{1}{5}x^2$ $(x \geq 0)$

Example 22 *Finding a change in the value of an antiderivative*

Suppose that the rate of change of a quantity is given by the function $f(x) = \frac{1}{5}x^2$ $(x \geq 0)$, as shown in Figure 35. By what amount does the quantity change from $x = 3$ to $x = 6$?

Solution

Find any antiderivative F of f and calculate $F(6) - F(3)$.

An antiderivative of f is

$$F(x) = \tfrac{1}{5} \times \tfrac{1}{3}x^3 = \tfrac{1}{15}x^3.$$

So the change in the quantity from $x = 3$ to $x = 6$ is

$$\begin{aligned} F(6) - F(3) &= (\tfrac{1}{15} \times 6^3) - (\tfrac{1}{15} \times 3^3) \\ &= \tfrac{72}{5} - \tfrac{9}{5} \\ &= \tfrac{63}{5} = 12.6. \end{aligned}$$

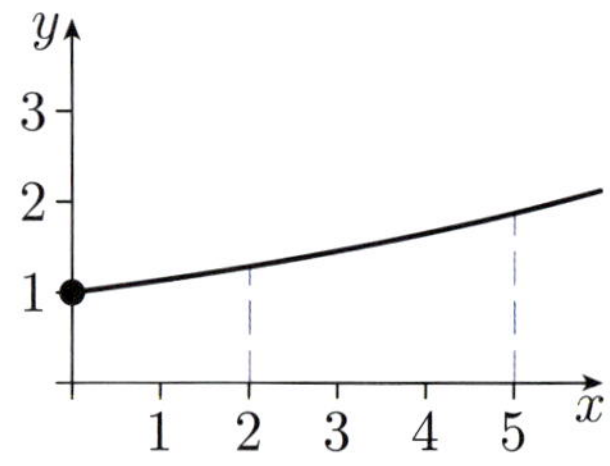

Figure 36 The graph of $f(x) = e^{x/8}$ $(x \geq 0)$

Here's a similar activity for you to try.

Activity 35 *Finding a change in the value of an antiderivative*

Suppose that the rate of change of a quantity is given by the function $f(x) = e^{x/8}$ $(x \geq 0)$, as shown in Figure 36. Find the amount by which the quantity changes from $x = 2$ to $x = 5$, to three significant figures.

In the next example and activity, the changing quantity is the displacement of a moving object, and hence the rate of change of the quantity is the velocity of the object.

The example is about a car whose velocity is increasing, as shown in Figure 37.

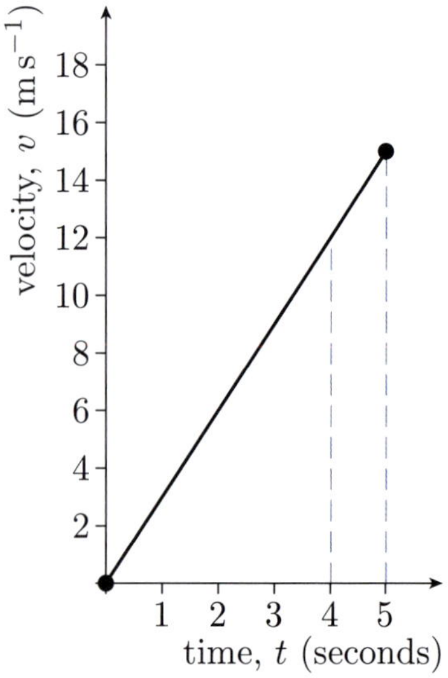

Figure 37 The graph of $v = 3t$ $(0 \leq t \leq 5)$

Example 23 *Finding a change in displacement*

Suppose that a car begins to move along a straight road, and its velocity v (in $\mathrm{m\,s^{-1}}$) during the first five seconds of its journey is given by $v = f(t)$, where

$$f(t) = 3t.$$

Here t is the time in seconds since the car began moving. How far does the car travel from the end of the fourth second to the end of the fifth second?

Solution

Find any antiderivative F of f and calculate $F(5) - F(4)$.

An antiderivative of f is

$$F(t) = \tfrac{3}{2}t^2.$$

So the change in the displacement of the car from $t = 4$ to $t = 5$ is

$$\begin{aligned} F(5) - F(4) &= (\tfrac{3}{2} \times 5^2) - (\tfrac{3}{2} \times 4^2) \\ &= 37.5 - 24 \\ &= 13.5. \end{aligned}$$

That is, the car travels 13.5 m (in the positive direction) during the fifth second.

The activity below is about an object whose velocity is decreasing, and then becomes negative, as shown in Figure 38; that is, it starts to move in the opposite direction, as shown in Figure 39.

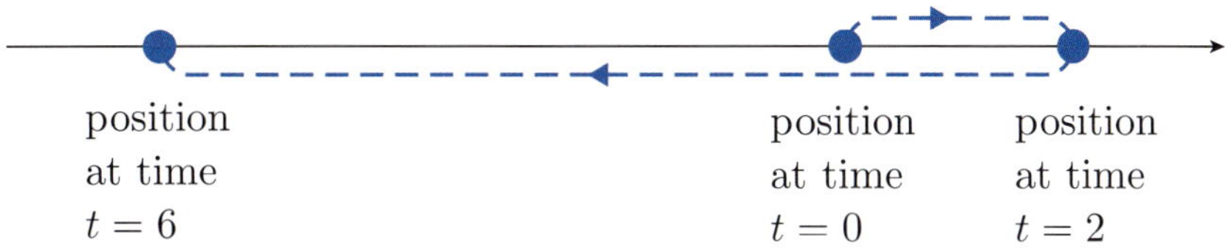

Figure 39 The object whose motion is given by Figure 38 is moving in the positive direction at time $t = 0$, then turns at time $t = 2$ to move in the negative direction

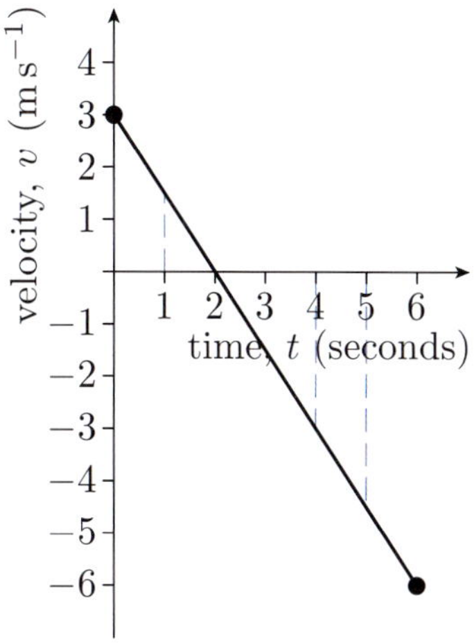

Figure 38 The graph of $v = 3 - \frac{3}{2}t$ $(0 \le t \le 6)$

Activity 36 *Finding changes in displacement*

Suppose that an object moves along a straight line, and its velocity v (in m s^{-1}) at time t (in seconds) is given by $v = f(t)$, where $f(t) = 3 - \frac{3}{2}t$ $(0 \le t \le 6)$, as shown in Figure 38.

(a) By what amount does the displacement of the object change from time $t = 0$ to time $t = 1$? What is the *distance* that the object travels in that time?

(b) Repeat part (a) for the times $t = 4$ and $t = 5$.

The ideas that you've met in this subsection will be important in Unit 8.

6 Antiderivatives of further standard functions

In this section you'll meet and use formulas for antiderivatives of some further standard functions. These formulas are all found by using formulas for derivatives that you met earlier in this unit.

6.1 An antiderivative of the reciprocal function

We'll begin by finding a formula for an antiderivative of the reciprocal function $f(x) = 1/x$, which is the only power function that isn't covered by the general formula for the indefinite integral of a power function that you met in Subsection 4.2.

Here's how you can find an antiderivative of this function, by using what you already know about differentiation.

Remember that you saw earlier in this unit that

$$\frac{\mathrm{d}}{\mathrm{d}x}(\ln x) = \frac{1}{x}. \quad (1)$$

Unfortunately this formula doesn't immediately give you a formula for an antiderivative of the function $f(x) = 1/x$, because the function $f(x) = 1/x$ has domain $(-\infty, 0) \cup (0, \infty)$, but the formula applies only when x is positive (because $\ln x$ is defined only when x is positive). However, the formula does tell you that an antiderivative of the function

$$g(x) = \frac{1}{x} \quad (x > 0)$$

is

$$G(x) = \ln x.$$

This fact is illustrated in Figure 40. The graph of $g(x) = 1/x$ $(x > 0)$, in Figure 40(a), gives the gradients of the graph of $G(x) = \ln x$, in Figure 40(b).

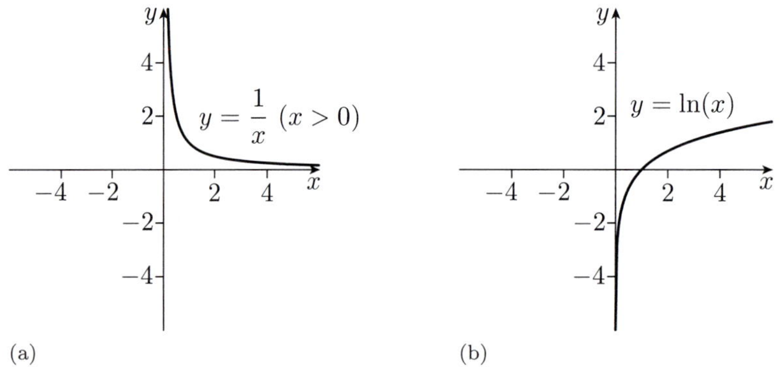

Figure 40 The graphs of (a) $g(x) = 1/x$ $(x > 0)$ and (b) $G(x) = \ln x$

To deal with the 'other half' of the function $f(x) = 1/x$, you can use the fact that equation (1), together with the chain rule, gives

$$\frac{d}{dx}(\ln(-x)) = \frac{1}{-x} \times (-1) = \frac{1}{x}.$$

This formula applies for *negative* values of x, because $\ln(-x)$ is defined when $-x$ is positive, that is, when x is negative. So it tells you that an antiderivative of the function

$$h(x) = 1/x \quad (x < 0)$$

is

$$H(x) = \ln(-x).$$

This fact is illustrated in Figure 41. The graph of $h(x) = 1/x$ $(x < 0)$, in Figure 41(a), gives the gradients of the graph of $H(x) = \ln(-x)$, in Figure 41(b).

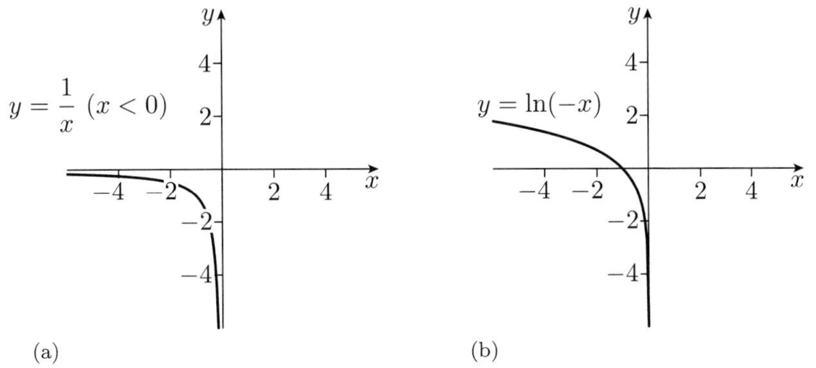

Figure 41 The graphs of (a) $h(x) = 1/x$ $(x < 0)$ and (b) $H(x) = \ln(-x)$

You can obtain an antiderivative for the function $f(x) = 1/x$, with its whole domain $(-\infty, 0) \cup (0, \infty)$, by putting together the antiderivatives found for the two 'halves' of f. The graph of this antiderivative is shown in Figure 42(b). The graph of $f(x) = 1/x$ is shown in Figure 42(a): it gives the gradients of the graph in Figure 42(b).

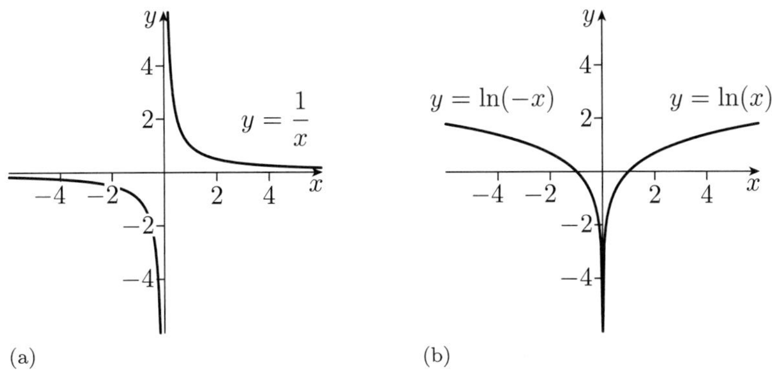

Figure 42 The graphs of (a) $f(x) = 1/x$ and (b) $F(x) = \ln x$ and $F(x) = \ln(-x)$

The formula for the antiderivative of $f(x) = 1/x$ found above can be written as

$$F(x) = \begin{cases} \ln(-x) & (x < 0) \\ \ln x & (x > 0). \end{cases} \qquad (2)$$

However, there's a more concise way to write it, using the notation $|\ |$ for the modulus of a number. Notice that if x is a negative number, then

$$-x = |x|$$

(for example, $-(-3) = 3 = |-3|$). Also, if x is a positive number, then

$$x = |x|$$

(for example, $3 = |3|$). So formula (2) can be written concisely as

$$F(x) = \ln |x|.$$

Hence the indefinite integral of $f(x) = 1/x$ is

$$F(x) = \ln |x| + c.$$

This formula is sometimes useful, but often when you're working with the function $f(x) = 1/x$, the variable x takes only *positive* values. If this is the case, then you can just use the antiderivative of the function $f(x) = 1/x$ $(x > 0)$, which has the simpler formula

$$F(x) = \ln x + c.$$

The facts that you've seen so far in this section are summarised below.

Indefinite integral of the reciprocal function

The indefinite integral of $\dfrac{1}{x}$ is $\ln |x| + c$.

(If x takes only positive values, then the indefinite integral of $\dfrac{1}{x}$ is simply $\ln x + c$.)

In the next activity you're asked to use this formula, together with the constant multiple rule and the sum rule for antiderivatives, to find the indefinite integrals of some further functions.

Activity 37 *Finding indefinite integrals of more functions*

Find the indefinite integrals of the following functions.

(a) $f(x) = \dfrac{4}{x}$ (b) $f(x) = \dfrac{x+1}{x^2}$ (c) $f(x) = \dfrac{1}{2x}$ $(x > 0)$

6.2 Table of standard indefinite integrals

You can use all of the derivatives of standard functions that you met in Unit 6 and in the first three sections of this unit to deduce antiderivatives, and hence indefinite integrals, of further functions.

You saw in the last subsection that it isn't straightforward to use the fact that the derivative of $\ln x$ is $1/x$ to deduce an antiderivative of $1/x$, because the domain of the function $f(x) = 1/x$ contains numbers that aren't in the domain of the function $F(x) = \ln x$. Luckily there are no similar problems with any of the other derivatives of standard functions that you've met, so it's straightforward to use them to deduce antiderivatives.

For example, you can immediately deduce antiderivatives of the standard functions sin, cos and exp, as follows. Since

the derivative of $\sin x$ is $\cos x$,

it follows that

an antiderivative of $\cos x$ is $\sin x$.

Similarly, since

the derivative of $\cos x$ is $-\sin x$,

it follows that

an antiderivative of $-\sin x$ is $\cos x$,

and hence (by the constant multiple rule for antiderivatives)

an antiderivative of $\sin x$ is $-\cos x$.

Finally, since the derivative of e^x is e^x, it follows that an antiderivative of e^x is e^x.

The table in the box below contains all the indefinite integrals of standard functions that you've met so far, and all the further indefinite integrals that can be deduced from the table of standard derivatives in Subsection 3.4, in the way just described. You should try to memorise at least the ones above the line in the middle of the table. For the others, you should try to make sure that you can at least recognise the functions on the left as functions that can be integrated easily, even if you continue to refer to the table for their indefinite integrals. This table of standard indefinite integrals is also included in the *Handbook*.

Standard indefinite integrals

Function	Indefinite integral
a (constant)	$ax + c$
x^n $(n \neq -1)$	$\dfrac{1}{n+1}x^{n+1} + c$
$\dfrac{1}{x}$	$\ln\lvert x\rvert + c$ or $\ln x + c$, for $x > 0$
e^x	$e^x + c$
$\sin x$	$-\cos x + c$
$\cos x$	$\sin x + c$
$\sec^2 x$	$\tan x + c$
$\operatorname{cosec}^2 x$	$-\cot x + c$
$\sec x \tan x$	$\sec x + c$
$\operatorname{cosec} x \cot x$	$-\operatorname{cosec} x + c$
$\dfrac{1}{\sqrt{1-x^2}}$	$\sin^{-1} x + c$ or $-\cos^{-1} x + c$
$\dfrac{1}{1+x^2}$	$\tan^{-1} x + c$

In the next activity you're asked to use these formulas, together with the constant multiple rule and the sum rule for antiderivatives, to find the indefinite integrals of some further functions. In some of the parts you have to start by using an algebraic technique to express the function in a form that you can integrate.

Activity 38 *Finding indefinite integrals of more functions*

Find the indefinite integrals of the following functions.

(a) $h(\theta) = 3\cos\theta + 4\sin\theta$ (b) $g(\phi) = 6 - 5\operatorname{cosec}^2\phi$ (c) $f(x) = \dfrac{7}{x}$

(d) $g(t) = \dfrac{1}{3t}$ (e) $f(x) = \dfrac{1}{\pi x}$ $(x > 0)$ (f) $f(x) = \dfrac{3}{1+x^2}$

(g) $h(t) = \dfrac{1}{4\sqrt{1-t^2}}$ (h) $p(x) = \dfrac{1}{5+5x^2}$ (i) $g(x) = \dfrac{1}{\sqrt{4-4x^2}}$

(j) $q(x) = 5(x-3)(2x-1)$ (k) $f(x) = \dfrac{x-2}{x}$ (l) $g(x) = \dfrac{x-2}{x^3}$

(m) $f(x) = 8e^x$ (n) $f(x) = e^{1+x}$ (o) $r(\phi) = -\operatorname{cosec}\phi\cot\phi$

(p) $r(\theta) = \sec\theta(\sec\theta + \tan\theta)$

You'll learn much more about integration in Unit 8. In particular you'll meet a surprising and useful alternative way to understand it. You'll also meet some further methods for combining antiderivatives of functions to obtain antiderivatives of more functions.

Learning outcomes

After studying this unit, you should be able to:

- remember the formulas for the derivatives of some standard functions
- differentiate a wide variety of functions, including products, quotients and composites
- use differentiation to solve simple optimisation problems
- find and use derivatives using a computer
- understand and use the relationship between the derivative of an invertible function and the derivative of its inverse function
- understand that integration reverses differentiation
- understand what's meant by an antiderivative and the indefinite integral of a function
- remember the formulas for the indefinite integrals of some standard functions
- use the constant multiple rule and the sum rule for antiderivatives
- use integration to work with the relationship between a changing quantity and its rate of change
- in particular, use integration to work with the relationships between displacement, velocity and acceleration.

Solutions to activities

Solution to Activity 1

The graph of the derivative of the sine function for x between 0 and 2π must look something like the sketch below.

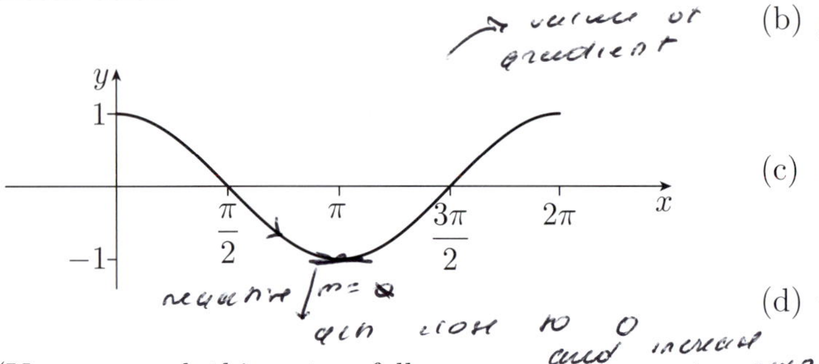

(You can work this out as follows.

When $x = \pi$, the gradient of the graph of the sine function seems to be about the same as the gradient of the line $y = -x$; that is, it seems to be about -1. As x increases, the graph gradually gets less steep – that is, its gradient takes negative values of smaller and smaller magnitude, until $x = 3\pi/2$, when the gradient seems to be zero. The gradient changes slowly when x is only a little larger than π, but changes more rapidly as x gets closer to $3\pi/2$.

In other words, for x between π and $3\pi/2$ the value of the derivative seems to increase from -1 to zero, slowly at first, but then more rapidly.

When x increases from $3\pi/2$, the gradient starts at zero, then it becomes positive. At first the graph is not very steep – that is, the gradient has small positive values – but it becomes steeper and steeper – that is, the gradient takes larger and larger positive values. Eventually, when $x = 2\pi$, the gradient seems to be about the same as the gradient of the line $y = x$; that is, it seems to be about 1. The gradient increases fairly rapidly when x is only a little larger than $3\pi/2$, but increases more slowly as x gets closer to 2π.

So, for x between $3\pi/2$ and 2π, the value of the derivative seems to increase from 0 to 1, fairly rapidly at first, but then more slowly.)

Solution to Activity 2

(a) $f(x) = \sin x + \cos x$, so
$$f'(x) = \cos x - \sin x.$$

(b) $g(u) = u^2 - \cos u$, so
$$g'(u) = 2u - (-\sin u) = 2u + \sin u.$$

(c) $P = 6\tan\theta$, so
$$\frac{\mathrm{d}P}{\mathrm{d}\theta} = 6\sec^2\theta.$$

(d) $r = -2(1 + \sin\phi)$, so
$$\frac{\mathrm{d}r}{\mathrm{d}\phi} = -2(0 + \cos\phi) = -2\cos\phi.$$

Solution to Activity 3

(a) $f(x) = e^x + \ln x$, so
$$f'(x) = e^x + \frac{1}{x}.$$

(b) $h(r) = r - \cos r - 3\ln r$, so
$$h'(r) = 1 - (-\sin r) - 3 \times \frac{1}{r} = 1 + \sin r - \frac{3}{r}.$$

(c) $v = \dfrac{1}{t} + \ln t$, so
$$\frac{\mathrm{d}v}{\mathrm{d}t} = -\frac{1}{t^2} + \frac{1}{t} = -\frac{1}{t^2} + \frac{t}{t^2} = \frac{t-1}{t^2}.$$

(d) $w = 5 - 3e^u$, so
$$\frac{\mathrm{d}w}{\mathrm{d}u} = -3e^u.$$

(e) $k = 4(\ln v - \tan v)$, so
$$\frac{\mathrm{d}k}{\mathrm{d}v} = 4\left(\frac{1}{v} - \sec^2 v\right).$$

Solution to Activity 4

(a) The function is $k(x) = x^3 \tan x$.
Let $f(x) = x^3$ and $g(x) = \tan x$.
Then $f'(x) = 3x^2$ and $g'(x) = \sec^2 x$.

By the product rule,

$$\begin{aligned} k'(x) &= f(x)g'(x) + g(x)f'(x) \\ &= x^3 \sec^2 x + (\tan x) \times 3x^2 \\ &= x^3 \sec^2 x + 3x^2 \tan x \\ &= x^2(x \sec^2 x + 3 \tan x). \end{aligned}$$

(b) The function is $k(x) = xe^x$.
Let $f(x) = x$ and $g(x) = e^x$.
Then $f'(x) = 1$ and $g'(x) = e^x$.
By the product rule,

$$\begin{aligned} k'(x) &= f(x)g'(x) + g(x)f'(x) \\ &= xe^x + e^x \times 1 \\ &= xe^x + e^x \\ &= (x+1)e^x. \end{aligned}$$

Solution to Activity 5

The function is $k(x) = x^2 \times x^3$.
Let $f(x) = x^2$ and $g(x) = x^3$.
Then $f'(x) = 2x$ and $g'(x) = 3x^2$.
By the product rule,

$$\begin{aligned} k'(x) &= f(x)g'(x) + g(x)f'(x) \\ &= x^2 \times 3x^2 + x^3 \times 2x \\ &= 3x^4 + 2x^4 \\ &= 5x^4. \end{aligned}$$

This is the same answer that you obtain by differentiating $k(x) = x^5$ directly, using the usual rule for the derivative of a power function.

Solution to Activity 6

(a) $k(x) = (3x^2 + 2x + 1)e^x$, so, by the product rule,

$$\begin{aligned} k'(x) &= (3x^2 + 2x + 1)e^x + e^x(6x + 2) \\ &= (3x^2 + 2x + 1 + 6x + 2)e^x \\ &= (3x^2 + 8x + 3)e^x. \end{aligned}$$

(b) $k(x) = \sin x \cos x$, so, by the product rule,

$$\begin{aligned} k'(x) &= (\sin x)(-\sin x) + (\cos x)(\cos x) \\ &= \cos^2 x - \sin^2 x. \end{aligned}$$

(c) $k(z) = z \sin z$, so, by the product rule,

$$\begin{aligned} k'(z) &= z \cos z + (\sin z) \times 1 \\ &= z \cos z + \sin z. \end{aligned}$$

(d) $v = (2t^2 - 1)\cos t$, so, by the product rule,

$$\begin{aligned} \frac{dv}{dt} &= (2t^2 - 1)(-\sin t) + (\cos t) \times 4t \\ &= (1 - 2t^2)\sin t + 4t \cos t. \end{aligned}$$

(e) $m = (u^2 + 3)\ln u$, so, by the product rule,

$$\begin{aligned} \frac{dm}{du} &= (u^2 + 3) \times \frac{1}{u} + (\ln u) \times 2u \\ &= \frac{u^2 + 3}{u} + 2u \ln u \\ &= \frac{u^2 + 3}{u} + \frac{2u^2 \ln u}{u} \\ &= \frac{u^2 + 3 + 2u^2 \ln u}{u} \\ &= \frac{u^2(1 + 2\ln u) + 3}{u}. \end{aligned}$$

(f) $y = \sqrt{x} \sin x = x^{1/2} \sin x$, so, by the product rule,

$$\begin{aligned} \frac{dy}{dx} &= x^{1/2} \cos x + (\sin x) \times \frac{1}{2} x^{-1/2} \\ &= \sqrt{x} \cos x + \frac{\sin x}{2\sqrt{x}} \\ &= \frac{2x \cos x + \sin x}{2\sqrt{x}}. \end{aligned}$$

Solution to Activity 7

If $y = x^2 e^x$, then, by the product rule,

$$\begin{aligned} \frac{dy}{dx} &= x^2 e^x + e^x \times 2x \\ &= x(x+2)e^x. \end{aligned}$$

Substituting $x = -1$ gives

$$\frac{dy}{dx} = (-1)(-1+2)e^{-1} = -\frac{1}{e}.$$

So the gradient is $-1/e$, which is approximately -0.37.

(The graph of $y = x^2 e^x$ is shown below. You can see that the gradient at the point with x-coordinate -1 does seem to be approximately -0.37.)

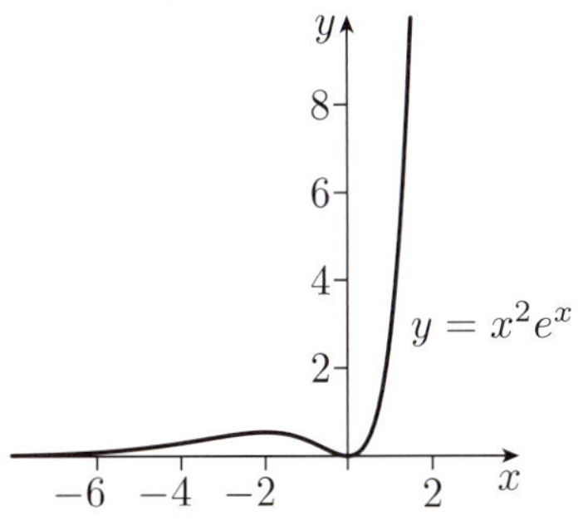

Solution to Activity 8

(a) The function is $k(x) = \dfrac{e^x}{x}$.

Let $f(x) = e^x$ and $g(x) = x$.
Then $f'(x) = e^x$ and $g'(x) = 1$.
So, by the quotient rule,

$$\begin{aligned} k'(x) = &= \frac{g(x)f'(x) - f(x)g'(x)}{(g(x))^2} \\ &= \frac{xe^x - e^x \times 1}{x^2} \\ &= \frac{(x-1)e^x}{x^2}. \end{aligned}$$

(b) The function is $k(x) = \dfrac{x^3}{2x+1}$.

Let $f(x) = x^3$ and $g(x) = 2x + 1$.
Then $f'(x) = 3x^2$ and $g'(x) = 2$.
So, by the quotient rule,

$$\begin{aligned} k'(x) = &= \frac{g(x)f'(x) - f(x)g'(x)}{(g(x))^2} \\ &= \frac{(2x+1) \times 3x^2 - x^3 \times 2}{(2x+1)^2} \\ &= \frac{(3(2x+1) - 2x)x^2}{(2x+1)^2} \\ &= \frac{(6x+3-2x)x^2}{(2x+1)^2} \\ &= \frac{(4x+3)x^2}{(2x+1)^2}. \end{aligned}$$

Solution to Activity 9

(a) $f(x) = \dfrac{e^x}{x^3}$, so, by the quotient rule,

$$\begin{aligned} f'(x) &= \frac{x^3e^x - e^x(3x^2)}{(x^3)^2} \\ &= \frac{x^2(x-3)e^x}{x^6} \\ &= \frac{(x-3)e^x}{x^4}. \end{aligned}$$

(b) $g(u) = \dfrac{u-1}{e^u}$, so, by the quotient rule,

$$\begin{aligned} g'(u) &= \frac{(e^u) \times 1 - (u-1)e^u}{(e^u)^2} \\ &= \frac{(1-u+1)e^u}{(e^u)^2} \\ &= \frac{2-u}{e^u}. \end{aligned}$$

(c) $z = \dfrac{r+1}{r-1}$, so, by the quotient rule,

$$\begin{aligned} \frac{dz}{dr} &= \frac{(r-1) \times 1 - (r+1) \times 1}{(r-1)^2} \\ &= \frac{r-1-r-1}{(r-1)^2} \\ &= -\frac{2}{(r-1)^2}. \end{aligned}$$

(d) $m = \dfrac{\theta}{\cos\theta}$, so, by the quotient rule,

$$\begin{aligned} \frac{dm}{d\theta} &= \frac{(\cos\theta) \times 1 - \theta(-\sin\theta)}{(\cos\theta)^2} \\ &= \frac{\cos\theta + \theta\sin\theta}{\cos^2\theta} \\ &= \frac{1 + \theta\tan\theta}{\cos\theta}. \end{aligned}$$

(The final expression here is obtained by dividing the top and bottom of the previous expression by $\cos\theta$. The previous expression would also be an acceptable final answer.)

(e) $p(t) = \dfrac{\ln t}{t^2}$, so, by the quotient rule,

$$\begin{aligned} p'(t) &= \frac{t^2 \times \frac{1}{t} - (\ln t) \times 2t}{(t^2)^2} \\ &= \frac{t - 2t\ln t}{t^4} \\ &= \frac{1 - 2\ln t}{t^3}. \end{aligned}$$

Solution to Activity 10

(a) By the quotient rule,

$$\begin{aligned} \frac{d}{dx}(\cot x) &= \frac{d}{dx}\left(\frac{\cos x}{\sin x}\right) \\ &= \frac{(\sin x)(-\sin x) - (\cos x)(\cos x)}{(\sin x)^2} \\ &= -\frac{\sin^2 x + \cos^2 x}{\sin^2 x} \\ &= -\frac{1}{\sin^2 x} \\ &= -\operatorname{cosec}^2 x. \end{aligned}$$

(b) By the quotient rule,

$$\begin{aligned}\frac{d}{dx}(\sec x) &= \frac{d}{dx}\left(\frac{1}{\cos x}\right)\\ &= \frac{(\cos x)\times 0 - 1\times(-\sin x)}{(\cos x)^2}\\ &= \frac{\sin x}{\cos^2 x}\\ &= \left(\frac{1}{\cos x}\right)\left(\frac{\sin x}{\cos x}\right)\\ &= \sec x \tan x.\end{aligned}$$

(c) By the quotient rule,

$$\begin{aligned}\frac{d}{dx}(\operatorname{cosec} x) &= \frac{d}{dx}\left(\frac{1}{\sin x}\right)\\ &= \frac{(\sin x)\times 0 - 1\times\cos x}{(\sin x)^2}\\ &= \frac{-\cos x}{\sin^2 x}\\ &= -\left(\frac{1}{\sin x}\right)\left(\frac{\cos x}{\sin x}\right)\\ &= -\operatorname{cosec} x \cot x.\end{aligned}$$

Solution to Activity 11

(a) $y = \ln(x^4)$
so $y = \ln u$ where $u = x^4$.

(b) $y = \sin(x^2 - 1)$
so $y = \sin u$ where $u = x^2 - 1$.

(c) $y = \cos(e^x)$
so $y = \cos u$ where $u = e^x$.

(d) $y = \cos^3 x$
so $y = u^3$ where $u = \cos x$.

(e) $y = e^{\tan x}$
so $y = e^u$ where $u = \tan x$.

(f) $y = \sqrt{\ln x}$
so $y = \sqrt{u}$ where $u = \ln x$.

(g) $y = (x + \sqrt{x})^9$
so $y = u^9$ where $u = x + \sqrt{x}$.

(h) $y = \sqrt[3]{x^2 - x - 1}$
so $y = \sqrt[3]{u}$ where $u = x^2 - x - 1$.

(i) $y = e^{\sqrt{x}}$
so $y = e^u$ where $u = \sqrt{x}$.

Solution to Activity 12

(a) $y = e^{5x}$, so

$$y = e^u \text{ where } u = 5x.$$

This gives

$$\frac{dy}{du} = e^u, \quad \frac{du}{dx} = 5.$$

By the chain rule,

$$\begin{aligned}\frac{dy}{dx} &= \frac{dy}{du}\frac{du}{dx}\\ &= e^u \times 5\\ &= e^{5x}\times 5\\ &= 5e^{5x}.\end{aligned}$$

(b) $y = \sin(2x^2 - 3)$, so

$$y = \sin u \text{ where } u = 2x^2 - 3.$$

This gives

$$\frac{dy}{du} = \cos u, \quad \frac{du}{dx} = 4x.$$

By the chain rule,

$$\begin{aligned}\frac{dy}{dx} &= \frac{dy}{du}\frac{du}{dx}\\ &= (\cos u)\times 4x\\ &= (\cos(2x^2 - 3))\times 4x\\ &= 4x\cos(2x^2 - 3).\end{aligned}$$

(c) $y = \sin\left(\frac{x}{4}\right)$, so

$$y = \sin u \text{ where } u = \frac{x}{4}.$$

This gives

$$\frac{dy}{du} = \cos u, \quad \frac{du}{dx} = \frac{1}{4}.$$

By the chain rule,

$$\begin{aligned}\frac{dy}{dx} &= \frac{dy}{du}\frac{du}{dx}\\ &= (\cos u)\times\frac{1}{4}\\ &= \frac{1}{4}\cos\left(\frac{x}{4}\right).\end{aligned}$$

(d) $y = \cos(e^x)$, so

$$y = \cos u \text{ where } u = e^x.$$

This gives

$$\frac{dy}{du} = -\sin u, \quad \frac{du}{dx} = e^x.$$

By the chain rule,

$$\begin{aligned}\frac{dy}{dx} &= \frac{dy}{du}\frac{du}{dx}\\ &= (-\sin u)\times e^x\\ &= (-\sin(e^x))\times e^x\\ &= -e^x\sin(e^x)\end{aligned}$$

(e) $y = \ln(\sin x)$, so

$$y = \ln u \text{ where } u = \sin x.$$

This gives

$$\frac{dy}{du} = \frac{1}{u},\quad \frac{du}{dx} = \cos x.$$

By the chain rule,

$$\begin{aligned}\frac{dy}{dx} &= \frac{dy}{du}\frac{du}{dx}\\ &= \frac{1}{u}\times\cos x\\ &= \frac{1}{\sin x}\times\cos x\\ &= \frac{\cos x}{\sin x}\\ &= \cot x.\end{aligned}$$

(f) $y = \sin^3 x = (\sin x)^3$, so

$$y = u^3 \text{ where } u = \sin x.$$

This gives

$$\frac{dy}{du} = 3u^2,\quad \frac{du}{dx} = \cos x.$$

By the chain rule,

$$\begin{aligned}\frac{dy}{dx} &= \frac{dy}{du}\frac{du}{dx}\\ &= 3u^2\cos x\\ &= 3(\sin x)^2\cos x\\ &= 3\sin^2 x\cos x.\end{aligned}$$

Solution to Activity 13

(a) $f(x) = \cos(4x)$, so, by the chain rule,

$$\begin{aligned}f'(x) &= -\sin(4x)\times 4\\ &= -4\sin(4x).\end{aligned}$$

(b) $f(x) = e^{2x+1}$, so, by the chain rule,

$$\begin{aligned}f'(x) &= e^{2x+1}\times 2\\ &= 2e^{2x+1}.\end{aligned}$$

(c) $r(x) = \cos^5 x$, so, by the chain rule,

$$\begin{aligned}r'(x) &= (5\cos^4 x)(-\sin x)\\ &= -5\sin x\cos^4 x.\end{aligned}$$

(Remember that $\cos^5 x$ means $(\cos x)^5$.)

(d) $g(x) = \ln(x^2+x+1)$, so, by the chain rule,

$$\begin{aligned}g'(x) &= \frac{1}{x^2+x+1}\times(2x+1)\\ &= \frac{2x+1}{x^2+x+1}.\end{aligned}$$

(e) $v(t) = \sin(e^t)$, so, by the chain rule,

$$\begin{aligned}v'(t) &= \cos(e^t)\times e^t\\ &= e^t\cos(e^t).\end{aligned}$$

(f) $v = \ln(r^5)$, so, by the chain rule,

$$\begin{aligned}\frac{dv}{dr} &= \frac{1}{r^5}\times 5r^4\\ &= \frac{5}{r}.\end{aligned}$$

(Here's an alternative way to differentiate this function, without using the chain rule:

$v = \ln(r^5) = 5\ln r$, so, by the constant multiple rule,

$$\frac{dv}{dr} = 5\times\frac{1}{r} = \frac{5}{r}.)$$

(g) $r = \tan^2\theta$, so, by the chain rule,

$$\begin{aligned}\frac{dr}{d\theta} &= (2\tan\theta)\times\sec^2\theta\\ &= 2\tan\theta\sec^2\theta.\end{aligned}$$

(h) $R = e^{\tan q}$, so, by the chain rule,

$$\frac{dR}{dq} = e^{\tan q}\sec^2 q.$$

(i) $A = (1+p^2)^9$ so, by the chain rule,

$$\begin{aligned}\frac{dA}{dp} &= 9(1+p^2)^8\times 2p\\ &= 18p(1+p^2)^8.\end{aligned}$$

Solution to Activity 14

(a) $f(x) = \sqrt{\ln x} = (\ln x)^{1/2}$, so, by the chain rule,

$$\begin{aligned}f'(x) &= \tfrac{1}{2}(\ln x)^{-1/2}\times\frac{1}{x}\\ &= \frac{1}{2x\sqrt{\ln x}}.\end{aligned}$$

(b) $p(w) = \dfrac{1}{(w^3+7)^4} = (w^3+7)^{-4}$, so, by the chain rule,

$$\begin{aligned}p'(w) &= -4(w^3+7)^{-5}\times 3w^2\\ &= -\frac{12w^2}{(w^3+7)^5}.\end{aligned}$$

(c) $f(x) = \cos(\sqrt{x}) = \cos(x^{1/2})$, so, by the chain rule,

$$\begin{aligned} f'(x) &= -\sin(x^{1/2}) \times \tfrac{1}{2}x^{-1/2} \\ &= -\frac{\sin(\sqrt{x})}{2\sqrt{x}}. \end{aligned}$$

(d) $f(x) = \dfrac{1}{\sqrt{e^x}} = (e^x)^{-1/2} = e^{-x/2}$, so, by the chain rule,

$$\begin{aligned} f'(x) &= e^{-x/2} \times (-\tfrac{1}{2}) \\ &= -\tfrac{1}{2}e^{-x/2} \\ &= -\frac{1}{2\sqrt{e^x}}. \end{aligned}$$

(This solution uses the index laws $(a^m)^n = a^{mn}$ and $a^{m/n} = \sqrt[n]{a^m}$.)

(e) $a(t) = \dfrac{1}{(2-3t-3t^2)^{1/3}} = (2-3t-3t^2)^{-1/3}$, so, by the chain rule,

$$\begin{aligned} a'(t) &= -\tfrac{1}{3}(2-3t-3t^2)^{-4/3} \times (-3-6t) \\ &= \frac{1+2t}{(2-3t-3t^2)^{4/3}}. \end{aligned}$$

(f) $C(p) = e^{1/p^2} = e^{(p^{-2})}$, so, by the chain rule,

$$\begin{aligned} C'(p) &= e^{(p^{-2})} \times (-2p^{-3}) \\ &= -2p^{-3}e^{1/p^2} \\ &= -\frac{2e^{1/p^2}}{p^3}. \end{aligned}$$

Solution to Activity 15

(a) $k(x) = \cos(7x)$, so

$$k'(x) = -7\sin(7x).$$

(b) $r(\theta) = \sin\left(\dfrac{\theta}{2}\right)$, so

$$r'(\theta) = \tfrac{1}{2}\cos\left(\frac{\theta}{2}\right).$$

(c) $g(u) = e^{3u/2}$, so

$$g'(u) = \tfrac{3}{2}e^{3u/2}.$$

(d) $h(x) = e^{-x}$, so

$$h'(x) = -e^{-x}.$$

(e) $p(\alpha) = \cos(-8\alpha)$, so

$$\begin{aligned} p'(\alpha) &= -8(-\sin(-8\alpha)) \\ &= 8\sin(-8\alpha). \end{aligned}$$

(f) $w(x) = \ln(7x)$, so

$$w'(x) = \frac{7}{7x} = \frac{1}{x}.$$

(Here's an alternative way to differentiate this function:

$w(x) = \ln(7x) = \ln 7 + \ln x$, so

$$w'(x) = 0 + \frac{1}{x} = \frac{1}{x}.)$$

(g) $f(t) = \sin\left(\dfrac{-2t}{3}\right)$, so

$$f'(t) = -\frac{2}{3}\cos\left(\frac{-2t}{3}\right).$$

(h) $k(\phi) = \tan(3\phi)$, so

$$k'(\phi) = 3\sec^2(3\phi).$$

Solution to Activity 16

(a) $k(x) = \cos(7x+4)$, so

$$k'(x) = -7\sin(7x+4).$$

(b) $h(x) = e^{-x-3}$, so

$$h'(x) = -e^{-x-3}.$$

(c) $r(\theta) = \sin\left(\dfrac{\theta-1}{2}\right) = \sin\left(\dfrac{\theta}{2} - \dfrac{1}{2}\right)$, so

$$r'(\theta) = \frac{1}{2}\cos\left(\frac{\theta}{2} - \frac{1}{2}\right) = \frac{1}{2}\cos\left(\frac{\theta-1}{2}\right).$$

(d) $s(\theta) = \sin\left(\dfrac{2-\theta}{3}\right) = \sin\left(\dfrac{2}{3} - \dfrac{\theta}{3}\right)$, so

$$s'(\theta) = -\frac{1}{3}\cos\left(\frac{2}{3} - \frac{\theta}{3}\right) = -\frac{1}{3}\cos\left(\frac{2-\theta}{3}\right).$$

Solution to Activity 17

(a)

$$\begin{aligned} g(x) &= (x^2+1)(x^3+1) \\ &= x^5 + x^3 + x^2 + 1, \end{aligned}$$

so

$$\begin{aligned} g'(x) &= 5x^4 + 3x^2 + 2x \\ &= x(5x^3 + 3x + 2). \end{aligned}$$

(An alternative method is to use the product rule.)

(b) $f(x) = \dfrac{x}{(x-2)^3}$, so

$$f'(x) = \frac{(x-2)^3 \times 1 - x\dfrac{d}{dx}\left((x-2)^3\right)}{(x-2)^6}$$
(by the quotient rule)

$$= \frac{(x-2)^3 \times 1 - x \times 3(x-2)^2 \times 1}{(x-2)^6}$$
(by the chain rule)

$$= \frac{(x-2)^3 - 3x(x-2)^2}{(x-2)^6}$$
$$= \frac{(x-2) - 3x}{(x-2)^4}$$
$$= \frac{-2x-2}{(x-2)^4}$$
$$= -\frac{2(x+1)}{(x-2)^4}.$$

(c) $g(x) = \cos(x \ln x)$, so

$$g'(x) = -\sin(x \ln x)\,\frac{d}{dx}(x \ln x)$$
(by the chain rule)

$$= -\sin(x \ln x)\left(x \times \frac{1}{x} + (\ln x) \times 1\right)$$
(by the product rule)

$$= -\sin(x \ln x)(1 + \ln x).$$

(d) $h(x) = e^{x/2}\sin(3x)$, so

$$h'(x) = e^{x/2}\frac{d}{dx}(\sin(3x)) + \sin(3x)\,\frac{d}{dx}\left(e^{x/2}\right)$$
(by the product rule)

$$= e^{x/2} \times 3\cos(3x) + \sin(3x) \times \tfrac{1}{2}e^{x/2}$$
(by the chain rule)

$$= 3e^{x/2}\cos(3x) + \tfrac{1}{2}e^{x/2}\sin(3x)$$
$$= \tfrac{1}{2}e^{x/2}(6\cos(3x) + \sin(3x)).$$

(e) $q(u) = \dfrac{\cos(4u)}{e^{3u}}$, so

$$q'(u) = \frac{e^{3u}\,\dfrac{d}{dx}(\cos(4u)) - \cos(4u)\,\dfrac{d}{dx}\left(e^{3u}\right)}{(e^{3u})^2}$$
(by the quotient rule)

$$= \frac{e^{3u} \times (-4\sin(4u)) - \cos(4u) \times 3e^{3u}}{(e^{3u})^2}$$
(by the chain rule)

$$= \frac{-4e^{3u}\sin(4u) - 3e^{3u}\cos(4u)}{e^{6u}}$$
$$= \frac{-e^{3u}(4\sin(4u) + 3\cos(4u))}{e^{6u}}$$
$$= -\frac{4\sin(4u) + 3\cos(4u)}{e^{3u}}$$

(f) $f(x) = e^{\sin(5x)}$, so

$$f'(x) = e^{\sin(5x)} \times 5\cos(5x)$$
(by the chain rule)

$$= 5e^{\sin(5x)}\cos(5x).$$

(g) $h(x) = \cos^3(x^2)$, so

$$h'(x) = 3\cos^2(x^2)\,\frac{d}{dx}\left(\cos(x^2)\right)$$
(by the chain rule)

$$= 3\cos^2(x^2) \times (-\sin(x^2)) \times 2x$$
(by the chain rule again)

$$= -6x\cos^2(x^2)\sin(x^2).$$

(h) $z(x) = \dfrac{e^{-2x}}{5} = \tfrac{1}{5}e^{-2x}$, so

$$z'(x) = \tfrac{1}{5} \times (-2e^{-2x})$$
$$= -\frac{2}{5}e^{-2x}.$$

(i) $G(x) = 3\sin x\cos x + 2\sin(3x)$, so

$$G'(x) = 3((\sin x)(-\sin x) + (\cos x)(\cos x)) + 2(\cos(3x)) \times 3$$
(by the product rule)

$$= -3\sin^2 x + 3\cos^2 x + 6\cos(3x).$$

(You can simplify this answer further by using the trigonometric identity $\cos(2\theta) = \cos^2\theta - \sin^2\theta$:

$$-3\sin^2 x + 3\cos^2 x + 6\cos(3x)$$
$$= 3(-\sin^2 x + \cos^2 x) + 6\cos(3x)$$
$$= 3\cos(2x) + 6\cos(3x).)$$

(j) $p(t) = te^t \sin t = t(e^t \sin t)$, so

$$p'(t) = t\frac{d}{dt}(e^t \sin t) + (e^t \sin t) \times 1$$

(by the product rule)

$$= t(e^t \cos t + (\sin t)e^t) + e^t \sin t$$

(by the product rule again)

$$= te^t \cos t + te^t \sin t + e^t \sin t$$
$$= e^t(t\cos t + t\sin t + \sin t).$$

(Alternatively, you could start by writing $p(t) = (te^t)\sin t$, and again apply the product rule twice.)

(k) $r = \sin(2\theta)\cos\theta$, so

$$\frac{dr}{d\theta} = \sin(2\theta)(-\sin\theta) + (\cos\theta) \times 2\cos(2\theta)$$

(by the product rule)

$$= 2\cos\theta\cos(2\theta) - \sin\theta\sin(2\theta).$$

(l) $h = ze^{3z}$, so

$$\frac{dh}{dz} = z \times 3e^{3z} + e^{3z} \times 1$$

(by the product rule)

$$= 3ze^{3z} + e^{3z}$$
$$= (3z+1)e^{3z}.$$

(m) $v = \dfrac{\sin\sqrt{u}}{\sqrt{u}} = \dfrac{\sin(u^{1/2})}{u^{1/2}}$, so

$$\frac{dv}{du} = \frac{u^{1/2}\frac{d}{du}(\sin(u^{1/2})) - \sin(u^{1/2}) \times \frac{1}{2}u^{-1/2}}{u}$$

(by the quotient rule)

$$= \frac{u^{1/2}\cos(u^{1/2}) \times \frac{1}{2}u^{-1/2} - \sin(u^{1/2}) \times \frac{1}{2}u^{-1/2}}{u}$$

(by the chain rule)

$$= \frac{\cos(u^{1/2}) - u^{-1/2}\sin(u^{1/2})}{2u}$$
$$= \frac{u^{1/2}\cos(u^{1/2}) - \sin(u^{1/2})}{2u^{3/2}}$$
$$= \frac{\sqrt{u}\cos\sqrt{u} - \sin\sqrt{u}}{2u\sqrt{u}}.$$

Solution to Activity 18

(a) $g(x) = e^{-kx}$, where k is a constant, so

$$g'(x) = -ke^{-kx}.$$

(b) $f(x) = x\cos(\pi x)$, so, by the product rule,

$$f'(x) = x \times (-\pi\sin(\pi x)) + (\cos(\pi x)) \times 1$$
$$= -\pi x\sin(\pi x) + \cos(\pi x)$$
$$= \cos(\pi x) - \pi x\sin(\pi x).$$

(c) $h = \sin^2(a\theta)$, where a is a constant, so, by the chain rule,

$$\frac{dh}{d\theta} = 2\sin(a\theta) \times a\cos(a\theta)$$
$$= 2a\sin(a\theta)\cos(a\theta).$$

(You can use the trigonometric identity $\sin(2\theta) = 2\sin\theta\cos\theta$ to write this answer above in a neater form, namely $a\sin(2a\theta)$.)

(d) $y = \cos(3x) + c$, where c is a constant, so

$$\frac{dy}{dx} = (-\sin(3x)) \times 3 + 0$$
$$= -3\sin(3x).$$

Solution to Activity 19

The function is $p(t) = \dfrac{\ln t}{t^2} = t^{-2}\ln t$.

So, by the product rule,

$$p'(t) = t^{-2} \times \frac{1}{t} + (\ln t)(-2t^{-3})$$
$$= \frac{1}{t^3} - \frac{2\ln t}{t^3}$$
$$= \frac{1 - 2\ln t}{t^3}.$$

Solution to Activity 20

Let the length of the sides of the squares be x (in m), as shown below. The value of x must be between 0 and $\frac{1}{2}$.

Let the volume of the box be V (in m^3).

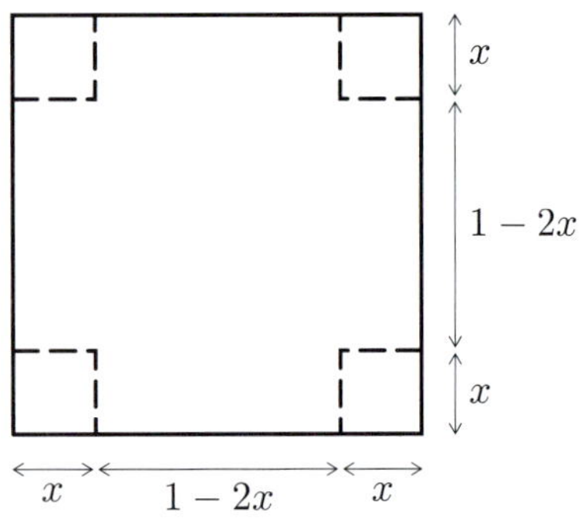

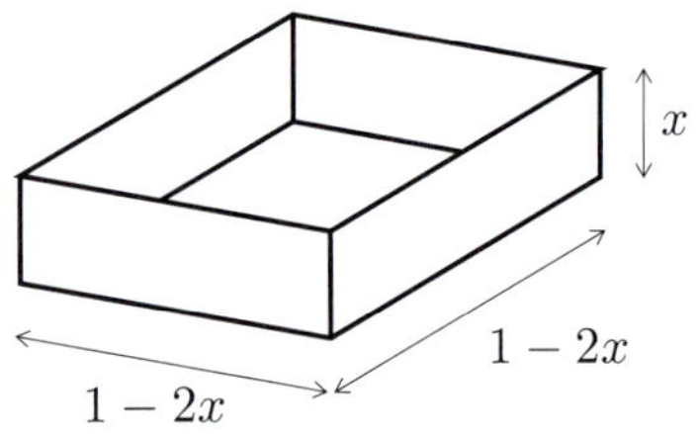

Then the height of the box will be x, and the length and width of the box will each be $1 - 2x$. Hence

$$V = x(1-2x)^2.$$

We have to find the value of x, between 0 and $\frac{1}{2}$, that gives the maximum value of V.

V is a polynomial function of x, so it is continuous on its whole domain and differentiable at every value of x. So the maximum value of V must occur either at an endpoint of the interval $[0, \frac{1}{2}]$, or at a stationary point.

The formula for V is

$$\begin{aligned} V &= x(1-2x)^2 \\ &= x(1-4x+4x^2) \\ &= x - 4x^2 + 4x^3, \end{aligned}$$

so

$$\frac{dV}{dx} = 1 - 8x + 12x^2.$$

At a stationary point, $dV/dx = 0$, which gives

$$\begin{aligned} &1 - 8x + 12x^2 = 0 \\ &12x^2 - 8x + 1 = 0 \\ &(6x-1)(2x-1) = 0 \\ &x = \tfrac{1}{6} \text{ or } x = \tfrac{1}{2}. \end{aligned}$$

So the stationary points occur when $x = \frac{1}{6}$ and $x = \frac{1}{2}$.

The endpoints of the interval are 0 and $\frac{1}{2}$, so the maximum value of V occurs when $x = 0$, $x = \frac{1}{6}$ or $x = \frac{1}{2}$.

When $x = 0$ the height of the box is zero, and when $x = \frac{1}{2}$ its length and width are zero, so in both these cases the volume V will be zero. When $x = \frac{1}{6}$, the length, width and height of the box are all positive, so the volume V will also be positive.

So the maximum value of V is achieved when $x = \frac{1}{6}$.

Hence the sides of the squares should be $\frac{1}{6}$ m; that is, about 16.7 cm.

(A graph of V against x, for x between 0 and $\frac{1}{2}$, is shown below. You can see that the maximum value of V does indeed seem to occur when x is approximately $\frac{1}{6} = 0.166\ldots$.)

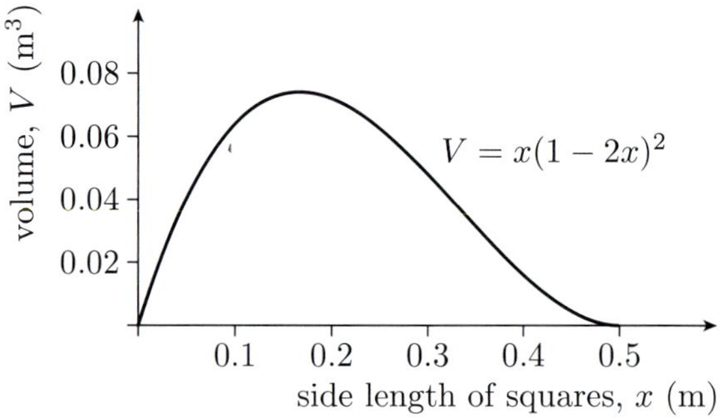

(Note also that an alternative way to differentiate the formula for V, rather than first multiplying out the brackets, is to use a combination of the product rule and the chain rule, as follows:

$$\begin{aligned} \frac{dV}{dx} &= \frac{d}{dx}(x(1-2x)^2) \\ &= x\frac{d}{dx}((1-2x)^2) + (1-2x)^2 \times 1 \\ &= x \times 2(1-2x) \times (-2) + (1-2x)^2 \\ &= -4x(1-2x) + (1-2x)^2 \\ &= (1-2x)(-4x+1-2x) \\ &= (1-2x)(1-6x).) \end{aligned}$$

Solution to Activity 21

The volume of the channel is the area of its cross-section times its (fixed) length. So to achieve a channel of maximum volume, we have to achieve the maximum area of the cross-section.

Let the angle between the two sides of the v-shape be θ, as shown below. The value of θ must be between 0 and π.

Let the area of the cross-section of the channel be A (in m^2).

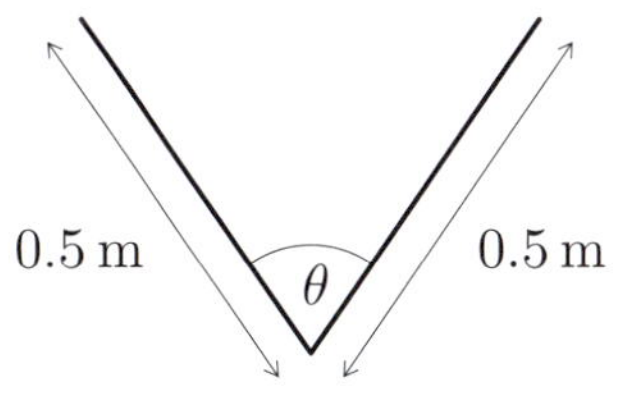

Each side of the v-shape has length $\frac{1}{2}$ m, so, by the formula $A = \frac{1}{2}ab\sin\theta$ for the area of a triangle,

$$A = \tfrac{1}{2} \times \tfrac{1}{2} \times \tfrac{1}{2}\sin\theta = \tfrac{1}{8}\sin\theta.$$

We have to find the value of θ, between 0 and π, that gives the maximum value of A.

Since the sine function is continuous on its whole domain and differentiable everywhere, so is the variable A as a function of θ. So the maximum value of A must occur either at an endpoint of the interval $[0, \pi]$, or at a stationary point.

The formula stated above for A gives

$$\frac{dA}{d\theta} = \tfrac{1}{8}\cos\theta.$$

At a stationary point, $dA/d\theta = 0$, which gives

$$\tfrac{1}{8}\cos\theta = 0$$
$$\cos\theta = 0$$
$$\theta = \frac{\pi}{2}$$

(since θ is in the interval $[0, \pi]$).

So the stationary point occurs when $\theta = \pi/2$.

The endpoints of the interval are 0 and π, so the maximum value of A occurs when $\theta = 0$, $\theta = \pi/2$ or $\theta = \pi$.

When $\theta = 0$ the width of the channel is zero, and when $\theta = \pi$ its height is zero, so in both these cases the area A will be zero. When $\theta = \pi/2$, the width and height of the channel are both positive, so the area A will also be positive.

So the maximum value of A is achieved when $\theta = \pi/2$.

That is, to give a channel with maximum volume, the angle between the sides of the v-shape should be $\pi/2$, which is 90°.

(A graph of A against θ, for θ between 0 and π, is shown below. You can see that the maximum value of A does indeed seem to occur when θ is approximately $\pi/2$.)

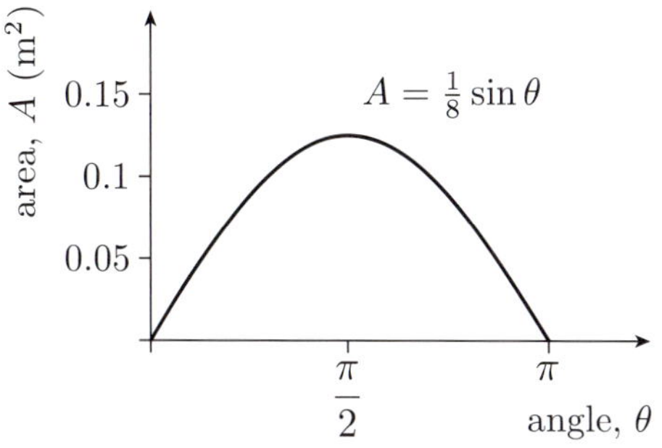

(Notice that in fact you could have answered this question without using calculus, because, from what you saw in Unit 4, the maximum value of $\sin\theta$ for values of θ between 0 and π occurs when $\theta = \pi/2$. It follows that the maximum value of $A = \frac{1}{8}\sin\theta$ for values of θ between 0 and π also occurs when $\theta = \pi/2$.)

Solution to Activity 23

(a) (You need to do this on a computer.)

(b)

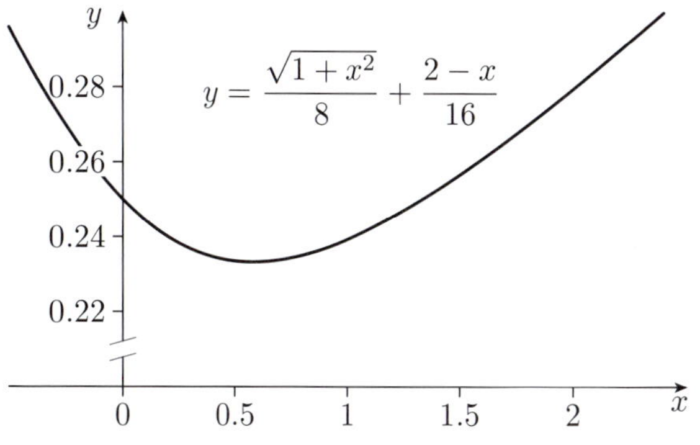

(The symbol consisting of two short slant lines on the y-axis of this graph indicates that part of the axis has been omitted.)

(c) The value of x that gives the minimum value of $f(x)$ seems to be about 0.6.

(d) The derivative of f is

$$f'(x) = \frac{x}{8\sqrt{1+x^2}} - \frac{1}{16}.$$

(e) The solution of the equation $f'(x) = 0$ is $x = 0.577$ (to 3 s.f.).

(f) The man should join the road 0.58 kilometres (to the nearest 10 metres) along from the point that is closest to his initial position.

(Because the graph of f indicates that the minimum value of $f(x)$ occurs at a local minimum rather than at one of the endpoints of the interval $[0, 2]$, in the solution above we don't consider the values of $f(0)$ and $f(2)$.)

(Details of how to use the CAS for this question are given in the *Computer algebra guide*, in the section 'Computer methods for CAS activities in Books A–D'.)

By-hand solution

The problem in this question can be solved by hand as follows.

The function to be minimised has the rule

$$f(x) = \frac{\sqrt{1+x^2}}{8} + \frac{2-x}{16}$$
$$= \tfrac{1}{8}(1+x^2)^{1/2} + \tfrac{1}{16}(2-x).$$

The variable x can take values between 0 and 2.

The function f is continuous on its domain $[0, 2]$, and differentiable at every value of x in this interval, except at the endpoints 0 and 2. (Remember that a function can't be differentiable at an endpoint of its domain.) (These facts can be deduced from the fact that f is a simple combination of functions that have these properties. You can learn more about making such deductions in later modules. For now, you can see from the graph of f that it appears to have these properties.)

Hence the minimum value of f on the interval $[0, 2]$ occurs at a stationary point of f or at an endpoint of the interval.

Differentiating f gives

$$f'(x) = \tfrac{1}{8} \times \tfrac{1}{2}(1+x^2)^{-1/2} \times 2x + \tfrac{1}{16} \times (-1)$$
$$= \frac{x}{8\sqrt{1+x^2}} - \frac{1}{16}.$$

At a stationary point, $f'(x) = 0$, which gives

$$\frac{x}{8\sqrt{1+x^2}} - \frac{1}{16} = 0$$
$$\frac{x}{8\sqrt{1+x^2}} = \frac{1}{16}$$
$$16x = 8\sqrt{1+x^2}$$
$$2x = \sqrt{1+x^2}$$
$$(2x)^2 = 1+x^2$$
$$4x^2 = 1+x^2$$
$$3x^2 = 1$$
$$x^2 = \frac{1}{3}$$
$$x = \frac{1}{\sqrt{3}}.$$

(Here we have used the fact that $x \geq 0$.)

So the only stationary point of f is $x = 1/\sqrt{3}$. The value of f at this stationary point is

$$f\left(\frac{1}{\sqrt{3}}\right) = \frac{\sqrt{1+\left(\frac{1}{\sqrt{3}}\right)^2}}{8} + \frac{2-\left(\frac{1}{\sqrt{3}}\right)}{16}$$
$$= \frac{\sqrt{1+\frac{1}{3}}}{8} + \frac{2\sqrt{3}-1}{16\sqrt{3}}$$
$$= \frac{\sqrt{\frac{4}{3}}}{8} + \frac{2\sqrt{3}-1}{16\sqrt{3}}$$
$$= \frac{2}{8\sqrt{3}} + \frac{2\sqrt{3}-1}{16\sqrt{3}}$$
$$= \frac{2\sqrt{3}}{8\sqrt{3}\sqrt{3}} + \frac{2\sqrt{3}\sqrt{3}-\sqrt{3}}{16\sqrt{3}\sqrt{3}}$$
$$= \frac{2\sqrt{3}}{24} + \frac{6-\sqrt{3}}{48}$$
$$= \frac{4\sqrt{3}}{48} + \frac{6-\sqrt{3}}{48}$$
$$= \frac{6+3\sqrt{3}}{48}$$
$$= \frac{2+\sqrt{3}}{16} \approx 0.23.$$

The values of f at the endpoints of the interval $[0, 2]$ are as follows:

$$f(0) = \frac{\sqrt{1+0^2}}{8} + \frac{2-0}{16}$$
$$= \frac{1}{8} + \frac{1}{8} = \frac{1}{4} = 0.25$$

$$f(2) = \frac{\sqrt{1+2^2}}{8} + \frac{2-2}{16}$$
$$= \frac{\sqrt{5}}{8} + 0 = \frac{\sqrt{5}}{8} \approx 0.28.$$

So the minimum value of $f(x)$ is achieved when $x = 1/\sqrt{3} = 0.577$ (to 3 s.f.). Hence, as was found above using a computer, the man should join the road 0.58 kilometres (to the nearest 10 metres) along from the point that is closest to his initial position.

Solution to Activity 24

(a) We have $y = \cos^{-1} x$, so

$$x = \cos y.$$

Therefore

$$\frac{dx}{dy} = -\sin y.$$

By the inverse function rule,

$$\frac{dy}{dx} = -\frac{1}{\sin y},$$

provided that $\sin y \neq 0$.

The identity $\sin^2 y + \cos^2 y = 1$ gives

$$\sin y = \pm\sqrt{1 - \cos^2 y} = \pm\sqrt{1 - x^2}.$$

The $+$ sign applies here, because y takes values only in the interval $[0, \pi]$ (since $y = \cos^{-1}(x)$) and so $\sin y$ is always non-negative. Hence

$$\sin y = \sqrt{1 - x^2}.$$

Therefore

$$\frac{dy}{dx} = -\frac{1}{\sqrt{1-x^2}},$$

provided that $x \neq \pm 1$.

(b) We have $y = \tan^{-1} x$, so

$$x = \tan y.$$

Therefore

$$\frac{dx}{dy} = \sec^2 y.$$

By the inverse function rule,

$$\frac{dy}{dx} = \frac{1}{\sec^2 y}.$$

The identity $\tan^2 y + 1 = \sec^2 y$ gives

$$\sec^2 y = 1 + \tan^2 y = 1 + x^2.$$

Therefore

$$\frac{dy}{dx} = \frac{1}{1+x^2}.$$

Solution to Activity 25

(a) $f(x) = \sin^{-1}(3x)$, so, by the chain rule,

$$f'(x) = \frac{1}{\sqrt{1-(3x)^2}} \times 3$$
$$= \frac{3}{\sqrt{1-9x^2}}.$$

(b) $f(x) = e^x \cos^{-1} x$, so, by the product rule,

$$f'(x) = e^x\left(-\frac{1}{\sqrt{1-x^2}}\right) + (\cos^{-1} x)e^x$$
$$= \left(\cos^{-1} x - \frac{1}{\sqrt{1-x^2}}\right)e^x.$$

Solution to Activity 26

(a) $F(x) = x^2$, so $F'(x) = 2x$.

(b) $F(x) = x^2 + 3$, so $F'(x) = 2x$.

(c) $F(x) = x^2 - \frac{5}{7}$, so $F'(x) = 2x$.

Solution to Activity 27

(a) Since

$$\frac{d}{dx}\left(\tfrac{1}{2}\sin(2x)\right) = \tfrac{1}{2} \times 2\cos(2x) = \cos(2x),$$

it follows that $F(x) = \frac{1}{2}\sin(2x)$ is an antiderivative of the function $f(x) = \cos(2x)$.

(b) The indefinite integral of $f(x) = \cos(2x)$ is $F(x) = \frac{1}{2}\sin(2x) + c$.

(c) An antiderivative of $f(x) = \cos(2x)$, other than the antiderivative in part (a), is $F(x) = \frac{1}{2}\sin(2x) + 1$.

(You probably chose a different antiderivative. Any function obtained by setting the constant c in the indefinite integral $F(x) = \frac{1}{2}\sin(2x) + c$ equal to a particular number will do.)

Solution to Activity 28

(a) The indefinite integral of $f(x) = x^9$ is

$$F(x) = \tfrac{1}{10}x^{10} + c.$$

(b) The indefinite integral of $p(u) = \dfrac{1}{u^3} = u^{-3}$ is

$$P(u) = \frac{1}{-2}u^{-2} + c = -\frac{1}{2u^2} + c.$$

(c) The indefinite integral of $f(x) = x^{2/3}$ is

$$F(x) = \tfrac{3}{5}x^{5/3} + c.$$

(d) The indefinite integral of $f(x) = \sqrt[4]{x} = x^{1/4}$ is

$$F(x) = \tfrac{4}{5}x^{5/4} + c.$$

(e) The indefinite integral of $g(t) = t^{-2/3}$ is

$$G(t) = 3t^{1/3} + c.$$

(f) The indefinite integral of $b(v) = \dfrac{1}{\sqrt{v}} = v^{-1/2}$ is

$$B(v) = 2v^{1/2} + c = 2\sqrt{v} + c.$$

Solution to Activity 29

(a) The indefinite integral of $f(x) = x^5$ is

$$F(x) = \tfrac{1}{6}x^6 + c.$$

(b) Three different antiderivatives of f are

$$F(x) = \tfrac{1}{6}x^6,$$
$$F(x) = \tfrac{1}{6}x^6 + 1,$$
$$F(x) = \tfrac{1}{6}x^6 - 2.$$

(You probably chose different examples.)

Solution to Activity 30

(a) The indefinite integral of $f(x) = 8x^3$ is

$$\begin{aligned} F(x) &= 8 \times \tfrac{1}{4}x^4 + c \\ &= 2x^4 + c. \end{aligned}$$

(b) The indefinite integral of $f(x) = 7$ is

$$F(x) = 7x + c.$$

(c) The indefinite integral of $f(x) = \sqrt{x} - x^2 = x^{1/2} - x^2$ is

$$\begin{aligned} F(x) &= \tfrac{2}{3}x^{3/2} - \tfrac{1}{3}x^3 + c \\ &= \tfrac{1}{3}(2x^{3/2} - x^3) + c. \end{aligned}$$

(d) We have

$$\begin{aligned} f(x) &= (x+2)(x-5) \\ &= x^2 - 3x - 10, \end{aligned}$$

which has indefinite integral

$$F(x) = \tfrac{1}{3}x^3 - \tfrac{3}{2}x^2 - 10x + c.$$

(e) We have

$$\begin{aligned} g(t) &= (t+1)^2 \\ &= t^2 + 2t + 1, \end{aligned}$$

which has indefinite integral

$$G(t) = \tfrac{1}{3}t^3 + t^2 + t + c.$$

(f) We have

$$g(x) = \frac{x^2+1}{x^4} = \frac{1}{x^2} + \frac{1}{x^4} = x^{-2} + x^{-4},$$

which has indefinite integral

$$\begin{aligned} G(x) &= -x^{-1} - \tfrac{1}{3}x^{-3} + c \\ &= -\frac{1}{x} - \frac{1}{3x^3} + c. \end{aligned}$$

(g) We have

$$\begin{aligned} f(r) &= \sqrt[3]{r}\left(10r^2 - \frac{1}{r^2}\right) \\ &= r^{1/3}(10r^2 - r^{-2}) \\ &= 10r^{7/3} - r^{-5/3} \end{aligned}$$

which has indefinite integral

$$\begin{aligned} F(r) &= 10(\tfrac{3}{10}r^{10/3}) - (-\tfrac{3}{2}r^{-2/3}) + c \\ &= 3r^{10/3} + \tfrac{3}{2}r^{-2/3} + c \\ &= \tfrac{3}{2}r^{-2/3}(1 + 2r^4) + c. \end{aligned}$$

Solution to Activity 31

(a) The function is $f(x) = x + 2$. Its indefinite integral is

$$F(x) = \tfrac{1}{2}x^2 + 2x + c.$$

Using the fact that $F(1) = \frac{9}{2}$ gives

$$\begin{aligned} \tfrac{1}{2} \times 1^2 + 2 \times 1 + c &= \tfrac{9}{2} \\ \tfrac{5}{2} + c &= \tfrac{9}{2} \\ c &= \tfrac{4}{2} = 2. \end{aligned}$$

Hence the required antiderivative is

$$F(x) = \tfrac{1}{2}x^2 + 2x + 2.$$

(b) The function is $f(x) = 1/x^3 = x^{-3}$ with domain $(0, \infty)$. Its indefinite integral is

$$F(x) = \frac{1}{-2}x^{-2} + c = -\frac{1}{2x^2} + c.$$

Using the fact that $F(3) = -\frac{1}{9}$ gives

$$\begin{aligned} -\frac{1}{2 \times 3^2} + c &= -\frac{1}{9} \\ -\frac{1}{18} + c &= -\frac{1}{9} \\ c = -\frac{1}{9} + \frac{1}{18} &= -\frac{1}{18}. \end{aligned}$$

Hence the required antiderivative is

$$\begin{aligned} F(x) &= -\frac{1}{2x^2} - \frac{1}{18} \\ &= -\frac{9 + x^2}{18x^2} \quad (x \in (0, \infty)). \end{aligned}$$

Solution to Activity 32

(a) The given equation for v in terms of t is

$$v = \tfrac{1}{10}t.$$

Hence the equation for s in terms of t is

$$s = \tfrac{1}{10} \times \tfrac{1}{2}t^2 + c,$$

that is,

$$s = \tfrac{1}{20}t^2 + c,$$

where c is a constant.

The marble starts rolling from a position 0.3 metres down the slope, so when $t = 0$, $s = 0.3$.

Substituting these values of s and t into the equation above gives

$$0.3 = \tfrac{1}{20} \times 0^2 + c$$
$$c = 0.3.$$

So the equation for s in terms of t is

$$s = \tfrac{1}{20}t^2 + 0.3.$$

(b) Substituting $t = 4$ into the equation from part (a) gives

$$s = \tfrac{1}{20} \times 4^2 + 0.3 = 0.8 + 0.3 = 1.1.$$

That is, the displacement of the marble from the top of the slope after it has been rolling for 4 seconds is 1.1 metres.

Solution to Activity 33

(a) The equation for a in terms of t is

$$a = -9.8.$$

Integrating this equation gives the following equation for v in terms of t:

$$v = -9.8t + c,$$

where c is a constant.

At the start of the motion, the velocity of the ball is $12\,\mathrm{m\,s^{-1}}$. That is, when $t = 0$, $v = 12$.

Substituting these values into the equation for v above gives

$$12 = -9.8 \times 0 + c, \quad \text{that is,} \quad c = 12.$$

So the equation for v in terms of t is

$$v = 12 - 9.8t.$$

(b) Integrating the equation found in part (a) gives the following equation for s in terms of t:

$$s = 12t - 9.8 \times \tfrac{1}{2}t^2 + b,$$

that is,

$$s = 12t - 4.9t^2 + b,$$

where again b is a constant.

At the start of the motion, the displacement of the ball is 0 m. That is, when $t = 0$, $s = 0$.

Substituting these values into the equation for s above gives

$$0 = 12 \times 0 - 4.9 \times 0^2 + b, \quad \text{that is,} \quad b = 0.$$

So the equation for s in terms of t is

$$s = 12t - 4.9t^2.$$

(c) When $t = 1$,

$$v = 12 - 9.8 \times 1 = 12 - 9.8 = 2.2,$$

and

$$s = 12 \times 1 - 4.9 \times 1^2 = 12 - 4.9 = 7.1.$$

So the velocity and the displacement of the ball 1 second after it was thrown are $2.2\,\mathrm{m\,s^{-1}}$ and 7.1 m, respectively.

(d) When the ball has fallen back to the point from which it was thrown, $s = 0$. Substituting this value of s into the equation found in part (b) gives

$$0 = 12t - 4.9t^2$$
$$12t - 4.9t^2 = 0$$
$$t(12 - 4.9t) = 0$$
$$t = 0 \text{ or } t = 12/4.9 = 2.4 \text{ (to 1 d.p.).}$$

So the ball takes about 2.4 seconds to fall back to the point from which it was thrown.

(The displacement–time graph for the ball is shown below. You can see that the ball first rises and then falls, as expected. You can also see that it reaches a height of about 7 m after 1 second, and that it returns to the point from which it was thrown after about 2.4 seconds, as you would expect from the answers found in parts (c) and (d).)

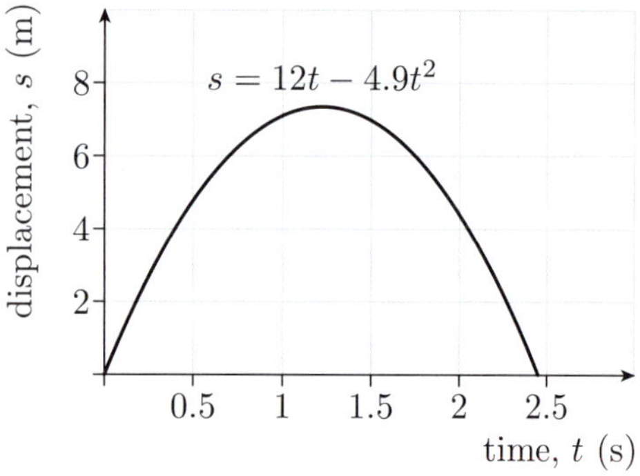

Solution to Activity 34

(a) The given equation for m in terms of q is

$$m = 250 - 0.5q.$$

Since marginal cost is the derivative of total cost, the equation for t in terms of q is

$$t = 250q - 0.5 \times \tfrac{1}{2}q^2 + c,$$

that is,

$$t = 250q - 0.25q^2 + c,$$

where c is a constant.

According to the information given in the question, when $q = 80$, $t = 30\,400$.

Substituting these values into the equation found for t above gives

$$30\,400 = 250 \times 80 - 0.25 \times 80^2 + c$$
$$30\,400 = 18\,400 + c$$
$$c = 12\,000.$$

Hence the equation for t in terms of q is

$$t = 250q - 0.25q^2 + 12\,000.$$

(b) The total weekly cost of producing 160 tonnes of fertiliser each week is

$$£(250 \times 160 - 0.25 \times 160^2 + 12\,000)$$
$$= £45\,600.$$

(c) The question states that the total weekly cost of producing 80 tonnes of fertiliser each week is £30 400. Hence the cost per tonne of producing 80 tonnes of fertiliser each week is

$$\frac{£30400}{80} = £380.$$

It was calculated in part (b) that the total weekly cost of producing 160 tonnes of fertiliser each week is £45 600. Hence the cost per tonne of producing 160 tonnes of fertiliser each week is

$$\frac{£45\,600}{160} = £285.$$

(The graph of t against q is shown below. You can see that the total weekly cost of producing 160 tonnes of fertiliser each week is about £45 000, as expected.)

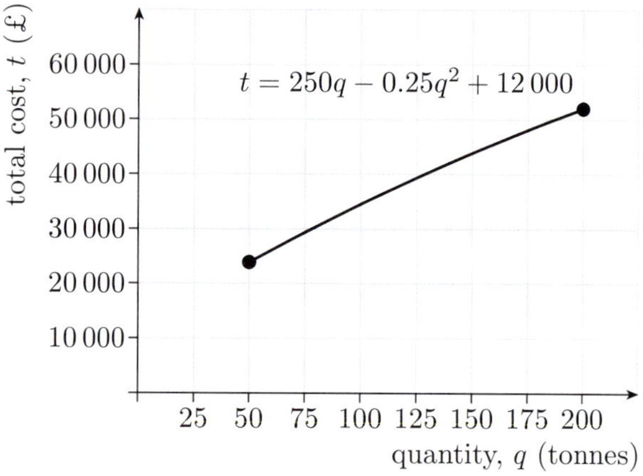

Solution to Activity 35

An antiderivative of f is

$$F(x) = 8e^{x/8}.$$

So the change in the quantity from $x = 2$ to $x = 5$ is

$$F(5) - F(2) = 8e^{5/8} - 8e^{2/8}$$
$$= 4.67 \text{ (to 3 s.f.)}.$$

Solution to Activity 36

An antiderivative of f is

$$F(t) = 3t - \tfrac{3}{2} \times \tfrac{1}{2}t^2 = 3t - \tfrac{3}{4}t^2.$$

(a) The change in the displacement (in metres) of the object from $t = 0$ to $t = 1$ is

$$F(1) - F(0) = (3 \times 1 - \tfrac{3}{4} \times 1^2) - (3 \times 0 - \tfrac{3}{4} \times 0^2)$$
$$= 2.25.$$

That is, the object travels 2.25 m in that time (in the positive direction).

(b) The change in the displacement (in metres) of the object from $t = 4$ to $t = 5$ is

$$\begin{aligned} F(5) - F(4) &= (3 \times 5 - \tfrac{3}{4} \times 5^2) \\ &\quad - (3 \times 4 - \tfrac{3}{4} \times 4^2) \\ &= (15 - \tfrac{75}{4}) - (12 - 12) \\ &= -3.75. \end{aligned}$$

That is, the object travels 3.75 m in that time (in the negative direction).

Solution to Activity 37

(a) The function $f(x) = \dfrac{4}{x}$ has indefinite integral

$$F(x) = 4 \ln |x| + c.$$

(b) The function

$$f(x) = \frac{x+1}{x^2} = \frac{1}{x} + \frac{1}{x^2} = \frac{1}{x} + x^{-2}$$

has indefinite integral

$$F(x) = \ln |x| - x^{-1} + c = \ln |x| - \frac{1}{x} + c.$$

(c) The function

$$f(x) = \frac{1}{2x} \quad (x > 0)$$

has indefinite integral

$$F(x) = \tfrac{1}{2} \ln x + c.$$

(The answer $F(x) = \frac{1}{2} \ln |x| + c$ is also correct, but can be simplified, as x takes only positive values.)

Solution to Activity 38

(a) The indefinite integral of $h(\theta) = 3\cos\theta + 4\sin\theta$ is

$$\begin{aligned} H(\theta) &= 3\sin\theta + 4(-\cos\theta) + c \\ &= 3\sin\theta - 4\cos\theta + c. \end{aligned}$$

(b) The indefinite integral of $g(\phi) = 6 - 5\,\mathrm{cosec}^2\,\phi$ is

$$\begin{aligned} G(\phi) &= 6\phi - 5(-\cot\phi) + c \\ &= 6\phi + 5\cot\phi + c. \end{aligned}$$

(c) The indefinite integral of $f(x) = \dfrac{7}{x}$ is

$$F(x) = 7 \ln |x| + c.$$

(d) The function is

$$g(t) = \frac{1}{3t},$$

which can be written as

$$g(t) = \frac{1}{3} \times \frac{1}{t}.$$

Hence its indefinite integral is

$$G(t) = \frac{1}{3} \ln |t| + c.$$

(e) The function is

$$f(x) = \frac{1}{\pi x} \quad (x > 0),$$

which can be written as

$$f(x) = \frac{1}{\pi} \times \frac{1}{x} \quad (x > 0).$$

Hence its indefinite integral is

$$F(x) = \frac{1}{\pi} \ln x + c.$$

(f) The indefinite integral of $f(x) = \dfrac{3}{1+x^2}$ is

$$F(x) = 3\tan^{-1} x + c.$$

(g) The indefinite integral of $h(t) = \dfrac{1}{4\sqrt{1-t^2}}$ is

$$H(t) = \tfrac{1}{4} \sin^{-1} t + c.$$

(Or, alternatively, $H(t) = -\frac{1}{4} \cos^{-1} t + c$.)

(h) The function is

$$p(x) = \frac{1}{5 + 5x^2},$$

which can be written as

$$p(x) = \frac{1}{5} \times \frac{1}{1+x^2}.$$

Hence its indefinite integral is

$$P(x) = \tfrac{1}{5} \tan^{-1} x + c.$$

(i) The function is

$$g(x) = \frac{1}{\sqrt{4 - 4x^2}},$$

which can be written as

$$g(x) = \frac{1}{\sqrt{4}\sqrt{1-x^2}};$$

that is,

$$g(x) = \frac{1}{2\sqrt{1-x^2}}.$$

Hence its indefinite integral is

$$G(x) = \tfrac{1}{2} \sin^{-1} x + c.$$

(Or, alternatively, $G(x) = -\frac{1}{2} \cos^{-1} x + c$.)

(j) The function is

$$q(x) = 5(x-3)(2x-1),$$

which can be written as

$$q(x) = 5(2x^2 - 7x + 3).$$

Hence its indefinite integral is

$$Q(x) = 5(2 \times \tfrac{1}{3}x^3 - 7 \times \tfrac{1}{2}x^2 + 3x) + c$$
$$= \tfrac{10}{3}x^3 - \tfrac{35}{2}x^2 + 15x + c.$$

(It could also be written as

$$Q(x) = \tfrac{5}{6}x(4x^2 - 21x + 18) + c,$$

for example.)

(k) The function is

$$f(x) = \frac{x-2}{x},$$

which can be written as

$$f(x) = 1 - \frac{2}{x}.$$

Hence its indefinite integral is

$$F(x) = x - 2\ln|x| + c.$$

(l) The function is

$$g(x) = \frac{x-2}{x^3},$$

which can be written as

$$g(x) = x^{-2} - 2x^{-3}.$$

Hence its indefinite integral is

$$G(x) = -x^{-1} - 2 \times \frac{1}{-2}x^{-2} + c$$
$$= -\frac{1}{x} + \frac{1}{x^2} + c$$
$$= -\frac{x}{x^2} + \frac{1}{x^2} + c$$
$$= \frac{1-x}{x^2} + c.$$

(m) The indefinite integral of $f(x) = 8e^x$ is

$$F(x) = 8e^x + c.$$

(n) The function is

$$f(x) = e^{1+x},$$

which can be written as

$$f(x) = e \times e^x.$$

Hence its indefinite integral is

$$F(x) = e \times e^x + c$$
$$= e^{1+x} + c.$$

(You can integrate the expression $e \times e^x$ by using the constant multiple rule for antiderivatives, since e is a constant.)

(o) The indefinite integral of $r(\phi) = -\operatorname{cosec}\phi \cot\phi$ is

$$R(\phi) = -(-\operatorname{cosec}\phi) + c$$
$$= \operatorname{cosec}\phi + c.$$

(p) The function is

$$r(\theta) = \sec\theta(\sec\theta + \tan\theta)$$

which can be written as

$$r(\theta) = \sec^2\theta + \sec\theta\tan\theta.$$

Hence its indefinite integral is

$$R(\theta) = \tan\theta + \sec\theta + c.$$

Acknowledgements

Grateful acknowledgement is made to the following sources:

Page 30: Mark Hobbs, for the idea for the cartoon

Page 55: © gallica.bnf.fr / Bibliothèque nationale de France

Every effort has been made to contact copyright holders. If any have been inadvertently overlooked the publishers will be pleased to make the necessary arrangements at the first opportunity.

Unit 8

Integration methods

Introduction

In this unit you'll continue your study of integration. In Unit 7 you saw that integration is the reverse of differentiation. You saw that you can use it to solve problems in which you know the values taken by the rate of change of a continuously changing quantity, and you want to work out the values taken by the quantity, or how much the quantity changes over some period. For example, you saw that if you have a formula for the velocity of an object in terms of time, then you can use integration to find a formula for the displacement of the object in terms of time, or work out the change in the displacement of the object over some period of time.

In Sections 1 and 2 of this unit, you'll meet a different way to think about integration. This way of thinking about it involves areas of regions of the plane, rather than rates of change. You'll see that the link between the two ways of thinking about integration helps you to solve problems both about areas and about rates of change.

In Sections 3 to 5 of the unit, you'll learn some more methods for integrating functions, which you can use in addition to the methods that you learned in Unit 7. These further methods will allow you to integrate a much wider variety of functions than you've learned to integrate so far.

Many students find some of the material in this unit quite challenging when they first meet it, so don't worry if that's the case for you too. As with most mathematics, it will seem more straightforward once you're more familiar with it. If you find a particular technique difficult, then make sure that you watch the associated tutorial clips, and use the practice quiz and exercise booklet for this unit for more practice.

This unit is likely to take you more time to study than most of the other units, so the study planner allows extra time for it.

1 Areas, signed areas and definite integrals

In this section you'll look at a type of mathematical problem that you might think has little connection with the mathematics of rates of change that you studied in Units 6 and 7. In Section 2 you'll see that in fact the two areas of mathematics are closely related.

Throughout Sections 1 and 2, we'll work with *continuous* functions. Remember that a **continuous** function is a function whose graph has no breaks in it, over its whole domain. Informally, it's a function whose graph you can draw without taking your pen off the paper (though you might need an infinitely large piece of paper, as the graph might be infinitely long!).

1.1 Areas and signed areas

Sometimes it's useful to calculate the area of a flat shape with a curved boundary. For example, suppose that an architect is designing a large building shaped as shown in Figure 1(a). The curve of the roof is given by the graph of the function

$$f(x) = 9 - \tfrac{1}{36}x^2 \quad (-18 \le x \le 18),$$

where x and $f(x)$ are measured in metres. This graph is shown in Figure 1(b).

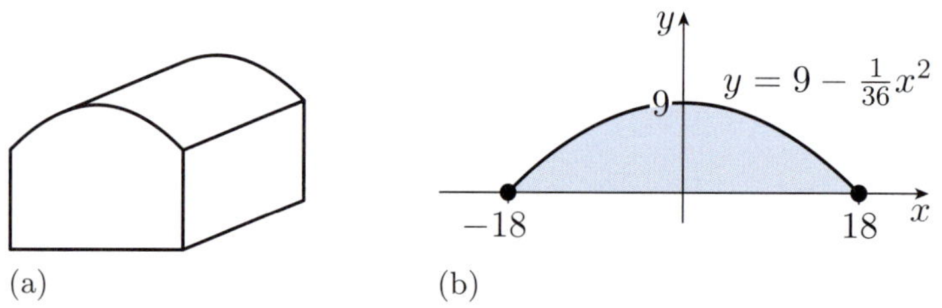

Figure 1 (a) The shape of a building (b) the graph that gives the curve of the roof

Suppose that the architect wants to calculate the area shaded in Figure 1(b), so that she can determine the amounts of materials needed to build the end wall of the building.

Here's a way of calculating an approximate value for this area, which you can make as accurate as you wish. You start by dividing the interval $[-18, 18]$, which is the interval of x-values that corresponds to the curve, into a number of subintervals of equal width.

For example, in Figure 2 the interval has been divided into twelve subintervals. The right endpoint of the first subinterval is equal to the left endpoint of the next subinterval, and so on.

For each subinterval, you approximate the shape of the curve on that subinterval by a horizontal line segment whose height is the value of the function at the left endpoint of the subinterval, as shown in Figure 2. You calculate the areas of the rectangles between these line segments and the x-axis (simply by multiplying their heights by their widths in the usual way), and add up all these areas to give you an approximate value for the required area. The more subintervals you use, the better will be the approximation.

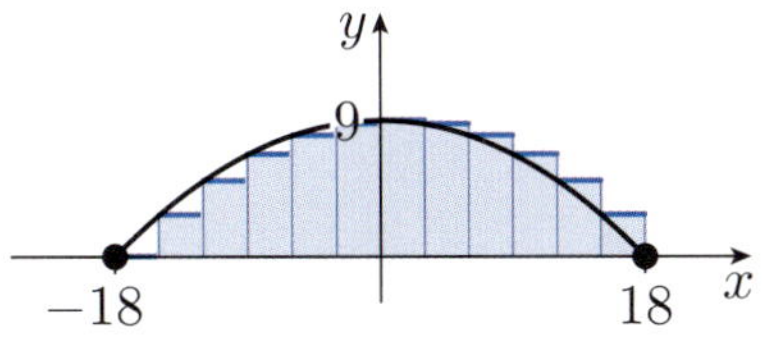

Figure 2 A collection of rectangles whose total area is approximately the area in Figure 1(b)

In the next example this method is used, with four subintervals, to find an approximate value for the area discussed above. Notice that, in both the example and in Figure 2 above, the rectangle corresponding to the first subinterval has zero height and therefore zero area – this happens because the value of the function at the left endpoint of this subinterval is zero.

Example 1 *Calculating an approximate value for an area*

Use the method described above, with four subintervals as shown below, to find an approximate value for the area between the graph of the function $f(x) = 9 - \frac{1}{36}x^2$ and the x-axis.

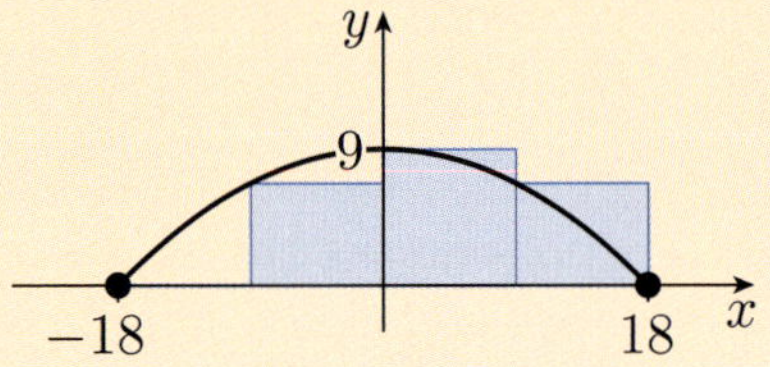

Solution

We divide the interval $[-18, 18]$ into four subintervals of equal width. The whole interval has width 36, so each subinterval has width $36/4 = 9$.

The left endpoints of the four subintervals are

$$-18, \quad -18+9, \quad -18+2\times 9, \quad -18+3\times 9,$$

that is

$$-18, \quad -9, \quad 0, \quad 9.$$

So the heights of the rectangles are

$$f(-18), \quad f(-9), \quad f(0), \quad f(9),$$

that is,

$$9 - \tfrac{1}{36}\times(-18)^2, \quad 9 - \tfrac{1}{36}\times(-9)^2, \quad 9 - \tfrac{1}{36}\times 0^2, \quad 9 - \tfrac{1}{36}\times 9^2,$$

which evaluate to

$$0, \quad \tfrac{27}{4}, \quad 9, \quad \tfrac{27}{4}.$$

So we obtain the following approximate value for the area:

$$(0\times 9) + (\tfrac{27}{4}\times 9) + (9\times 9) + (\tfrac{27}{4}\times 9) = \frac{405}{2} = 202.5.$$

The calculation in Example 1 shows that a (rather crude) approximate value for the area of the cross-section of the roof discussed at the beginning of this subsection is $202.5\,\text{m}^2$.

Note that the units for the area of a region on a graph are, as you'd expect, the units on the vertical axis times the units on the horizontal axis. For example, if the units on both axes are metres, then the units for area are square metres (m^2). If a graph has no specific units on the axes, then the units for area are simply 'square units', and we usually omit them when we state an area.

In the next activity you're asked to find another approximation for the area of the cross-section of the roof discussed earlier, by using the subinterval method with six subintervals instead of four.

Activity 1 *Calculating an approximate value for an area*

Use the method described above, with six subintervals as shown below, to find an approximate value for the area between the graph of the function $f(x) = 9 - \frac{1}{36}x^2$ and the x-axis.

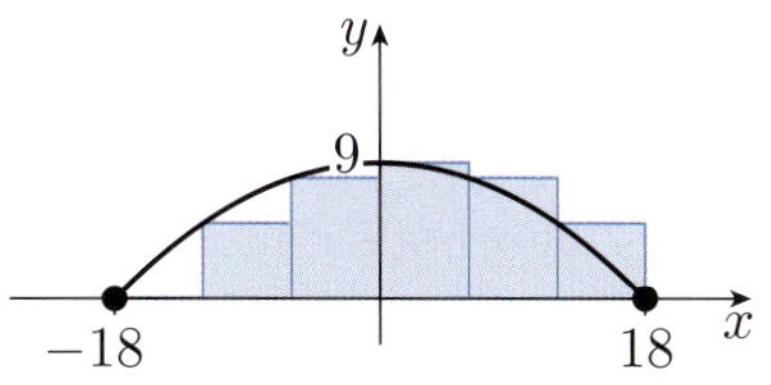

Table 1 shows the approximate values for the area between the graph of the function $f(x) = 9 - \frac{1}{36}x^2$ and the x-axis that were found in Example 1 and in the solution to Activity 1. It also shows, to three decimal places, some further approximate values that were found in the same way, but using larger numbers of subintervals. The calculations were carried out using a computer.

Table 1 Approximate values for the area in Figure 1(b)

Number of subintervals	4	6	12	50	100	500
Approximation obtained	202.5	210	214.5	215.914	215.978	215.999

You can see that as the number of subintervals gets larger and larger, the approximation obtained seems to be getting closer and closer to a particular number. We say that this number is the **limit** of the approximations as the number of subintervals **tends to infinity**. It's the *exact value* of the area between the graph of the function $f(x) = 9 - \frac{1}{36}x^2$ and the x-axis. From Table 1, it looks as if the exact value of this area is 216, or perhaps a number very close to 216.

In general, if you have a continuous function f whose graph lies on or above the x-axis throughout an interval $[a, b]$, then you can use the method demonstrated above to find, as accurately as you want, the area between the graph of f and the x-axis, from $x = a$ to $x = b$. Figure 3(a) illustrates an area of this type, and Figure 3(b) illustrates the approximation to the area that's obtained by using 8 subintervals.

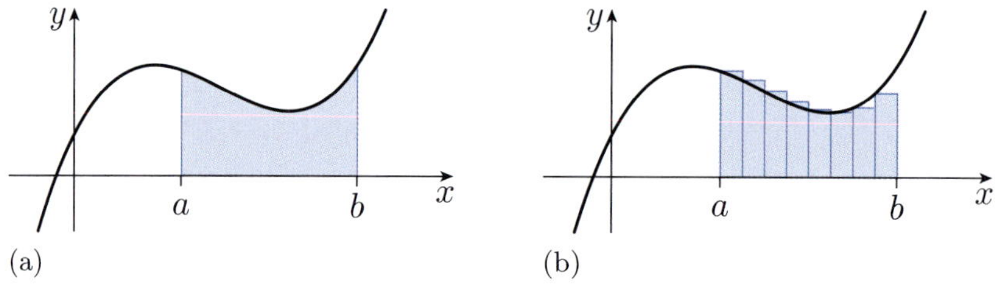

Figure 3 (a) The area between a graph and the x-axis, from $x = a$ to $x = b$ (b) A collection of rectangles whose total area is approximately this area

Here's a summary of the method that you've met for finding an approximate value for an area of this type. You start by dividing the interval $[a, b]$ into a number, say n, of subintervals of equal width. The width of each subinterval is then $(b - a)/n$. For each subinterval you calculate the product

$$f\begin{pmatrix}\text{left endpoint}\\ \text{of subinterval}\end{pmatrix} \times \frac{b-a}{n}, \tag{1}$$

and you add up all these products. In general, the larger the number n of subintervals, the closer your answer will be to the area between the graph of f and the x-axis, from $x = a$ to $x = b$.

Now suppose that you have a continuous function f whose graph lies on or *below* the x-axis throughout an interval $[a, b]$, and you want to calculate the area between the graph of f and the x-axis, from $x = a$ to $x = b$, as illustrated in Figure 4(a).

You can use the method above, with a small adjustment at the end, to calculate an approximate value for an area like this. Consider what happens when you apply the method to a graph like the one in Figure 4(a). You start by dividing the interval $[a, b]$ into n subintervals of equal width, then you calculate all the products of form (1) and add them all up. For a graph like this, because the value of f at the left endpoint of each subinterval is *negative* (or possibly zero), each product of form (1) will also be negative (or zero). In fact, each product of form (1) will be the *negative* of the area between the line segment that approximates the curve and the x-axis, as illustrated in Figure 4(b). So when you add up all the products of form (1), you'll obtain an approximate value for the *negative* of the area between the curve and the x-axis, from $x = a$ to $x = b$.

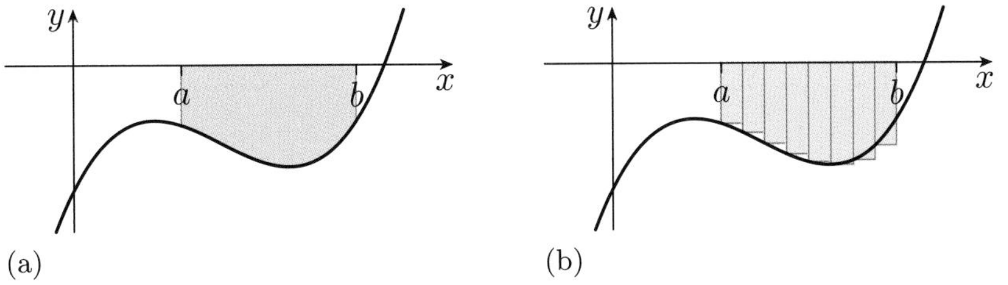

Figure 4 (a) The area between another graph and the x-axis, from $x = a$ to $x = b$ (b) A collection of rectangles whose total area is approximately this area

That's not a problem, because you can simply remove the minus sign to obtain the approximate value for the area that you want. However, to help us deal with situations like this, it's useful to make the following definitions. These definitions will be important throughout the unit.

Consider any region on a graph that lies either entirely above or entirely below the x-axis. The **signed area** of the region is its area with a plus or minus sign according to whether it lies above or below the x-axis, respectively. For example, in Figure 5 the two shaded regions above the x-axis have signed areas $+4$ and $+6$, respectively, which you can write simply as 4 and 6, and the shaded region below the x-axis has signed area -3.

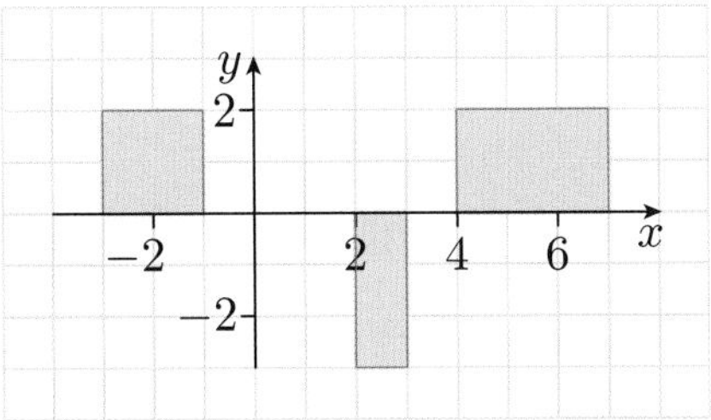

Figure 5 Regions on a graph

If you have a collection of regions on a graph, where each region in the collection lies either entirely above or entirely below the x-axis, then the total signed area of the collection is the sum of the signed areas of the individual regions. For example, the total signed area of the collection of three regions in Figure 5 is $4 + (-3) + 6 = 7$.

The units for signed area on a graph are, as you'd expect, the same as the units for area on the graph. That is, they're the units on the vertical axis times the units on the horizontal axis. If there are no specific units on the axes, then we don't use any specific units for signed area.

Positive signed area

Negative signed area

Example 2 *Understanding signed areas*

The areas of some regions on a graph are marked below.

In each part (a)–(c), use these areas to find the signed area between the graph and the x-axis, from the first value of x to the second value of x.

(a) From $x = -3$ to $x = -2$.

(b) From $x = -2$ to $x = 1$.

(c) From $x = -3$ to $x = 1$.

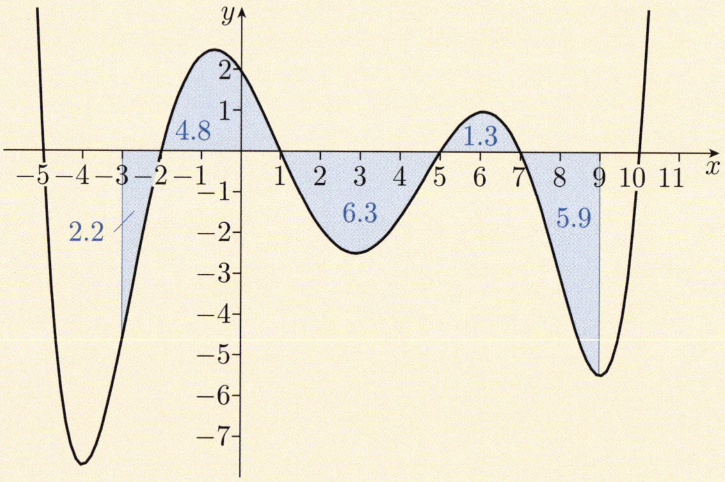

Solution

(a) The signed area from $x = -3$ to $x = -2$ is -2.2.

(b) The signed area from $x = -2$ to $x = 1$ is $+4.8 = 4.8$.

(c) The signed area from $x = -3$ to $x = 1$ is $-2.2 + 4.8 = 2.6$.

Of course, the signed area value found in Example 2(c) doesn't correspond to any actual area on the graph, as it's the sum of a positive signed area and a negative signed area.

Activity 2 *Understanding signed areas*

Consider again the graph in Example 2. In each of parts (a)–(d) below, use the given areas to find the signed area between the graph and the x-axis, from the first value of x to the second value of x.

(a) From $x = 1$ to $x = 5$.

(b) From $x = 5$ to $x = 7$.

(c) From $x = 1$ to $x = 7$.

(d) From $x = 1$ to $x = 9$.

With the definition of signed area that you've now seen, the method that you've met in this subsection can be described concisely as follows.

Strategy:
To find an approximate value for the signed area between the graph of a continuous function f and the x-axis, from $x = a$ to $x = b$

Divide the interval between a and b into n subintervals, each of width $(b - a)/n$. For each subinterval, calculate the product

$$f\left(\begin{array}{c}\text{endpoint of subinterval}\\ \text{nearest } a\end{array}\right) \times \frac{b-a}{n},$$

and add up all these products.

In general, the larger the number n of subintervals, the closer your answer will be to the required signed area.

Notice that the box above uses the phrase 'endpoint of subinterval nearest a', rather than 'left endpoint of subinterval', which means the same thing. This will be convenient later in this subsection. (Note that, in this phrase, it's the endpoint of the subinterval, not the subinterval itself, that's nearest a!)

Figures 6 and 7 illustrate the strategy above. If you add up the signed areas of all the rectangles in Figure 6(a), then you'll obtain an approximate value for the signed area in Figure 6(b).

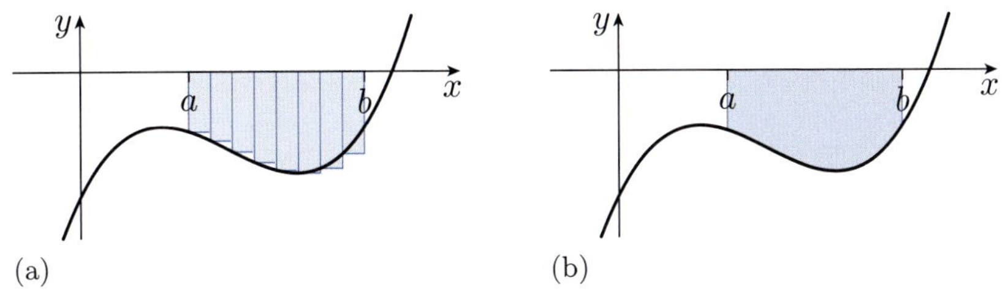

Figure 6 (a) A collection of rectangles whose total signed area is approximately the signed area shown in (b)

Similarly, if you add up the signed areas of all the rectangles in Figure 7(a), then you'll obtain an approximate value for the signed area in Figure 7(b).

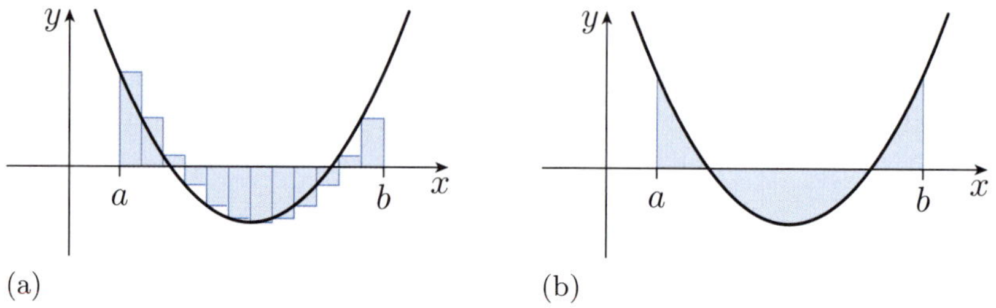

Figure 7 (a) Another instance of a collection of rectangles whose total signed area is approximately the signed area shown in (b)

Activity 3 *Calculating an approximate value for a signed area*

Use the method described in the box above, with six subintervals as shown on the left below, to find an approximate value for the signed area between the graph of the function $f(x) = 3 - x^2$ and the x-axis from $x = -3$ to $x = 3$, as shown on the right below.

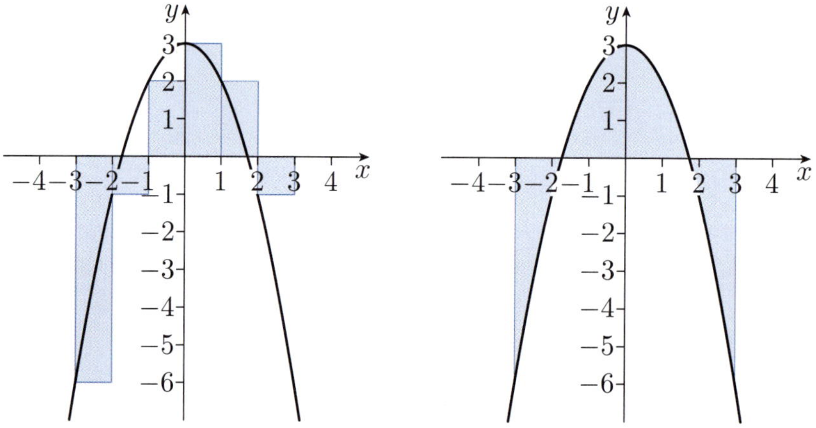

In the next activity you can use a computer to explore approximations to signed areas found by using subintervals in the way that you've seen.

Activity 4 *Calculating more approximate values for signed areas*

Open the applet *Approximations for signed areas*. Check that the function is set to $f(x) = 3 - x^2$, the values of a and b are set to -3 and 3 respectively, and the number of subintervals is set to 6. The resulting approximation for the signed area between the graph of f and the x-axis, from $x = a$ to $x = b$, should then be the value found in the solution to Activity 3.

Now increase the number of subintervals, and observe the effect on the approximation. What do you think is the exact value of the signed area?

Experiment by changing the function, the values of a and b, and the number of subintervals, to find approximate values for some other signed areas.

So far in this subsection, whenever a signed area from $x = a$ to $x = b$ has been discussed, the value of b has been greater than the value of a.

The value of b can also be *equal* to the value of a, and this gives a signed area of zero, as illustrated in Figure 8. In cases like this, the method that you've seen for using subintervals to find approximate values for signed areas gives the answer zero, since the width $(b - a)/n$ of each subinterval is zero.

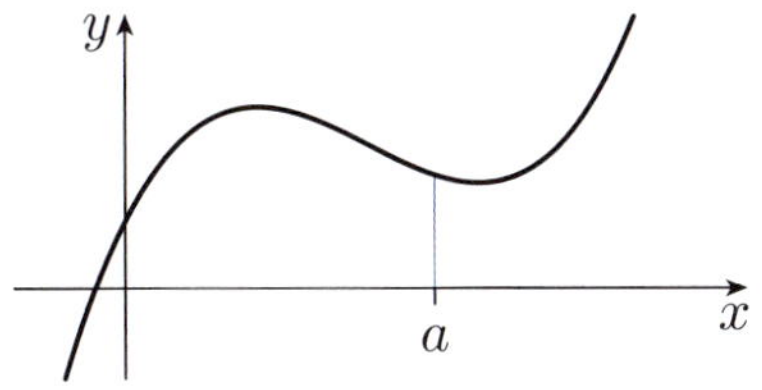

Figure 8 A signed area equal to zero

In fact, for reasons that you'll see later in the unit, it's useful to extend the definition of signed area a little, to give a meaning to the phrase 'the signed area between the graph of f and the x-axis from $x = a$ to $x = b$' when b is *less* than a. It's not immediately obvious what the phrase means in this case, but it turns out that the natural meaning is as follows. This definition will be important throughout the unit.

Suppose that f is a continuous function whose domain includes the interval $[b, a]$, as illustrated in Figure 9. Then the **signed area** between the graph of f and the x-axis from $x = a$ to $x = b$ is defined to be the *negative* of the signed area between the graph of f and the x-axis from $x = b$ to $x = a$.

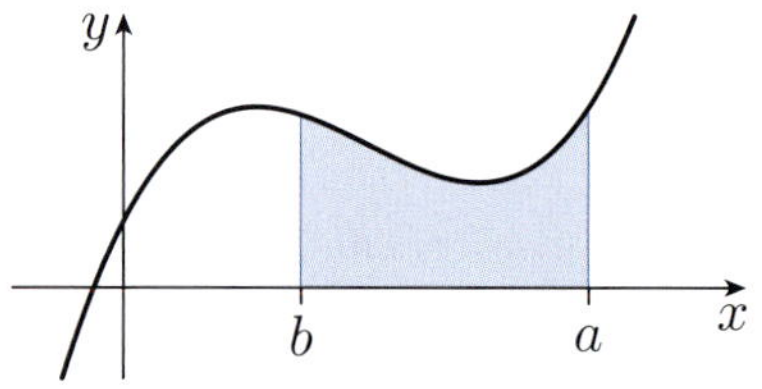

Figure 9 The graph of a continuous function f whose domain includes the interval $[b, a]$

For example, consider the graph in Figure 10. The signed area from $x = 1$ to $x = 3$ is 4, so the signed area from $x = 3$ to $x = 1$ is -4. Similarly, the signed area from $x = 6$ to $x = 8$ is -5, so the signed area from $x = 8$ to $x = 6$ is $-(-5) = 5$.

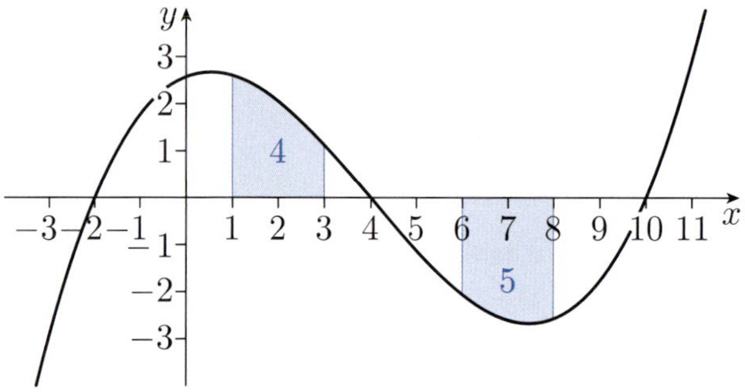

Figure 10 Areas on a graph

Because of this extended definition of signed area, you now have to think a little more carefully when you read or use the phrase 'signed area' (and the signed area isn't zero). If there's no mention of 'from $x = a$ to $x = b$', then to decide whether the signed area is positive or negative you just need to consider whether it lies above or below the x-axis. However, if the text is of the form 'the signed area from $x = a$ to $x = b$', then to decide whether the signed area is positive or negative you also have to consider whether b is greater than or less than a.

This is illustrated in the next example. A helpful way to think about whether b is greater than or less than a is to think about whether x moves forward or backward as it changes from $x = a$ to $x = b$.

Example 3 *Understanding the extended definition of signed area*

The areas of some regions on a graph are marked below. In each of parts (a)–(d), use the given areas to find the signed area between the graph and the x-axis, from the first value of x to the second value of x.

(a) From $x = 3$ to $x = 3$. (b) From $x = 6$ to $x = 10$.

(c) From $x = 10$ to $x = 6$. (d) From $x = 6$ to $x = -1$.

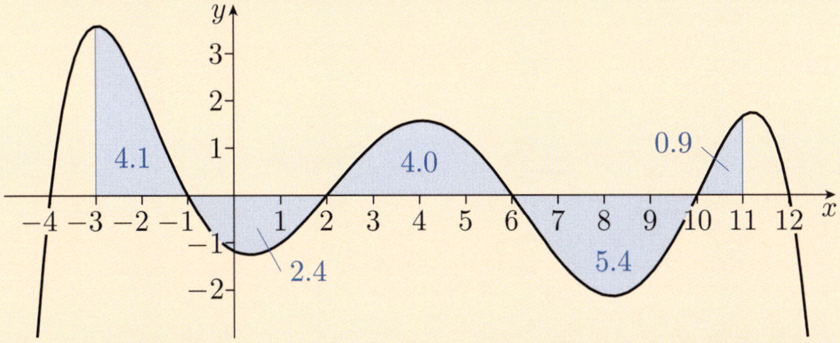

Solution

(a) From 3 to 3 is from a number to the same number.

The signed area from $x = 3$ to $x = 3$ is 0.

(b) From 6 to 10 is forward.

The signed area from $x = 6$ to $x = 10$ is -5.4.

(c) From 10 to 6 is backward.

The signed area from $x = 6$ to $x = 10$ is -5.4, so the signed area from $x = 10$ to $x = 6$ is 5.4.

(d) From 6 to -1 is backward.

The signed area from $x = -1$ to $x = 6$ is $-2.4 + 4.0 = 1.6$, so the signed area from $x = 6$ to $x = -1$ is -1.6.

In each part of the next activity, start by thinking about whether x moves forward or backward (or stays fixed) as it moves from the first value of x to the second value of x.

Activity 5 *Understanding the extended definition of signed area*

The graph in Example 3 is repeated below. In each of parts (a)–(g), use the given areas to find the signed area between the graph and the x-axis, from the first value of x to the second value of x.

(a) From $x = 2$ to $x = 6$.
(b) From $x = 6$ to $x = 2$.
(c) From $x = 2$ to $x = 10$.
(d) From $x = 10$ to $x = 2$.
(e) From $x = 8$ to $x = 8$.
(f) From $x = 2$ to $x = -3$.
(g) From $x = 10$ to $x = -1$.

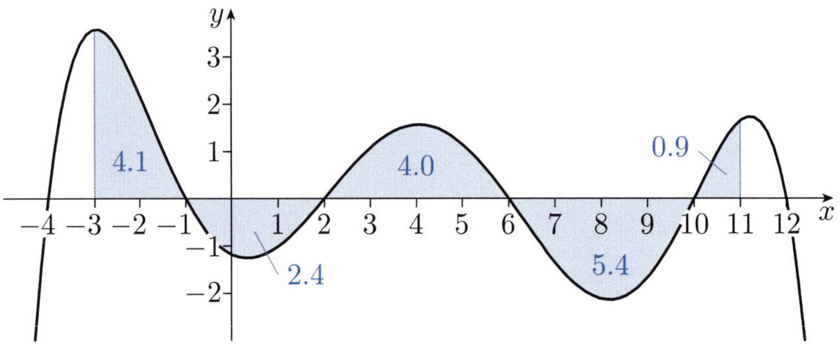

The subinterval method for finding approximate values for signed areas, which is summarised in the box on page 113, applies no matter whether the value of b is greater than, less than, or equal to the value of a. The reason why it works in cases where b is less than a is that in these cases the 'width' $(b - a)/n$ of each subinterval is *negative*.

You might like to use the applet *Approximations for signed areas* to see examples of the method giving approximate values for signed areas in cases where b is less than a.

Of course, if you want to work out an area between a curve and the x-axis over an interval $[a, b]$ by using the method that you've seen in this subsection, then normally you would work out the signed area from a to b, rather than the signed area from b to a. However, it's important to understand what's meant by the signed area from b to a, for reasons that you'll see later in the unit.

1.2 Definite integrals

Before you learn more about why signed areas are important, it's useful for you to learn some terminology and notation that are used when working with them.

If f is a continuous function and a and b are numbers in its domain, then the signed area between the graph of f and the x-axis from $x = a$ to $x = b$ is called the **definite integral** of f from a to b, and is denoted by

$$\int_a^b f(x)\,dx.$$

This notation is read as 'the integral from a to b of f of x, d x'. The numbers a and b are called the **lower** and **upper limits of integration**, respectively.

For example, for the function f in Figure 11,

$$\int_3^7 f(x)\,dx = -9, \quad \int_7^9 f(x)\,dx = 2,$$

$$\text{and} \quad \int_3^9 f(x)\,dx = -9 + 2 = -7.$$

Similarly, for the same function,

$$\int_4^4 f(x)\,dx = 0, \quad \text{and} \quad \int_7^3 f(x)\,dx = -(-9) = 9.$$

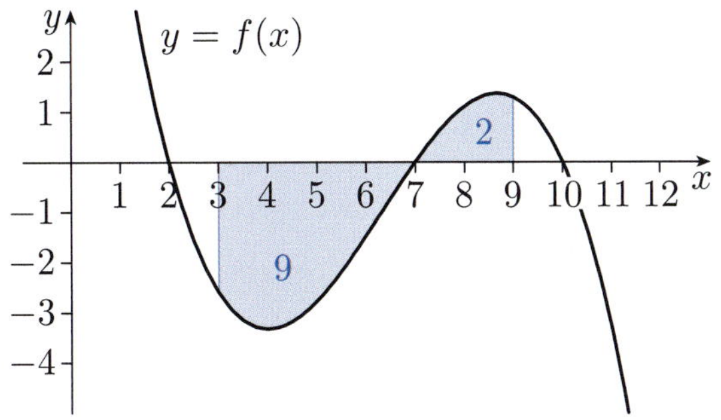

Figure 11 The graph of a function, and some areas

You'll see a little later in this subsection where this notation for a definite integral comes from.

The symbol $\displaystyle\int$ is called the **integral sign**. It sometimes appears in a smaller form, $\int$, when it's in a line of typed text.

As with Leibniz notation for derivatives, the 'd' in the notation for definite integrals has no independent meaning. In many texts, including this one, it's printed in upright type, rather than italic type, to emphasise that it's not a variable. You don't need to do anything special when you handwrite it.

When you use the notation $\int_a^b f(x)\,\mathrm{d}x$, remember that you must include not only the $\int_a^b$ at the beginning, but also the $\mathrm{d}x$ at the end. Try not to forget the $\mathrm{d}x$!

You've now seen that a *definite integral* is a type of signed area, and you saw in Unit 7 that an *indefinite integral* is a type of general antiderivative. These two seemingly unrelated concepts are, as you probably suspect, closely related. You'll see how in Section 2.

The box below lists some standard properties of definite integrals, which come from their definition as signed areas. These properties hold for all values of a, b and c in the domain of the continuous function f. The second property comes from the extended definition of signed area, to cases where b is less than a. The third property expresses the fact that the signed area from a to c is equal to the signed area from a to b plus the signed area from b to c. Figure 12 illustrates this property in a case where $a < b < c$, but the extended definition of signed area ensures that the property holds even when a, b and c aren't in this order. This is one reason why the definition of signed area is extended in the way that you've seen.

Standard properties of definite integrals

$$\int_a^a f(x)\,\mathrm{d}x = 0$$

$$\int_b^a f(x)\,\mathrm{d}x = -\int_a^b f(x)\,\mathrm{d}x$$

$$\int_a^c f(x)\,\mathrm{d}x = \int_a^b f(x)\,\mathrm{d}x + \int_b^c f(x)\,\mathrm{d}x$$

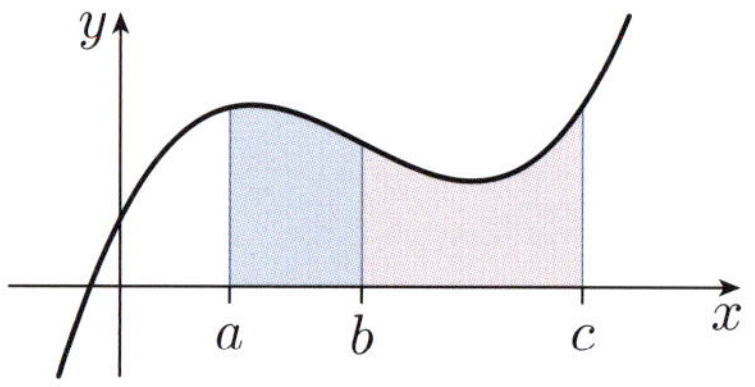

Figure 12 Two adjacent signed areas

Activity 6 *Understanding definite integrals*

Consider the function f whose graph is shown below. The areas of some regions are marked. By using these areas, write down the values of the following definite integrals.

Hint: notice that in some of these definite integrals the upper limit of integration is less than the lower limit of integration.

(a) $\int_{-5}^{-3} f(x)\,dx$ (b) $\int_{-3}^{2} f(x)\,dx$ (c) $\int_{2}^{7} f(x)\,dx$

(d) $\int_{-5}^{2} f(x)\,dx$ (e) $\int_{-3}^{7} f(x)\,dx$ (f) $\int_{-5}^{7} f(x)\,dx$

(g) $\int_{5}^{5} f(x)\,dx$ (h) $\int_{2}^{-3} f(x)\,dx$ (i) $\int_{7}^{2} f(x)\,dx$

(j) $\int_{-3}^{-5} f(x)\,dx$ (k) $\int_{7}^{-3} f(x)\,dx$ (l) $\int_{7}^{-5} f(x)\,dx$

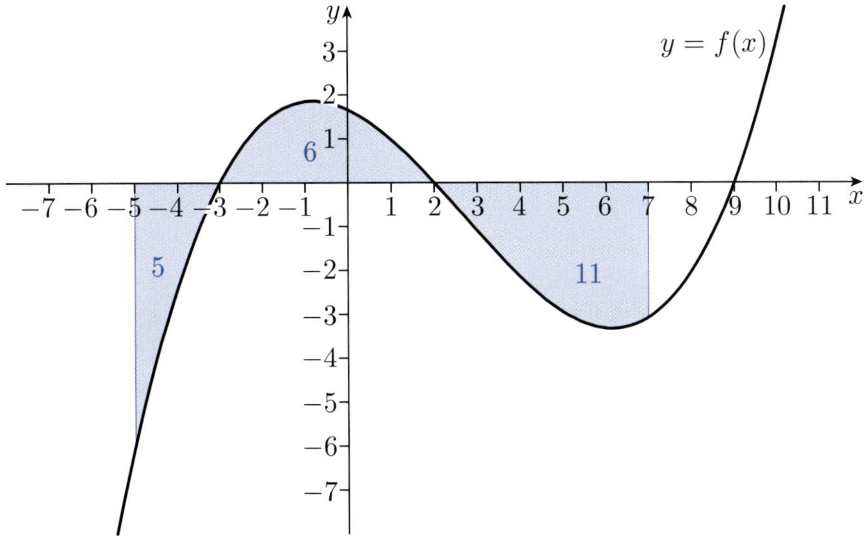

Activity 7 *Using a standard property of definite integrals*

Consider again the graph in Activity 6.

Given that $\int_{2}^{9} f(x)\,dx = -15$, find $\int_{7}^{9} f(x)\,dx$.

As you'd expect, you can replace the expression $f(x)$ in the notation for a definite integral by the formula for a particular function. For example, the signed areas in Figure 13 are denoted by

$$\int_{-1}^{1} x^2 \, dx \quad \text{and} \quad \int_{0}^{1} (x^3 - 1) \, dx,$$

respectively.

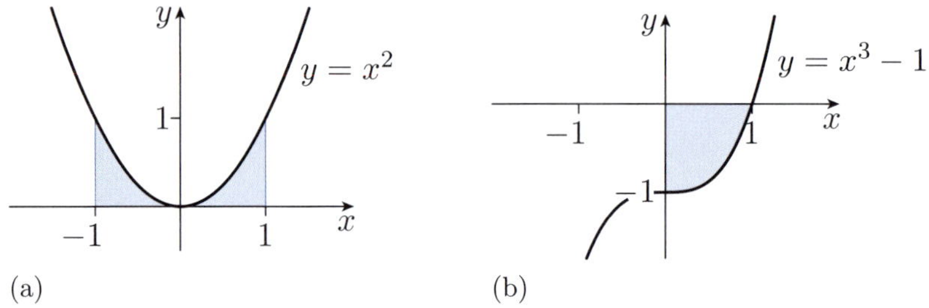

Figure 13 Signed areas between the graphs of particular functions and the x-axis

As always with algebraic notation, the notation for a definite integral can be used with letters other than the standard ones. For example, if g is a continuous function whose domain contains the numbers p and q, and you use t to denote the input variable of g, then the definite integral of g from $t = p$ to $t = q$ is denoted by

$$\int_{p}^{q} g(t) \, dt.$$

In fact, the input variable of the function in a definite integral is what's known as a **dummy variable** – you can change its name to any other variable name that you like, without affecting the value of the definite integral. For example, if f is a continuous function whose domain contains the numbers a and b, then

$$\int_{a}^{b} f(x) \, dx = \int_{a}^{b} f(t) \, dt = \int_{a}^{b} f(u) \, du,$$

and so on. As a particular example,

$$\int_{-1}^{1} x^2 \, dx = \int_{-1}^{1} t^2 \, dt = \int_{-1}^{1} u^2 \, du,$$

since all these definite integrals denote the signed area shown in Figure 13(a).

If f is any continuous function, and a and b are numbers in its domain, then you can find an approximate value for the definite integral $\int_a^b f(x) \, dx$, as accurately as you like, by using the method that you met in the last subsection. Here's the method expressed algebraically.

Suppose that you want to find $\int_a^b f(x)\,\mathrm{d}x$, as illustrated in Figure 14(a). You divide the interval between a and b into n subintervals, each of width $(b-a)/n$, as illustrated in Figure 14(b). We'll denote $(b-a)/n$ by w here, for conciseness.

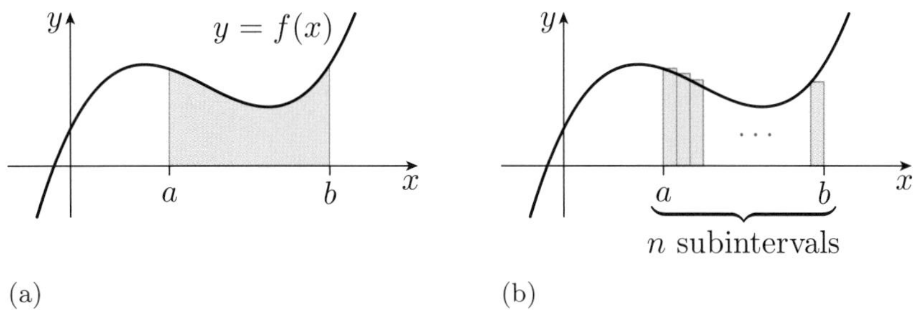

Figure 14 (a) A definite integral $\int_a^b f(x)\,\mathrm{d}x$ (b) A collection of n rectangles whose total signed area is approximately this definite integral

The endpoints nearest a of the subintervals are

$$a+0w, \quad a+1w, \quad a+2w, \quad \ldots, \quad a+(n-1)w.$$

(Remember that the strategy on page 113 specifies 'endpoint of subinterval nearest a', rather than 'left endpoint of subinterval'. The reason for this is that it ensures that the algebraic expressions for the endpoints stated above are correct when b is less than a, as well as when b is greater than a or equal to a.)

So an approximate value for the definite integral of f from $x=a$ to $x=b$ is given by

$$f(a+0w)\times w + f(a+1w)\times w + f(a+2w)\times w + \cdots$$
$$\cdots + f(a+(n-1)w)\times w.$$

This expression can be simplified slightly by taking out the common factor w. Doing this (and writing the common factor at the end) gives

$$\Big(f(a+0w)+f(a+1w)+f(a+2w)+\cdots+f(a+(n-1)w)\Big)w.$$

As the value of n gets larger and larger, the value of this expression gets closer and closer to the value of $\int_a^b f(x)\,\mathrm{d}x$.

The *limit* of the value of the expression, as the value of n tends to infinity, is the *exact* value of $\int_a^b f(x)\,\mathrm{d}x$. This idea is summarised below. The notation '$\lim\limits_{n\to\infty}$' means 'the limit, as n tends to infinity, of'.

Algebraic definition of a definite integral

Suppose that f is a continuous function and a and b are numbers in its domain. Then the **definite integral** of f from $x = a$ to $x = b$ is given by the equation

$$\int_a^b f(x)\,\mathrm{d}x = \lim_{n\to\infty}\Big(f(a+0w) + f(a+1w) + f(a+2w) + \cdots \\ \cdots + f(a+(n-1)w)\Big)w$$

where $w = (b-a)/n$.

If f is a particular continuous function, and a and b are numbers in its domain, then you can sometimes calculate an exact value for $\int_a^b f(x)\,\mathrm{d}x$ by using an algebraic method to find the limit in the box above. In some cases you might even be able to obtain a general formula for $\int_a^b f(x)\,\mathrm{d}x$, in terms of a and b. However, this is usually a complicated and difficult process, so normally we don't do it. Instead, if you want to evaluate a definite integral $\int_a^b f(x)\,\mathrm{d}x$, then you have two main options.

The first option is to obtain an approximate value by using subintervals in the way that you've seen. We usually use a computer to carry out the calculations. The computer algebra system that you're using in this module includes a command that instructs the computer to find an approximate value for a definite integral, to a particular level of precision, by using a version of the subinterval method that you've seen. You'll learn about this command in the final subsection of this unit.

The other option for evaluating a definite integral $\int_a^b f(x)\,\mathrm{d}x$ is to use an algebraic method that arises from the *fundamental theorem of calculus*, which you'll meet in Section 2. This algebraic method involves much less calculation than using subintervals, and it has the advantage of giving an exact answer. However, sometimes it's not possible or not practicable to use this method, as you'll see.

Now here's an explanation of where the notation and terminology for definite integrals come from. They arise directly from the method that you've seen for calculating approximate values for definite integrals. To see how, consider applying the method to a function f, as illustrated in Figure 15(a).

Consider any one of the subintervals. You could denote its endpoint nearest a by x, and its width by δx, as shown in Figure 15(b). As you saw in Unit 6, the notation δx represents a small change in x.

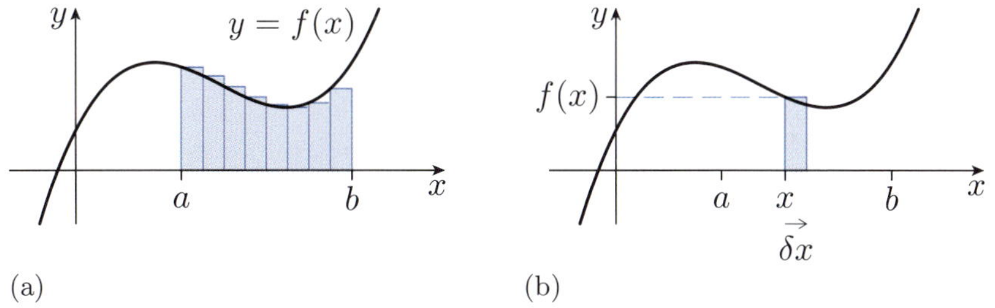

Figure 15 (a) The subinterval method applied to a function f from $x = a$ to $x = b$ (b) One of the subintervals and its corresponding rectangle

With this notation, the signed area of the rectangle corresponding to the subinterval, from x to $x + \delta x$, is

$$f(x)\,\delta x. \qquad (2)$$

So the total signed area of all the rectangles is the sum of a number of terms, each of form (2). As the number of subintervals gets larger and larger, the size of δx gets smaller and smaller, and the total signed area of the rectangles gets closer and closer to the definite integral of f from a to b. So, loosely, you can think of this definite integral as the sum of infinitely many terms of form (2), where the quantity δx is infinitely small. Historically, as you saw in Unit 6, the quantity δx was denoted by $\mathrm{d}x$ when it becomes infinitely small, so the sum described above was denoted by

$$\int_a^b f(x)\,\mathrm{d}x,$$

where the symbol $\int$ is an elongated 'S', which stands for 'sum'.

To see where the term 'integral' comes from, remember that the verb 'to integrate' means 'to join together'. Loosely, you can think of a definite integral as being obtained by joining together infinitely many signed areas, each infinitely narrow, into a single signed area.

Johann Bernoulli (1667–1748)

The name and principal symbol for integral calculus were discussed by Gottfried Wilhelm Leibniz and the Swiss mathematician Johann Bernoulli. Leibniz preferred 'calculus summatorius' as the name, and $\int$, the elongated S, as the symbol. Bernoulli preferred 'calculus integralis' as the name, and the capital letter I as the symbol. Eventually they agreed to compromise, adopting Bernoulli's name 'integral calculus' and Leibniz's symbol $\int$. The first appearance in print of the word 'integral' was in a work by Johann's brother, Jacob Bernoulli, in 1690, though Johann insisted that he had been the one to introduce the term.

Jacob Bernoulli (1654–1705)

Joseph Fourier (1768–1830)

The modern notation for a definite integral, with limits at the bottom and top of the integral sign, is due to the French mathematician Joseph Fourier, who introduced it in his pioneering book on heat diffusion *Théorie analytique de la chaleur* (The analytic theory of heat) of 1822. He first published the notation four years previously in an announcement for the book. Fourier accompanied Napoleon Bonaparte on his Egyptian expedition of 1798, and later supported Jean-François Champollion, the decipherer of the Rosetta Stone. He is remembered today for initiating the investigation of *Fourier series*. These are infinite trigonometric series, which are now named after him, and which arose in the context of his work on heat diffusion. You'll meet the idea of a *series* in Unit 10.

To finish this section, here's a summary of the main idea that you've met.

Definite integrals

Suppose that f is a continuous function, and a and b are numbers in its domain. The signed area between the graph of f and the x-axis from $x = a$ to $x = b$ is called the **definite integral** of f from a to b, and is denoted by

$$\int_a^b f(x)\,\mathrm{d}x.$$

2 The fundamental theorem of calculus

In this section you'll discover how the ideas about signed areas that you met in Section 1 are connected with the ideas about rates of change that you met in Units 6 and 7. The two areas of mathematics are linked by an important fact known as *the fundamental theorem of calculus.*

This theorem is stated in the first subsection of this section, preceded by an explanation of why it's true. It's a good idea to read the explanation, to increase your understanding, particularly if you intend to study more mathematics after this module. However, if you find the explanation hard to understand, then you can skip it, and go straight to the box headed 'Fundamental theorem of calculus', near the end of Subsection 2.1. Read the contents of this box, and continue reading to the end of the subsection. Make sure that you understand the fundamental theorem of calculus, and the other important facts stated there, even if you don't know why they're true. The two subsections that follow, Subsections 2.2 and 2.3, show you how to use the fundamental theorem of calculus. You might like to try reading Subsection 2.1 again later, as some of the ideas that you found difficult the first time might seem easier once you're more familiar with the ways in which you can use the fundamental theorem of calculus.

2.1 What is the fundamental theorem of calculus?

The first step in obtaining a link between signed areas and rates of change is to define, for any continuous function f, a new function whose values are signed areas between the graph of f and the x-axis. Here's how we do that.

Consider any continuous function f, as illustrated in Figure 16.

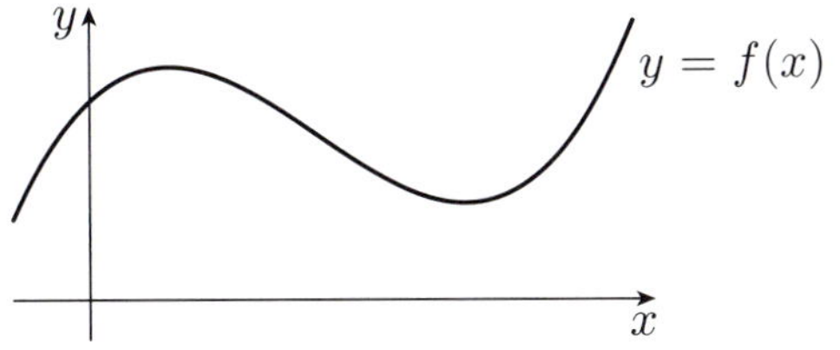

Figure 16 The graph of a continuous function f

Let's choose any number, say s, in the domain of f, and define a new function, A, to have the same domain as f and the following rule:

$A(x)$ = signed area between the graph of f and the x-axis, from s to x.

This definition is illustrated in Figure 17.

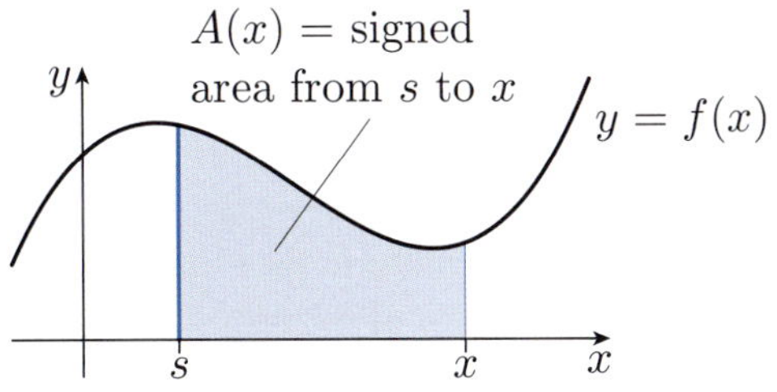

Figure 17 A continuous function f and a signed area $A(x)$

We call the function A the **signed-area-so-far function** for the function f, with starting point s. Its rule can be expressed concisely as follows:

$$A(x) = \int_s^x f(t)\,dt.$$

(Here the variable t has been used in the definite integral instead of the usual variable x, to avoid confusion with the input variable x of the function A.)

Note that if a and b are any two numbers in the domain of f, then the signed area between the graph of f and the x-axis, from $x = a$ to $x = b$, is given by

$$A(b) - A(a).$$

In other words,

$$\int_a^b f(x)\,dx = A(b) - A(a).$$

This fact is illustrated in Figure 18, in a case where $s < a < b$. The third property of definite integrals in the box on page 120 tells you that the fact holds even when s, a and b aren't in this order.

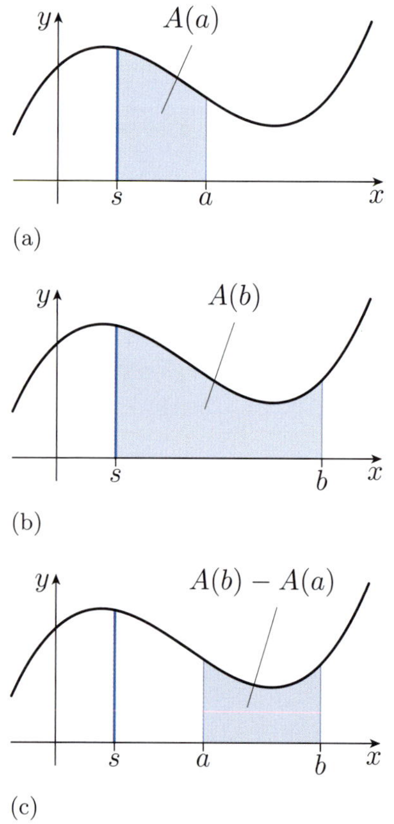

Figure 18 An illustration of the fact that $\int_a^b f(x)\,dx = A(b) - A(a)$

Now let's consider how the value of a signed-area-so-far function changes as the value of the input variable x changes. Let's start by looking at what happens when the function f is a *constant* function.

For example, consider the constant function $f(x) = 3$, whose graph is shown in Figure 19. Let's take the starting point s for the signed-area-so-far function A to be 1, as indicated in Figure 19.

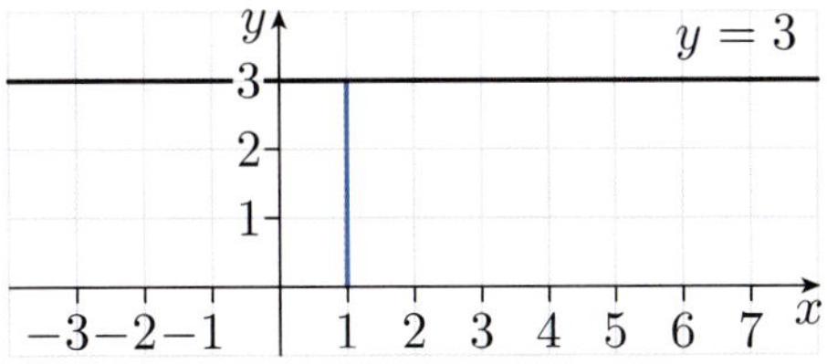

Figure 19 The graph of the constant function $f(x) = 3$

Figure 20 shows some values of $A(x)$.

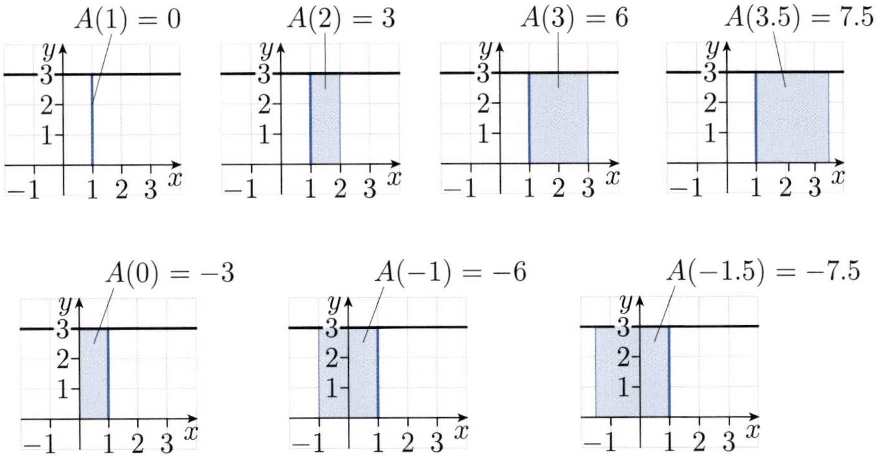

Figure 20 Some values of $A(x)$ when f is the constant function $f(x) = 3$ and the starting point s is 1

You can see that the signed area $A(x)$ changes at the rate of 3 square units for every unit by which x changes. For example,

if x increases by 1, then $A(x)$ increases by 3
if x increases by 0.5, then $A(x)$ increases by $0.5 \times 3 = 1.5$
if x increases by -1, then $A(x)$ increases by $-1 \times 3 = -3$.

(Remember that a negative increase is a decrease.)

In other words, the *rate of change* of $A(x)$ with respect to x is 3.

More generally, you can see that for any constant function $f(x) = k$, and any signed-area-so-far function A for f, the value of $A(x)$ changes at the rate of k square units for every unit by which x changes. In other words, the rate of change of $A(x)$ is k.

There are two more examples of this in Figure 21. In Figure 21(a), the function is $f(x) = 1.5$, so the signed area $A(x)$ changes at the rate of 1.5 square units for each unit by which x changes. In Figure 21(b), the function is $f(x) = -2$, so the signed area $A(x)$ changes at the rate of -2 square units for each unit by which x changes.

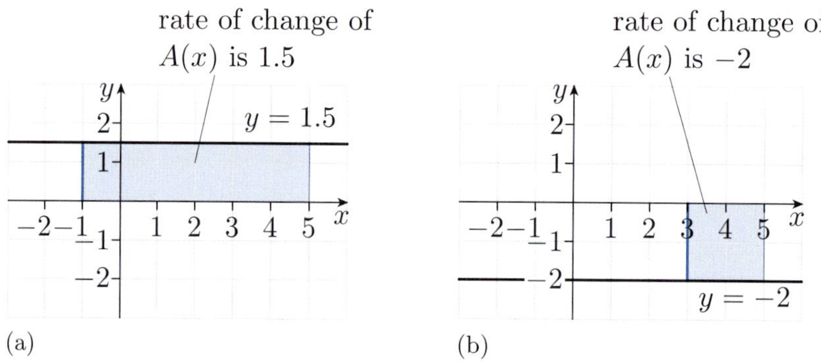

Figure 21 Signed-area-so-far functions for constant functions, with starting points $s = -1$ in (a) and $s = 3$ in (b)

Now let's look at what happens for a function f that isn't a constant function. Consider, for example, the function f whose graph is shown in Figure 22. The starting point s for the signed-area-so-far function A has been chosen to be 2.

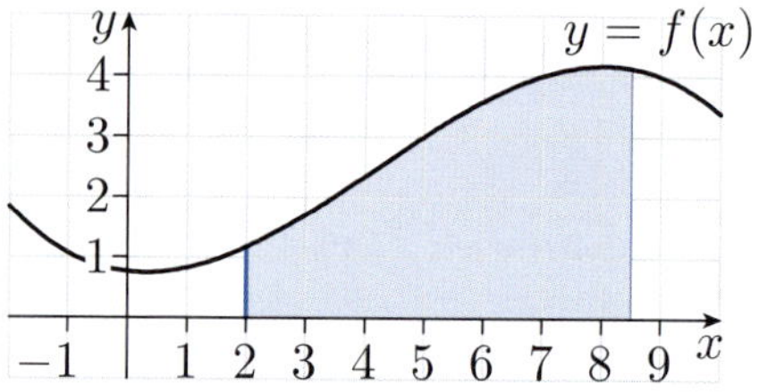

Figure 22 A signed-area-so-far function for a function f that isn't a constant function

For this function, as the value of x changes, the value of $A(x)$ doesn't change at a constant rate. For example, as illustrated in Figure 23(a), when x is close to 5 the value of $A(x)$ changes at the rate of about 3 square units for every unit by which x increases. Similarly, as illustrated in

Figure 23(b), when x is close to 7 the value of $A(x)$ changes at the rate of about 4 square units for every unit by which x increases.

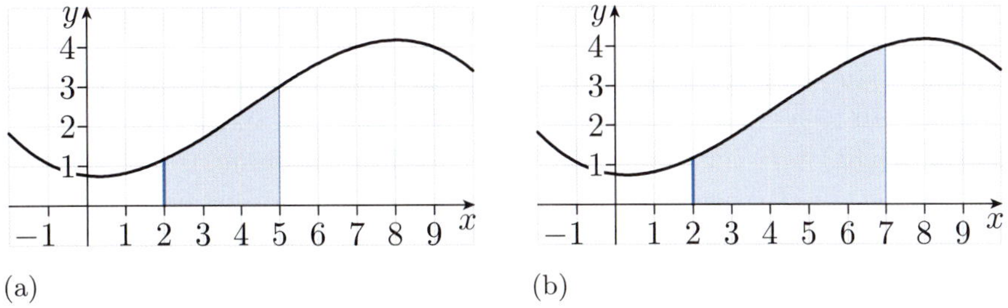

Figure 23 (a) When x is close to 5, the rate of change of $A(x)$ is about 3 (b) When x is close to 7, the rate of change of $A(x)$ is about 4

More precisely, when $x = 5$ the *instantaneous* rate of change of $A(x)$ is 3, and when $x = 7$ the *instantaneous* rate of change of $A(x)$ is 4. In fact – and this is the crucial thing to realise – at any value of x, the instantaneous rate of change of $A(x)$ is $f(x)$, as illustrated in Figure 24.

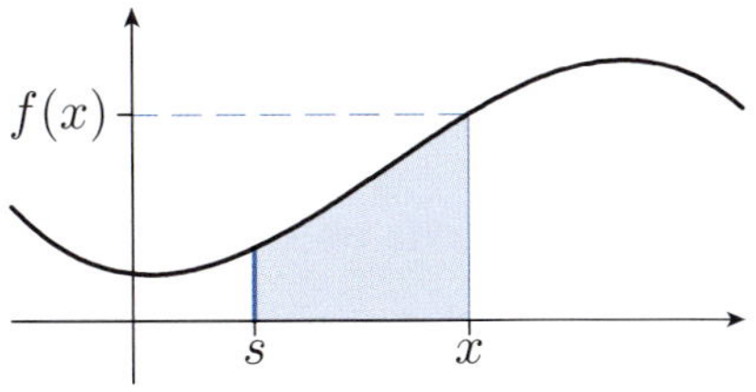

Figure 24 The instantaneous rate of change of $A(x)$ at x is $f(x)$

In other words, the derivative of the function A is the function f, and so the function A is an antiderivative of the function f.

You can see that this will be true for any continuous function f and any signed-area-so-far function A for f. That is, we have the important fact below. Because of its importance, it deserves to be called a *theorem*.

Theorem

Suppose that f is a continuous function, and s is any number in its domain. Let A be the signed-area-so-far function for f with starting point s. Then A is an antiderivative of f.

It's this fact that leads to the fundamental theorem of calculus. Consider any continuous function f. Let s be any number in its domain, and let A be the signed-area-so-far function for f with starting point s.

Suppose that a and b are any numbers in the domain of f. Then, as you saw earlier in this subsection, the value of $\int_a^b f(x)\,dx$, the signed area between the graph of f and the x-axis from $x = a$ to $x = b$, is given by

$$\int_a^b f(x)\,dx = A(b) - A(a). \qquad (3)$$

Now let F be *any* antiderivative of f. Then, since A and F are both antiderivatives of f, the value of $F(b) - F(a)$ is the same as the value of $A(b) - A(a)$. That's because, as you saw in Unit 7, the value of $F(b) - F(a)$ is the same no matter which antiderivative of f you take F to be.

So you can replace $A(b) - A(a)$ by $F(b) - F(a)$ in equation (3) above. This gives the crucial fact that's known as the **fundamental theorem of calculus**. It's stated concisely below.

Fundamental theorem of calculus

Suppose that f is a continuous function whose domain contains the numbers a and b, and that F is an antiderivative of f. Then

$$\int_a^b f(x)\,dx = F(b) - F(a).$$

This theorem can be stated in words as follows.

Fundamental theorem of calculus (in words)

Suppose that f is a continuous function whose domain contains the numbers a and b, and that F is an antiderivative of f. Then the signed area between the graph of f and the x-axis from $x = a$ to $x = b$ is equal to the change in the value of F as x changes from $x = a$ to $x = b$.

The equation in the fundamental theorem of calculus is true no matter whether the value a is less than, equal to or greater than the value b. This is the main reason why the definition of signed area is extended in the way that you saw earlier in this unit.

Here's another important fact that's come out of the discussion in this subsection. The theorem in the box on page 131 tells you immediately that the following fact is true.

Every continuous function has an antiderivative.

In fact there's some disagreement among mathematicians about exactly which theorem should be called 'the fundamental theorem of calculus'. Everyone agrees that this name should be given to a theorem that links signed areas with rates of change. However, while some mathematicians use the name for the theorem that's given this name in this module, other mathematicians use it for the theorem in the box on page 131. Still other mathematicians take the view that these two theorems are *two* fundamental theorems of calculus, or two parts of a single fundamental theorem of calculus. So if you search for 'fundamental theorem of calculus' on the internet, or look it up in books, then you're likely to find a variety of different results.

2.2 Using the fundamental theorem to find signed areas

The fundamental theorem of calculus gives you a quick way to find the signed area between the graph of a continuous function f and the x-axis, from $x = a$ to $x = b$, in cases *where you know a formula for an antiderivative F of f*. The theorem tells you that this signed area, which is the definite integral

$$\int_a^b f(x)\,\mathrm{d}x,$$

is given simply by $F(b) - F(a)$.

For example, suppose that you want to find the area between the graph of the function $f(x) = \cos x$ and the x-axis, from $x = 0$ to $x = \pi/2$, as illustrated in Figure 25. This area is the definite integral

$$\int_0^{\pi/2} \cos x\,\mathrm{d}x.$$

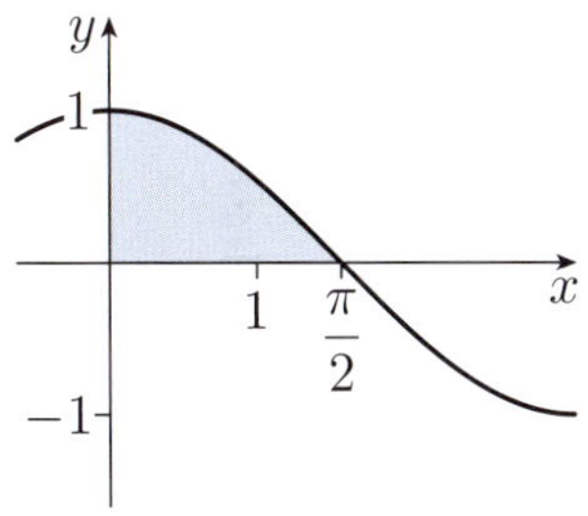

Figure 25 The area between the graph of $f(x) = \cos x$ and the x-axis, from $x = 0$ to $x = \pi/2$

Since an antiderivative of $f(x) = \cos x$ is $F(x) = \sin x$ (as you saw in Unit 7), the fundamental theorem of calculus tells you that

$$\int_0^{\pi/2} \cos x\,\mathrm{d}x = \sin\left(\frac{\pi}{2}\right) - \sin 0 = 1 - 0 = 1.$$

So the area in Figure 25 is 1 square unit.

You may find it magical that a complicated area like the one in Figure 25 can be found from such a simple calculation. In particular, it may seem amazing that although the area in Figure 25 depends on infinitely many values of $\cos x$ (namely all the values of $\cos x$ for $x = 0$ to $x = \pi/2$), it can be calculated from the values of an antiderivative of $\cos x$ at just *two* values of x, namely $x = 0$ and $x = \pi/2$.

Notice that in the calculation above it doesn't matter which of the infinitely many antiderivatives of $f(x) = \cos x$ you use, because they all differ by a constant and hence they all lead to the same final answer. For example, using the antiderivative $F(x) = \sin x + 7$ gives

$$\int_0^{\pi/2} \cos x \, dx = \left(\sin\left(\frac{\pi}{2}\right) + 7\right) - (\sin 0 + 7) = 1 + 7 - 0 - 7 = 1.$$

It's usually best to use the simplest antiderivative in calculations of this kind.

You'll see another example of the fundamental theorem of calculus used to find a signed area shortly, and you'll then have the opportunity to find some signed areas in this way yourself. First, however, it's helpful for you to learn a type of notation that's convenient in calculations of this kind.

For any function F, the expression

$$F(b) - F(a)$$

can be denoted by

$$\left[F(x)\right]_a^b.$$

For example,

$$\left[\sin x\right]_0^{\pi/2} = \sin\left(\frac{\pi}{2}\right) - \sin 0 = 1 - 0 = 1.$$

You can practise evaluating expressions of this type in the next activity.

Activity 8 *Evaluating expressions of the form* $[F(x)]_a^b$

Evaluate the following expressions. In part (c), give your answer to three significant figures.

(a) $\left[\frac{1}{2}x^2\right]_3^5$ (b) $\left[\cos x\right]_0^{2\pi}$ (c) $\left[e^x\right]_{-1}^1$

With the square bracket notation introduced above, the fundamental theorem of calculus can be restated as follows.

Fundamental theorem of calculus (square bracket notation)

Suppose that f is a continuous function whose domain contains the numbers a and b, and F is an antiderivative of f. Then

$$\int_a^b f(x) \, dx = \left[F(x)\right]_a^b.$$

In the next example, the fundamental theorem of calculus is used, together with the square bracket notation, to find a signed area – that is, to evaluate a definite integral.

Example 4 *Evaluating a definite integral*

Evaluate the definite integral

$$\int_1^2 x^3\,dx.$$

(This definite integral is shown in Figure 26.)

Solution

Use the fundamental theorem of calculus. By the general formula for the indefinite integral of a power function, an antiderivative of x^3 is $\frac{1}{4}x^4$.

$$\int_1^2 x^3\,dx = \left[\tfrac{1}{4}x^4\right]_1^2$$

Evaluate this expression.

$$= \left(\tfrac{1}{4}\times 2^4\right) - \left(\tfrac{1}{4}\times 1^4\right)$$
$$= 4 - \tfrac{1}{4} = \tfrac{15}{4}.$$

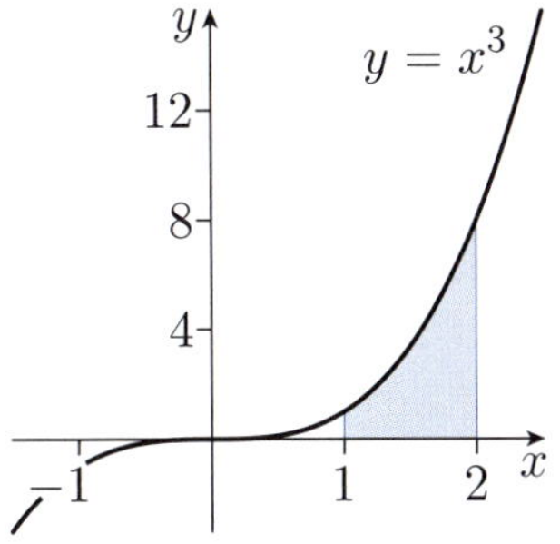

Figure 26 The definite integral in Example 4

You've now seen that, when you want to find a signed area, it's much quicker and simpler to use the fundamental theorem of calculus than to use the method involving subintervals that you saw in Subsection 1.1. Another advantage of using the fundamental theorem of calculus is that it gives you an *exact* answer, rather than an approximate one. However, remember that you can use the fundamental theorem of calculus only when you can find a formula for the antiderivative that you need. Sometimes it's difficult or impossible to find such a formula.

In the next activity, you can try evaluating some definite integrals by using the fundamental theorem of calculus. Use the square bracket notation, as in Example 4, since you need to get used to this notation. You can find the antiderivatives that you need by using the table of standard indefinite integrals given near the end of Unit 7. This table is also included in the *Handbook*.

Activity 9 *Evaluating definite integrals*

Evaluate the following definite integrals (which are shown in Figure 27). Give exact answers, and (except in parts (a) and (d), where the answer is a simple fraction or an integer) also give answers to three significant figures.

(a) $\displaystyle\int_{-1}^{1} x^2\,\mathrm{d}x$ (b) $\displaystyle\int_{0}^{2} e^t\,\mathrm{d}t$ (c) $\displaystyle\int_{1}^{4} \frac{1}{u}\,\mathrm{d}u$

(d) $\displaystyle\int_{-\pi/4}^{\pi/4} \sec^2\theta\,\mathrm{d}\theta$ (e) $\displaystyle\int_{-1}^{1} \frac{1}{1+u^2}\,\mathrm{d}u$

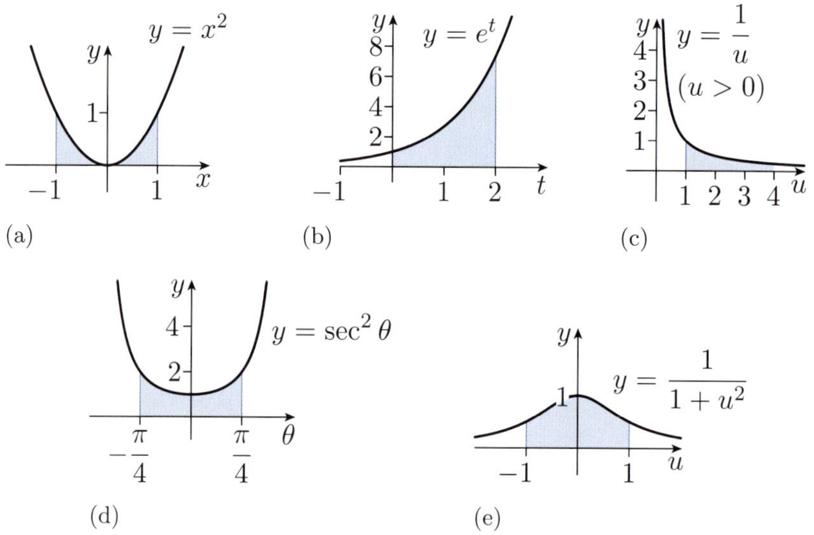

Figure 27 The definite integrals in Activity 9

The next example demonstrates how to evaluate another definite integral by using the fundamental theorem of calculus. The only difference from the examples that you've seen so far is that the antiderivative that's needed for the method can't be found directly from the table of standard indefinite integrals. Instead, we have to start by manipulating the given function to express it as a sum of constant multiples of functions that we can integrate, then use the constant multiple rule and sum rule for antiderivatives, together with the table of standard indefinite integrals. You practised finding antiderivatives in this way in Unit 7.

As shown in the example, a convenient way to carry out these sorts of manipulations is to manipulate the expression inside the notation $\int_a^b \ldots\,\mathrm{d}x$. This expression is known as the **integrand**. For example, the definite integral $\int_0^1 x^2\,\mathrm{d}x$ has integrand x^2.

Example 5 *Evaluating a more complicated definite integral*

Evaluate the definite integral

$$\int_2^4 \frac{(2x+1)(x-4)}{x}\,\mathrm{d}x.$$

(It is shown in Figure 28.)

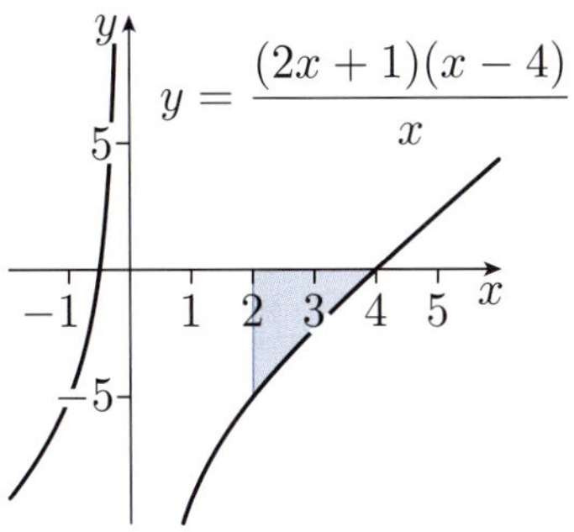

Figure 28 The definite integral in Example 5

Solution

Manipulate the integrand to get it into a form that you can integrate.

$$\begin{aligned}
&\int_2^4 \frac{(2x+1)(x-4)}{x}\,\mathrm{d}x \\
&= \int_2^4 \frac{2x^2-7x-4}{x}\,\mathrm{d}x \\
&= \int_2^4 \left(2x-7-\frac{4}{x}\right)\mathrm{d}x \\
&= \int_2^4 \left(2x-7-4\times\frac{1}{x}\right)\mathrm{d}x
\end{aligned}$$

Apply the fundamental theorem of calculus, and evaluate the result.

$$\begin{aligned}
&= \left[2\times\tfrac{1}{2}x^2-7x-4\ln x\right]_2^4 \\
&= \left[x^2-7x-4\ln x\right]_2^4 \\
&= (4^2-7\times 4-4\ln 4)-(2^2-7\times 2-4\ln 2) \\
&= 16-28-4\ln 4-4+14+4\ln 2 \\
&= -2-4(\ln 4-\ln 2) \\
&= -2-4\ln(4/2) \\
&= -2-4\ln 2
\end{aligned}$$

There's no need to find a decimal approximation, as the question didn't ask for one.

There are other acceptable ways to leave the final answer in Example 5. For example, you could write it as $-2(1+2\ln 2)$.

Activity 10 *Evaluating more complicated definite integrals*

Evaluate the following definite integrals (which are shown in Figure 29). Give exact answers.

(a) $\displaystyle\int_1^4 3\sqrt{x}\,\mathrm{d}x$ (b) $\displaystyle\int_{-1}^0 x(1+x)\,\mathrm{d}x$ (c) $\displaystyle\int_2^4 \frac{r+5}{r}\,\mathrm{d}r$

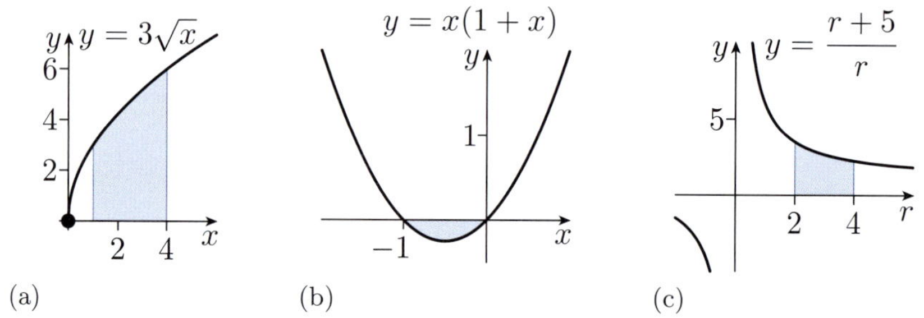

Figure 29 The definite integrals in Activity 10

When you use the square bracket notation, you can sometimes simplify your working by using the following rules.

Constant multiple rule and sum rule for the square bracket notation

$$\big[kF(x)\big]_a^b = k\big[F(x)\big]_a^b, \quad \text{where } k \text{ is a constant}$$

$$\big[F(x)+G(x)\big]_a^b = \big[F(x)\big]_a^b + \big[G(x)\big]_a^b$$

For example, the constant multiple rule for the square bracket notation tells you that

$$\big[5\sin x\big]_1^2 = 5\big[\sin x\big]_1^2$$

and the sum rule for the square bracket notation tells you that

$$\big[\sin x + \cos x\big]_1^2 = \big[\sin x\big]_1^2 + \big[\cos x\big]_1^2.$$

To see why these rules hold, you just need to use the definition of the square bracket notation. For example,

$$\begin{aligned}\big[5\sin x\big]_1^2 &= 5\sin 2 - 5\sin 1\\ &= 5(\sin 2 - \sin 1)\\ &= 5\big[\sin x\big]_1^2.\end{aligned}$$

Similarly,

$$\begin{aligned}\big[\sin x + \cos x\big]_1^2 &= (\sin 2 + \cos 2) - (\sin 1 + \cos 1)\\ &= (\sin 2 - \sin 1) + (\cos 2 - \cos 1)\\ &= \big[\sin x\big]_1^2 + \big[\cos x\big]_1^2.\end{aligned}$$

As always when a constant multiple rule and a sum rule hold, the sum rule for the square bracket notation also holds if you replace the plus sign on each side with a minus sign. To prove this, you combine the sum rule with the case $k = -1$ of the constant multiple rule. You saw this done for the constant multiple and sum rules for derivatives in Unit 6, Subsection 2.3.

The next example illustrates how you can use the constant multiple rule and the sum rule for the square bracket notation when you're evaluating a definite integral.

Example 6 *Using the constant multiple rule and sum rule for the square bracket notation*

Evaluate the definite integral

$$\int_1^2 (1 + x^3)\,dx.$$

(It is shown in Figure 30.)

Solution

Use the fundamental theorem of calculus.

$$\int_1^2 (1 + x^3)\,dx = \big[x + \tfrac{1}{4}x^4\big]_1^2$$

Use the sum rule.

$$= \big[x\big]_1^2 + \big[\tfrac{1}{4}x^4\big]_1^2$$

Use the constant multiple rule, and simplify the result.

$$\begin{aligned}&= \big[x\big]_1^2 + \tfrac{1}{4}\big[x^4\big]_1^2\\ &= (2 - 1) + \tfrac{1}{4}(2^4 - 1^4)\\ &= 1 + \tfrac{1}{4} \times 15\\ &= \tfrac{19}{4}.\end{aligned}$$

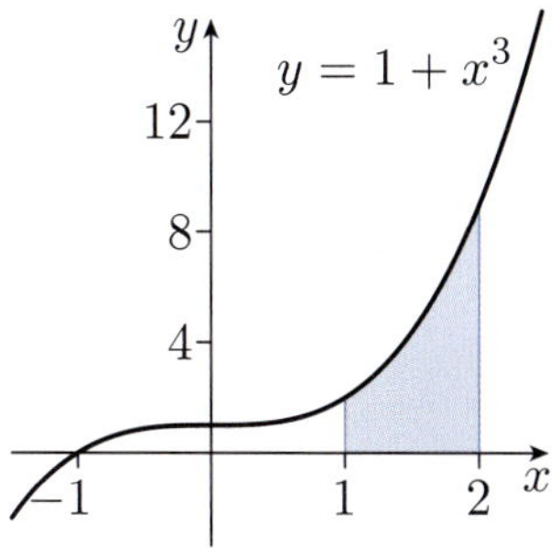

Figure 30 The definite integral in Example 6

You can practise using the constant multiple rule and sum rule for the square bracket notation in the next activity.

Activity 11 *Using the constant multiple rule and sum rule for the square bracket notation*

Evaluate the following definite integrals. (They're shown in Figure 31.) Give your answer to part (a) to three significant figures.

(a) $\displaystyle\int_3^5 x^{3/2}\,dx$ (b) $\displaystyle\int_{-1/2}^1 (u - 2u^2)\,du$

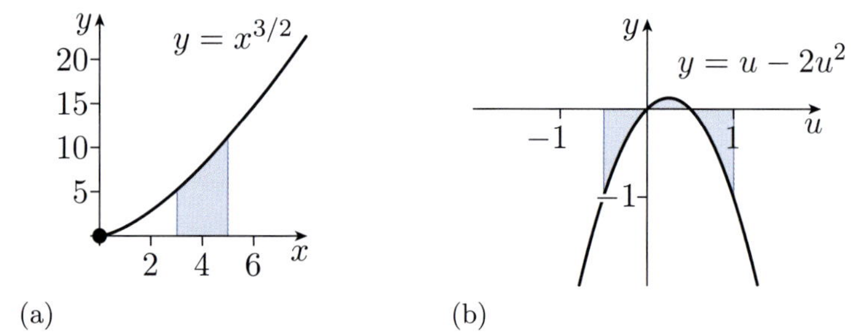

Figure 31 The definite integrals in Activity 11

Note that when you're evaluating a definite integral it's sometimes more convenient *not* to use the constant multiple rule and/or the sum rule for the square bracket notation, even if there's an opportunity to use them. You can choose whether or not to use them, as seems convenient.

A constant multiple rule and a sum rule also hold for definite integrals, as follows.

Constant multiple rule and sum rule for definite integrals

$$\int_a^b k\,f(x)\,dx = k\int_a^b f(x)\,dx, \quad \text{where } k \text{ is a constant}$$

$$\int_a^b (f(x) + g(x))\,dx = \int_a^b f(x)\,dx + \int_a^b g(x)\,dx$$

These rules follow from the fundamental theorem of calculus and the constant multiple rule and sum rule for antiderivatives, which you met in Unit 7. To see this for the constant multiple rule, suppose that f is a continuous function whose domain contains the numbers a and b, and that k is a constant. Let F be an antiderivative of f. Then, by the constant multiple rule for antiderivatives, $kF(x)$ is an antiderivative of $kf(x)$.

Hence, by the fundamental theorem of calculus and the constant multiple rule for the square bracket notation,

$$\int_a^b k\,f(x)\,\mathrm{d}x = \big[k\,F(x)\big]_a^b$$
$$= k\big[F(x)\big]_a^b$$
$$= k\int_a^b f(x)\,\mathrm{d}x.$$

You can prove the sum rule for definite integrals in a similar way.

Remember that the sum rule for definite integrals also holds if you replace the plus sign on each side with a minus sign, because, as mentioned earlier, this is always the case when a constant multiple rule and a sum rule hold.

The constant multiple rule for definite integrals tells you that, for example, the area in Figure 32(b) is twice the area in Figure 32(a).

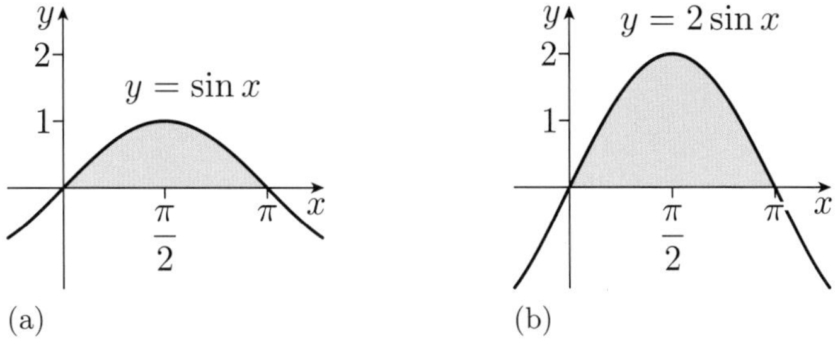

Figure 32 The area in (b) is twice the area in (a)

Similarly, the sum rule for definite integrals tells you that, for example, the area in Figure 33(c) is the sum of the areas in Figure 33(a) and (b).

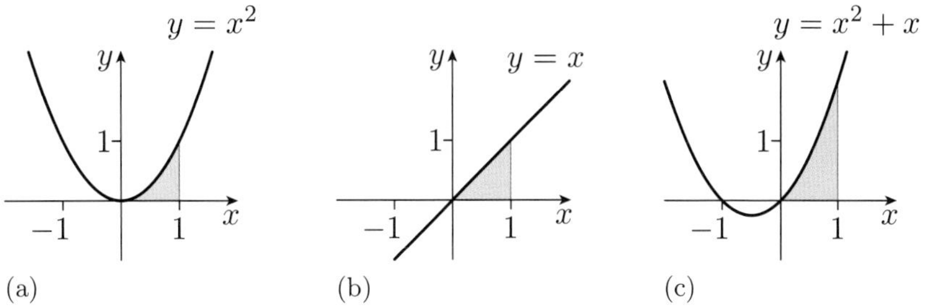

Figure 33 The area in (c) is the sum of the areas in (a) and (b)

The next example illustrates how you can use the constant multiple rule and the sum rule for definite integrals when you're evaluating a definite integral.

Example 7 *Using the constant multiple rule and sum rule for definite integrals*

Evaluate the definite integral

$$\int_0^{\pi} (x + 2\sin x)\,dx.$$

(It is shown in Figure 34.)

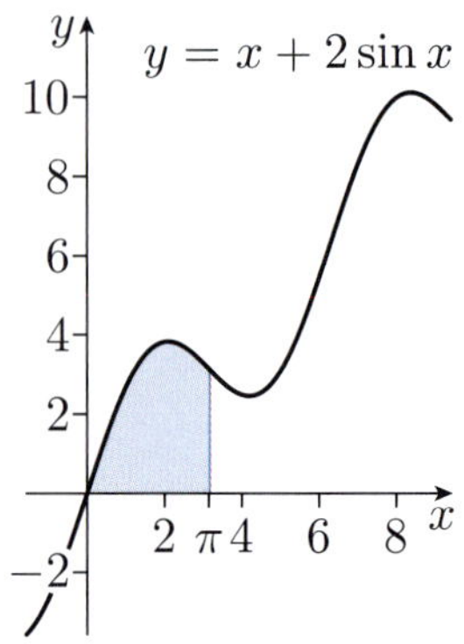

Figure 34 The definite integral in Example 7

Solution

Use the fundamental theorem of calculus. Use the constant multiple rule and sum rule, for definite integrals and for the square bracket notation, as convenient.

$$\begin{aligned}\int_0^{\pi} (x + 2\sin x)\,dx &= \int_0^{\pi} x\,dx + \int_0^{\pi} 2\sin x\,dx \\ &= \int_0^{\pi} x\,dx + 2\int_0^{\pi} \sin x\,dx \\ &= \left[\tfrac{1}{2}x^2\right]_0^{\pi} + 2\left[-\cos x\right]_0^{\pi} \\ &= \tfrac{1}{2}\left[x^2\right]_0^{\pi} - 2\left[\cos x\right]_0^{\pi} \\ &= \tfrac{1}{2}(\pi^2 - 0^2) - 2(\cos\pi - \cos 0) \\ &= \tfrac{1}{2}\pi^2 - 2(-1 - 1) \\ &= \tfrac{1}{2}\pi^2 + 4.\end{aligned}$$

You can practise using the constant multiple rule and sum rule for definite integrals in the next activity.

Activity 12 *Using the constant multiple rule and sum rule for definite integrals*

Evaluate the following definite integrals. (They are shown in Figure 35.) Give exact answers.

(a) $\displaystyle\int_{\sqrt{3}}^{\sqrt{7}} \tfrac{1}{2}x^3\,dx$ (b) $\displaystyle\int_{-\pi/4}^{\pi/4} (\sin x + \cos x)\,dx$

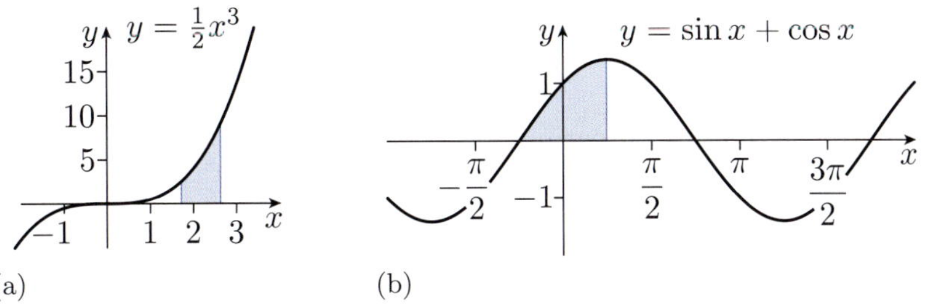

Figure 35 The definite integrals in Activity 12

As with the sum rule and constant multiple rule for the square bracket notation, you can choose whether or not to use the sum rule and the constant multiple rule for definite integrals in your working, as seems convenient. Often it's more convenient not to use them, even if there's an opportunity to do so.

In the next activity, you're asked to use the fundamental theorem of calculus to find the exact area of the cross-section of a roof that was discussed at the beginning of Subsection 1.1.

Activity 13 *Using the fundamental theorem for a practical problem*

The curve of the roof of a building is given by the graph of the function

$$f(x) = 9 - \tfrac{1}{36}x^2,$$

where x and $f(x)$ are measured in metres. Use the ideas that you've met in this section to find the area between this graph and the x-axis from $x = -18$ to $x = 18$, as shown below. Hence state the area of the cross-section of the roof.

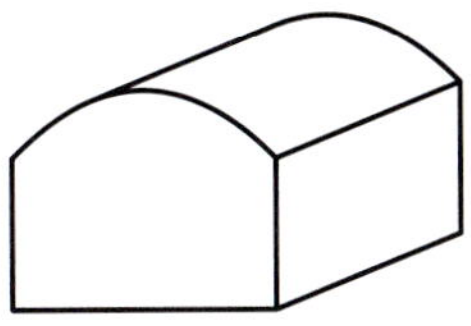

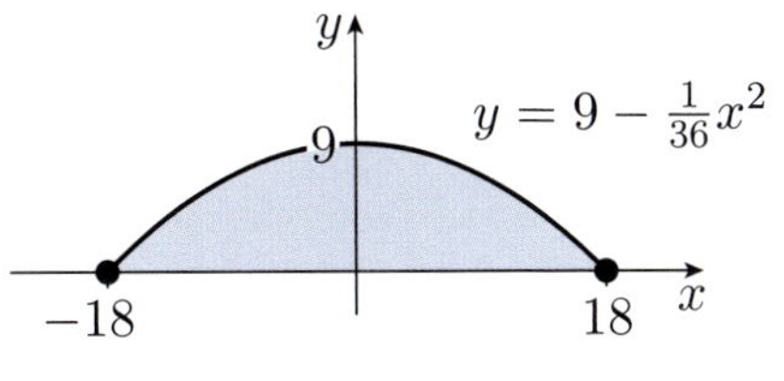

You may remember that in Subsection 1.1 we used subintervals to work out that the area in Activity 13 is approximately $216\,\text{m}^2$. The solution to Activity 13 confirms that the area is *exactly* $216\,\text{m}^2$.

Here's another practical example for you to try. In this example the graph lies below the x-axis, so the signed area that you'll obtain by using the fundamental theorem of calculus will be negative. You have to remove the minus sign to obtain the required area.

Activity 14 *Solving another practical problem about area*

An engineer is planning the construction of an open channel with a parabolic cross-section, as illustrated below. The shape of the cross-section is given by the equation

$$y = \tfrac{1}{32}x^2 - 200 \quad (-100 \le x \le 100),$$

where x and y are measured in centimetres, as shown in the graph below.

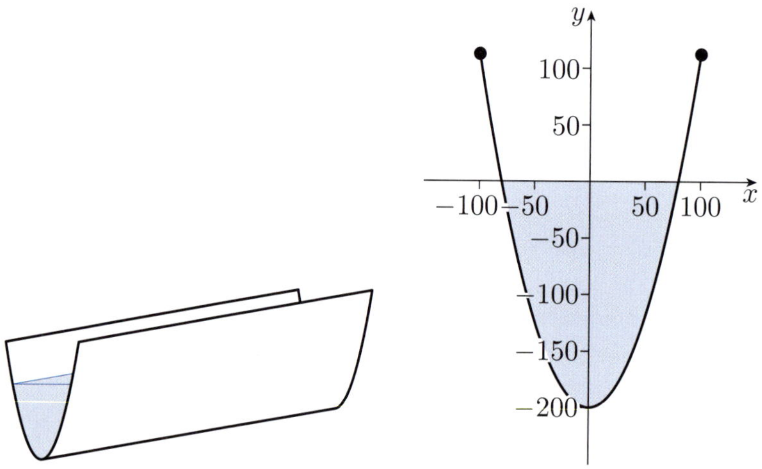

The maximum height of liquid permissible in the channel is the height of the x-axis on the graph; that is, 200 cm above the lowest point of the channel. To ensure that the channel is adequate for the expected flow of liquid, the engineer needs to know the area shaded on the graph.

(a) Find the x-intercepts of the graph.

(b) Hence calculate the shaded area, to the nearest square centimetre.

The history of the fundamental theorem of calculus

You've seen that the fundamental theorem of calculus relates rates of change – that is, gradients of tangents – to signed areas.

One of the earliest mathematicians to relate the problem of calculating tangents to the problem of calculating areas was Isaac Barrow, the first Lucasian professor at the University of Cambridge. However, his method was presented in the geometrical style of the day and was not computationally useful for solving problems. Isaac Newton was one of Barrow's students, and in 1669 Barrow resigned his professorship so that Newton could take it up.

Isaac Barrow (1630–1677)

Newton based his definition of quadrature (determining area) on the idea that it is the converse of finding tangents. So what for some people was, and for modern-day mathematicians is, a *theorem* that finding areas and tangents are inverse processes – the fundamental theorem of calculus – was taken by Newton as the very *definition* of quadrature. Newton's way of regarding area questions as the opposite of tangency questions is set down clearly in his *De analysi* of 1669.

Gottfried Wilhelm Leibniz published his first accounts of differential and integral calculus in the newly founded journal *Acta Eruditorum Lipsiensium* (Acts of the scholars of Leipzig) in 1684 and 1686 respectively, and his proof of the fundamental theorem of calculus followed in 1693.

The first rigorous proof of the fundamental theorem of calculus was provided in the early decades of the 19th century by the French mathematician Augustin Louis Cauchy, who included it in his course at the École Polytechnique in Paris. Cauchy was one of the most productive and influential mathematicians of his generation. His reformulation of the foundations of calculus, while very thorough, was also very hard to understand – so much so that not only his students but also his fellow professors protested!

Augustin Louis Cauchy (1789–1857)

2.3 Using the fundamental theorem to find changes in the values of antiderivatives

The fundamental theorem of calculus isn't useful just for solving problems related to finding areas. It's also useful for solving problems involving rates of change.

As you saw in Subsection 5.3 of Unit 7, in integral calculus you often want to solve problems of the following type. You have a continuous function f, and two numbers a and b in its domain, and you want to work out the amount by which an antiderivative of f changes from $x = a$ to $x = b$. (Remember that all the antiderivatives of f change by the same amount.)

If you can find a formula for an antiderivative F of f, then it's straightforward to do this: you just calculate $F(b) - F(a)$. You saw examples like this in Unit 7.

If you can't find a formula for an antiderivative, then you can use the fundamental theorem of calculus to help you find an approximate answer. The theorem tells you that, for any antiderivative F of f,

$$F(b) - F(a) = \int_a^b f(x)\,\mathrm{d}x.$$

So the answer that you want is equal to the signed area between the graph of f and the x-axis, from $x = a$ to $x = b$. You can use the method that you met in Subsection 1.1 to find an approximate value for this signed area. You would normally do that by using a computer, as mentioned earlier.

For example, suppose that you have a velocity-time graph for an object moving along a straight line, such as the graph in Example 8 below. Since displacement is an antiderivative of velocity, the fundamental theorem of calculus tells you that the change in the displacement of the object, from any point in time to any other point in time, is equal to the signed area between the graph and the time axis, from the first point in time to the second point in time. This fact is used in Example 8.

As with any graph, the units for signed area on a velocity-time graph are the units on the vertical axis times the units on the horizontal axis. So, for example, the units for signed area on the graph in Example 8 are

$$\mathrm{m\,s^{-1}} \times \mathrm{s} = \mathrm{m};$$

that is, metres. These are units for displacement, as you'd expect.

Example 8 *Using the fundamental theorem of calculus for a velocity-time graph*

The graph below is the velocity–time graph of an object moving along a straight line. The areas (in m) of some regions are marked on the graph.

In each of parts (a) to (c), use the given areas to find the amount by which the displacement of the object changes from the first time to the second time.

(a) From $t = 0$ to $t = 4$. (b) From $t = 4$ to $t = 10$.

(c) From $t = 0$ to $t = 10$.

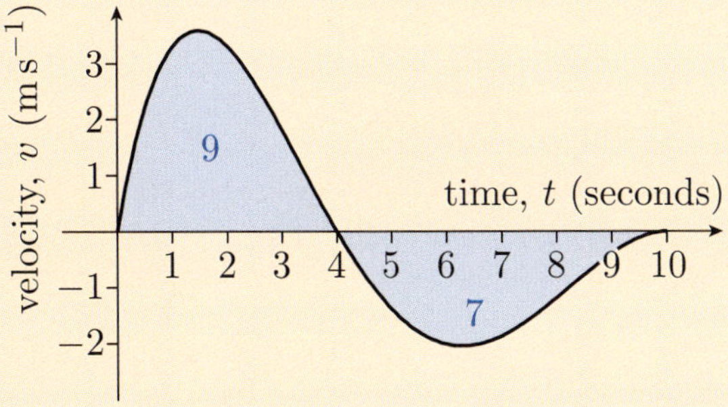

Solution

(a) The change in displacement from $t = 0$ to $t = 4$ is $9\,\mathrm{m}$.

(b) The change in displacement from $t = 4$ to $t = 10$ is $-7\,\mathrm{m}$.

(c) The change in displacement from $t = 0$ to $t = 10$ is $9\,\mathrm{m} + (-7)\,\mathrm{m} = 2\,\mathrm{m}$.

Notice that the velocity of the object in Example 8 changes from positive to negative at time $t = 4$. That is, at this time the object changes from moving in the positive direction to moving in the negative direction, as illustrated in Figure 36. The negative displacement that occurs after this time cancels out most of the positive displacement that occurred initially. Since the object was displaced by 9 m in the positive direction, and then by 7 m in the negative direction, its final displacement from its starting position is 2 m, as found in Example 8(c). The object travels a total distance of $9\,\mathrm{m} + 7\,\mathrm{m} = 16\,\mathrm{m}$ during the 10 seconds of its motion.

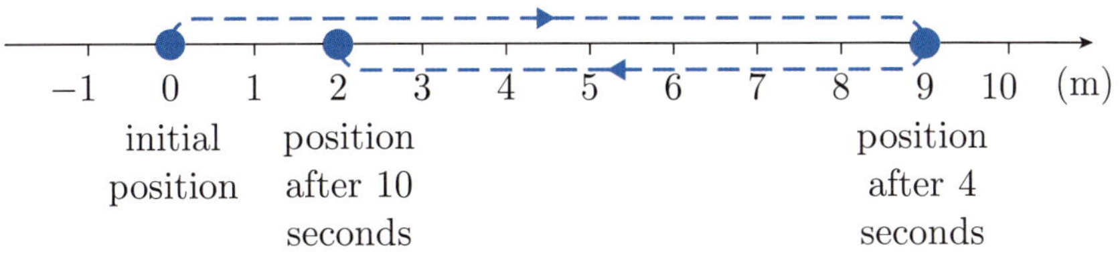

Figure 36 The object in Example 8 starts moving in the positive direction, then turns and moves in the negative direction

Activity 15 *Using the fundamental theorem of calculus for a velocity–time graph*

The velocity–time graph of an object moving along a straight line is shown below. The areas of some regions are marked on the graph. Use these areas to find the answers in parts (a) to (c) below.

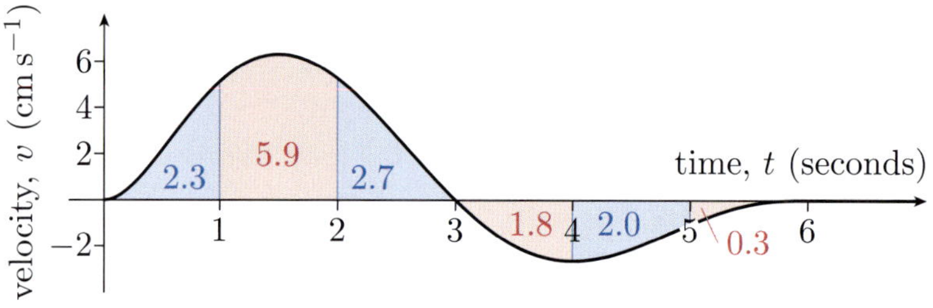

(a) In each of parts (i) to (iv), find the amount by which the displacement of the object changes from the first time to the second time.

(i) From $t = 0$ to $t = 1$. (ii) From $t = 3$ to $t = 4$.

(iii) From $t = 0$ to $t = 3$. (iv) From $t = 3$ to $t = 6$.

(b) What is the displacement of the object from its starting position at each of the following times?

(i) $t = 2$ (ii) $t = 6$

(c) What is the total distance travelled by the object during the 6 seconds of its motion?

If you want to solve a problem that involves finding a change in the value of an antiderivative, and you *can* find a formula for an antiderivative, then it's usually convenient to use the notation for a definite integral, and the square bracket notation, in your working. Here's Example 22 from Subsection 5.3 of Unit 7 again, but with the working carried out using this notation.

Example 9 *Finding a change in the value of an antiderivative*

Suppose that the rate of change of a quantity is given by the function $f(x) = \frac{1}{5}x^2$ $(x \geq 0)$, as shown in Figure 37. By what amount does the quantity change from $x = 3$ to $x = 6$?

Solution

The change in the quantity from $x = 3$ to $x = 6$ is

$$\begin{aligned}\int_3^6 \tfrac{1}{5}x^2\,\mathrm{d}x &= \left[\tfrac{1}{5} \times \tfrac{1}{3}x^3\right]_3^6\\ &= \left[\tfrac{1}{15}x^3\right]_3^6\\ &= \tfrac{1}{15}\left[x^3\right]_3^6\\ &= \tfrac{1}{15}(6^3 - 3^3)\\ &= \tfrac{1}{15} \times 189\\ &= \tfrac{63}{5}\\ &= 12.6.\end{aligned}$$

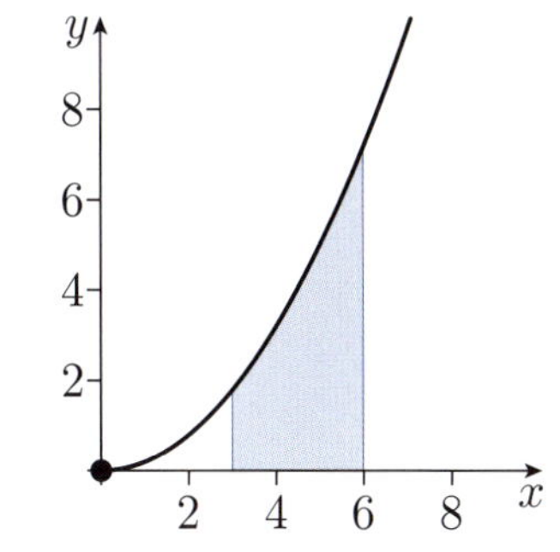

Figure 37 The graph of $y = \frac{1}{5}x^2$ $(x \geq 0)$

In the next activity, you're asked to repeat Activity 36 from Subsection 5.3 of Unit 7, but using the notation for a definite integral.

Activity 16 *Finding a change in displacement*

Suppose that an object moves along a straight line, and its velocity v (in $\mathrm{m\,s^{-1}}$) at time t (in seconds) is given by $v = 3 - \frac{3}{2}t$ $(0 \leq t \leq 6)$ (as shown in Figure 38).

(a) Write down a definite integral that gives the change in the displacement of the object from time $t = 0$ to time $t = 1$. Evaluate it, and hence state how far the object travels in that time.

(b) Repeat part (a), but from time $t = 4$ to time $t = 5$.

Try using the notation for a definite integral to solve the problem in the next activity.

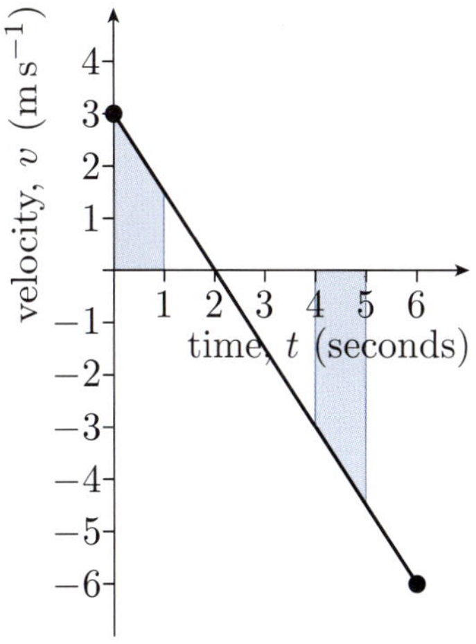

Figure 38 The graph of $v = 3 - \frac{3}{2}t$ $(0 \leq t \leq 6)$

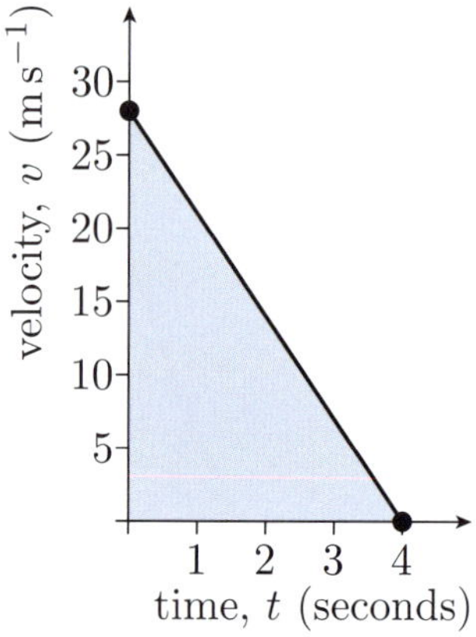

Figure 39 The graph of $v = 28 - 7t$ $(0 \leq t \leq 4)$

Activity 17 *Finding another change in displacement*

The driver of a car travelling at $28\,\mathrm{m\,s^{-1}}$ along a straight road applies the brakes. It takes four seconds for the car to stop, and its decreasing velocity v (in $\mathrm{m\,s^{-1}}$) during that time is given by the equation

$$v = 28 - 7t,$$

where t is the time in seconds since the brakes were applied (as shown in Figure 39). How far does the car travel during those four seconds?

In this subsection you've seen examples of the following two useful facts.

For the velocity-time graph of any object, and for any two points in time $t = a$ and $t = b$ in the time period covered by the graph, the following hold.

- The change in the displacement of the object from $t = a$ to $t = b$ is the total *signed area* between the graph and the time axis from $t = a$ to $t = b$.
- The total distance travelled by the object from $t = a$ to $t = b$ is the total *area* between the graph and the time axis from $t = a$ to $t = b$ (this applies when $a \leq b$).

Note that if a graph is a straight line, or consists of several line segments, then integration may not be the easiest way to work out an area or a signed area between the graph and the horizontal axis. For example, you can work out the area shaded in Figure 39 by using the formula for the area of a triangle, as shown at the end of the solution to Activity 17. Try using a similar geometric method in the next activity, which is a repeat of Example 23 in Unit 7.

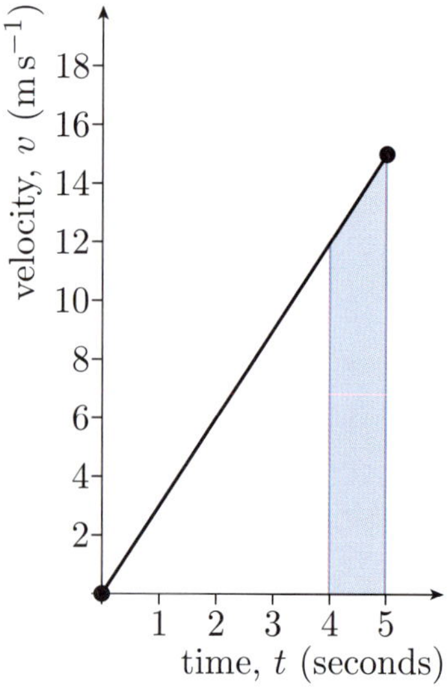

Figure 40 The graph of $v = 3t$ $(0 \leq t \leq 5)$

Activity 18 *Finding yet another change in displacement*

Suppose that a car begins to move along a straight road, and its velocity v (in $\mathrm{m\,s^{-1}}$) during the first five seconds of its journey is given by the equation

$$v = 3t,$$

where t is the time in seconds since it began moving, as shown in Figure 40. How far does the car travel from the end of the fourth second to the end of the fifth second?

2.4 Notation for indefinite integrals

Until now, when we've been working with antiderivatives and indefinite integrals, we've been using a lower-case letter, such as f, to denote the function that we start with, and the corresponding upper-case letter, such as F, to denote any of its antiderivatives, or its indefinite integral.

For example, some antiderivatives of the function $f(x) = x^2$ are

$$F(x) = \tfrac{1}{3}x^3, \quad F(x) = \tfrac{1}{3}x^3 + 7, \quad \text{and} \quad F(x) = \tfrac{1}{3}x^3 - \tfrac{2}{9},$$

and the indefinite integral of this function f is

$$F(x) = \tfrac{1}{3}x^3 + c.$$

However, there's a standard notation for indefinite integrals, which arises from the link between antiderivatives and definite integrals. For any function f that has an antiderivative, the indefinite integral of $f(x)$ is denoted by

$$\int f(x)\,\mathrm{d}x.$$

(This notation is read as 'the integral of f of x, d x'.)

For example,

$$\int x^2\,\mathrm{d}x = \tfrac{1}{3}x^3 + c,$$

and similarly

$$\int \sin x\,\mathrm{d}x = -\cos x + c.$$

(You saw in Unit 7 that the indefinite integral of $\sin x$ is $-\cos x + c$.)

Note that this notation is used only for indefinite integrals – *it's never used for antiderivatives*. So it's important to remember the following fact.

> If an equation has the notation $\int \ldots \mathrm{d}x$ on one side only, then there must be an arbitrary constant on the other side.

For example, the equation

$$\int x^2\,\mathrm{d}x = \tfrac{1}{3}x^3$$

is *incorrect*. The correct equation is

$$\int x^2\,\mathrm{d}x = \tfrac{1}{3}x^3 + c,$$

as you saw above.

We'll continue to use the capital letter notation for antiderivatives, when it's convenient to do so.

The notation for indefinite integrals introduced above, and the similar notation for definite integrals, are together known as **integral notation**. We refer to either an indefinite integral or a definite integral as an **integral**. In the indefinite integral

$$\int f(x)\,dx,$$

the expression $f(x)$ is called the **integrand**, just as in a definite integral.

The next example illustrates how integral notation is used for indefinite integrals in practice. It's similar to the examples and activities on finding indefinite integrals in the second half of Unit 7, except that it uses integral notation.

Example 10 *Finding indefinite integrals, using integral notation*

Find the following indefinite integrals.

(a) $\int \sec^2 x\,dx$ (b) $\int \frac{x-5}{x}\,dx$

Solution

(a) The indefinite integral of $\sec^2 x$ is given in the table of standard indefinite integrals, so you can simply write it down.

$$\int \sec^2 x\,dx = \tan x + c$$

(b) Manipulate the integrand to get it into a form that allows you to use the constant multiple rule and sum rule for antiderivatives, together with standard indefinite integrals.

$$\begin{aligned}\int \frac{x-5}{x}\,dx &= \int \left(1 - \frac{5}{x}\right) dx\\ &= \int \left(1 - 5 \times \frac{1}{x}\right) dx\\ &= x - 5\ln|x| + c\end{aligned}$$

You can practise using integral notation for indefinite integrals in the following activity. Don't forget to add the '$+c$' in each part.

Activity 19 *Finding indefinite integrals, using integral notation*

Find the following indefinite integrals.

(a) $\displaystyle\int e^x \,\mathrm{d}x$ (b) $\displaystyle\int (x+3)(x-2)\,\mathrm{d}x$ (c) $\displaystyle\int \frac{1}{x^3}\,\mathrm{d}x$

(d) $\displaystyle\int (\sin\theta - \cos\theta)\,\mathrm{d}\theta$ (e) $\displaystyle\int \frac{1}{1+x^2}\,\mathrm{d}x$ (f) $\displaystyle\int \frac{-3}{1+u^2}\,\mathrm{d}u$

(g) $\displaystyle\int \frac{1}{2(1+r^2)}\,\mathrm{d}r$ (h) $\displaystyle\int \frac{(1+x^2)}{2}\,\mathrm{d}x$

Since an indefinite integral is just an antiderivative with an arbitrary constant added on, the constant multiple rule and sum rule for antiderivatives, from Unit 7, can be written in terms of indefinite integrals. These forms of the rules are stated in the following box, in integral notation.

Constant multiple rule and sum rule for indefinite integrals

$$\int k\,f(x)\,\mathrm{d}x = k\int f(x)\,\mathrm{d}x, \quad \text{where } k \text{ is a constant}$$

$$\int (f(x)+g(x))\,\mathrm{d}x = \int f(x)\,\mathrm{d}x + \int g(x)\,\mathrm{d}x$$

As with the other constant multiple rules and sum rules, you can choose whether or not to use these rules in your working, as seems convenient. The next example repeats Example 10(b), but uses the constant multiple rule and sum rule for indefinite integrals in the forms stated above.

Example 11 *Using the constant multiple rule and sum rule for indefinite integrals*

Find the indefinite integral

$$\int \frac{x-5}{x}\,dx.$$

Solution

Manipulate the integrand to get it into a form that allows you to use the constant multiple rule and sum rule for indefinite integrals, together with standard indefinite integrals.

$$\begin{aligned}\int \frac{x-5}{x}\,dx &= \int \left(1-\frac{5}{x}\right)dx \\ &= \int 1\,dx - 5\int \frac{1}{x}\,dx \\ &= x - 5\ln|x| + c\end{aligned}$$

Here's something to notice about Example 11. You might have thought that each of the two indefinite integrals in the second line of the working should have been replaced by an expression containing an arbitrary constant. That would lead to a solution something like this:

$$\begin{aligned}\int \frac{x-5}{x}\,dx &= \int \left(1-\frac{5}{x}\right)dx \\ &= \int 1\,dx - 5\int \frac{1}{x}\,dx \\ &= (x+c) - 5(\ln|x| + d) \\ &= x - 5\ln|x| + c - 5d,\end{aligned}$$

where c and d are arbitrary constants.

However, since the arbitrary constants c and d can take any value, the expression $c - 5d$ can also take any value, so we may as well replace it by a single letter that can take any value. It's convenient to use the usual letter, c. In general, if an expression consists of a sum of constant multiples of indefinite integrals, then it needs only a single arbitrary constant.

You can practise using the constant multiple rule and sum rule for indefinite integrals in the next activity.

Activity 20 *Using the constant multiple rule and sum rule for indefinite integrals*

Find the following indefinite integrals.

(a) $\displaystyle\int \left(\frac{3}{x} - \frac{1}{1+x^2}\right) dx$ (b) $\displaystyle\int (\operatorname{cosec} x)(\operatorname{cosec} x + \cot x)\, dx$

You've now reached the end of the material in this module that tells you what integration is all about. In the rest of this unit you'll concentrate on actually finding indefinite and definite integrals.

Before you go on to that, you might like to watch the 25-minute video *The birth of calculus*, which is available on the module website. It tells you more about the history of the development of calculus.

3 Integration by substitution

To allow you to fully exploit all the ideas about integration that you've met, it's important that you're able to find formulas for the indefinite integrals of as wide a range of functions as possible. In the rest of this unit, you'll learn to integrate a much wider variety of functions than you learned to integrate in Unit 7. In the final subsection of the unit, you'll learn how to use the module computer algebra system to find indefinite and definite integrals.

Unfortunately, as mentioned in Unit 7, it's generally more tricky to integrate functions than it is to differentiate them. That's because there are no rules for antiderivatives that are similar to the product, quotient and chain rules for derivatives, so you can't usually integrate functions in the systematic way that you can differentiate them.

However, there are still many useful techniques that you can use to integrate functions. Each of these techniques applies to functions with particular characteristics. You'll meet some of these techniques in the rest of this unit. You'll learn how to recognise functions to which each technique can be applied, and how to use the techniques.

The technique that you'll learn in this section is *integration by substitution*. This technique might seem quite complicated when you first meet it, but you should find it more straightforward once you've had some practice with it. If you find it difficult, then make sure that you watch the tutorial clips, and try some extra practice. Remember that there are more practice exercises in the practice quiz and exercise booklet for this unit.

3.1 Basic integration by substitution

Integration by substitution is based on reversing the chain rule for differentiation.

Consider what happens when you differentiate an expression by using the chain rule. Remember that you start by recognising that the expression is a function of 'something'. You differentiate the expression with respect to the 'something', then you multiply the result by the derivative of the 'something' with respect to the input variable (usually x). For example, by the chain rule,

$$\frac{\mathrm{d}}{\mathrm{d}x}\left(\sin(x^2)\right) = \left(\cos(x^2)\right)(2x).$$

Since differentiation is the reverse of integration, this equation tells you that

$$\int \left(\cos(x^2)\right)(2x)\,\mathrm{d}x = \sin(x^2) + c.$$

Whenever you differentiate an expression by using the chain rule, you always obtain an expression of the form

f(something) $\times$ the derivative of the something,

where f is a function that you can integrate. The essential idea of integration by substitution is to recognise an expression that you want to integrate as having this form, and then perform the integration by reversing the chain rule.

To help you reverse the chain rule, it's helpful to denote the 'something' by an extra variable, in the way that you did when you first learned to use the chain rule in Unit 7. The usual choice of letter for the extra variable is u.

Here's how you can use this method to find the integral

$$\int \left(\cos(x^2)\right)(2x)\,\mathrm{d}x,$$

if you hadn't first seen the integrand obtained as an 'output' of the chain rule.

The first step is to recognise that the integrand has the form

f(something) $\times$ the derivative of the something,

where f is a function that you can integrate. You can see that it does, since it is

cos(something) $\times$ the derivative of the something,

where the 'something' is x^2.

Then, since the 'something' is x^2, you put $u = x^2$. This gives $\mathrm{d}u/\mathrm{d}x = 2x$. Hence, in the integral you can replace $\cos(x^2)$ by $\cos u$, and $2x$ by $\mathrm{d}u/\mathrm{d}x$, to give

$$\int \cos(u)\frac{\mathrm{d}u}{\mathrm{d}x}\,\mathrm{d}x.$$

Now here's the crucial step that you need to apply at this stage. Any integral of the form

$$\int f(u)\,\frac{\mathrm{d}u}{\mathrm{d}x}\,\mathrm{d}x$$

is equal to the simpler integral

$$\int f(u)\,\mathrm{d}u.$$

This is the step that uses the chain rule, and you'll see an explanation of why it's correct later in this subsection. Notice that it's easy to remember, because it looks like we've simply cancelled a '$\mathrm{d}x$' in a denominator with a '$\mathrm{d}x$' in a numerator. (Of course that's *not* what we've done, since $\mathrm{d}u/\mathrm{d}x$ isn't a fraction.)

So the integral that we're trying to find here can be expressed in terms of u as the simpler integral

$$\int \cos u\,\mathrm{d}u,$$

where $u = x^2$. You can now do the integration, which gives

$$\sin u + c.$$

The final step is to express this answer in terms of x, using the fact that $u = x^2$. This gives the final answer

$$\sin(x^2) + c.$$

So we've now worked out that

$$\int \left(\cos(x^2)\right)(2x)\,\mathrm{d}x = \sin(x^2) + c,$$

as expected.

Here's a summary of the method used above, which is the method known as **integration by substitution**.

Integration by substitution

1. Recognise that the integrand is of the form

   f(something) × the derivative of the something,

   where f is a function that you can integrate.
2. Set the something equal to u, and find $\mathrm{d}u/\mathrm{d}x$.
3. Hence write the integral in the form

   $$\int f(u)\,\mathrm{d}u,$$

   by using the fact that $\displaystyle\int f(u)\,\frac{\mathrm{d}u}{\mathrm{d}x}\,\mathrm{d}x = \int f(u)\,\mathrm{d}u.$
4. Do the integration.
5. Substitute back for u in terms of x.

As illustrated by the example above, the effect of substituting the variable u in place of the 'something' in the integral is to reduce the integral to a simpler integral in terms of u, which you may be able to find more easily.

You'll see another example of integration by substitution shortly, and you'll then have a chance to try it for yourself, but first it's useful for you to learn about a convenient way to carry out step 3 of the method.

Consider once more the integral above:

$$\int \left(\cos(x^2)\right)(2x)\,\mathrm{d}x.$$

Remember that we put $u = x^2$, which gave $\mathrm{d}u/\mathrm{d}x = 2x$, and this allowed us to write the integral in terms of u as

$$\int \cos u\,\mathrm{d}u.$$

It's convenient to think of this new form of the integral as being obtained from the original form by making two replacements, as shown below:

$$\int \underbrace{\left(\cos(x^2)\right)}_{\cos u}\ \underbrace{(2x)\,\mathrm{d}x}_{\mathrm{d}u}.$$

It's straightforward to work out that you can replace $\cos(x^2)$ by $\cos u$, as that just comes from the equation $u = x^2$.

A helpful way to work out that you can replace $(2x)\,\mathrm{d}x$ by $\mathrm{d}u$ is to imagine 'cross-multiplying' in the equation

$$\frac{\mathrm{d}u}{\mathrm{d}x} = 2x,$$

to obtain

$$\mathrm{d}u = (2x)\,\mathrm{d}x.$$

This isn't a real equation, of course, since $\mathrm{d}u$ and $\mathrm{d}x$ don't have independent meanings outside the notation for a derivative or integral. But it tells you immediately that you can replace $(2x)\,\mathrm{d}x$ by $\mathrm{d}u$. (In fact, it's possible to attach a meaning to $\mathrm{d}u$ and $\mathrm{d}x$, and hence to the equation above, but this is outside the scope of this module.)

Here's another example of integration by substitution, which illustrates all the ideas above.

Example 12 *Integrating by substitution*

Find the integral

$$\int e^{\tan x}\sec^2 x\,\mathrm{d}x.$$

Solution

Since the derivative of $\tan x$ is $\sec^2 x$, the integrand is of the form

$$e^{\text{something}} \times \text{the derivative of the something.}$$

So set the something equal to u.

Let $u = \tan x$; then $\dfrac{du}{dx} = \sec^2 x$.

Imagine 'cross-multiplying' in the second equation to obtain $du = (\sec^2 x)\,dx$.

So

$$\int e^{\tan x} \sec^2 x\,dx = \int e^u\,du$$

Do the integration.

$$= e^u + c$$

Substitute back for u in terms of x.

$$= e^{\tan x} + c.$$

Here are some examples for you to try. Make sure that you have your *Handbook* to hand, so you can consult the tables of standard derivatives and standard integrals.

Activity 21 *Integrating by substitution*

Find the following indefinite integrals.

(a) $\displaystyle\int e^{\cos x}(-\sin x)\,dx$ (b) $\displaystyle\int (\sin(x^3))\,(3x^2)\,dx$

(c) $\displaystyle\int \left(\frac{1}{\sin x}\right) \cos x\,dx$ (d) $\displaystyle\int \left(\frac{1}{\cos^2 x}\right) (-\sin x)\,dx$

(e) $\displaystyle\int \sin^4 x \cos x\,dx$ (f) $\displaystyle\int \left(\frac{1}{1+\sin^2 x}\right) \cos x\,dx$

In the examples that you've seen so far, it's been fairly straightforward to decide what the 'something' should be, so that the integrand is of the form

$$f(\text{something}) \times \text{the derivative of the something.}$$

However, sometimes you need to think about this more carefully, as illustrated in the next example.

Example 13 *Choosing the 'something' when you integrate by substitution*

Find the integral

$$\int \left(\frac{1}{2+\sin x}\right) \cos x \, dx.$$

Solution

The integrand is of the form

$$\frac{1}{2+\text{something}} \times \text{the derivative of the something,}$$

where the something is $\sin x$.

It's also of the form

$$\frac{1}{\text{something}} \times \text{the derivative of the something,}$$

where the something is $2 + \sin x$.

Choosing the first option would lead to the first part of the integrand being replaced by $1/(2+u)$, which isn't straightforward to integrate. Choosing the second option would lead to the first part of the integrand being replaced by $1/u$, which you can integrate using a result from the table of standard integrals.

So choose the second option: take the something to be $2 + \sin x$, and set it equal to u.

Let $u = 2 + \sin x$; then $\dfrac{du}{dx} = \cos x$.

Imagine 'cross-multiplying' to obtain $du = \cos x \, dx$.

So

$$\begin{aligned}\int \left(\frac{1}{2+\sin x}\right) \cos x \, dx &= \int \frac{1}{u} \, du \\ &= \ln |u| + c \\ &= \ln |2 + \sin x| + c\end{aligned}$$

In the particular case here you can remove the modulus signs, since $2 + \sin x$ is always positive.

$$= \ln(2 + \sin x) + c.$$

In each part of the next activity it's useful to take the 'something' to be an expression that includes a constant term, in a similar way to Example 13.

Activity 22 *Choosing the 'something' when you integrate by substitution*

Find the following indefinite integrals.

(a) $\displaystyle\int (4+\cos x)^7(-\sin x)\,\mathrm{d}x$ (b) $\displaystyle\int \sqrt{1+x^2}\,(2x)\,\mathrm{d}x$

(c) $\displaystyle\int (x^5-8)^{10}(5x^4)\,\mathrm{d}x$ (d) $\displaystyle\int \left(\frac{1}{e^x+5}\right) e^x\,\mathrm{d}x$

(e) $\displaystyle\int \left(\sin(5+2x^3)\right)(6x^2)\,\mathrm{d}x$

In the next activity, some of the integrals are similar to those in Activity 22, whereas others are similar to those in Activity 21. You have to decide in each case what the 'something' should be. Remember that the idea is that when you replace the something by u in the first part of the integrand, you obtain an expression in u that you can integrate.

Activity 23 *Choosing the 'something' when you integrate by substitution*

Find the following indefinite integrals.

(a) $\displaystyle\int (\cos(e^x))\,e^x\,\mathrm{d}x$ (b) $\displaystyle\int e^{1+\sin x}(\cos x)\,\mathrm{d}x$

(c) $\displaystyle\int \left(\frac{1}{x^{10}+6}\right)(10x^9)\,\mathrm{d}x$ (d) $\displaystyle\int (\cos^3 x)(-\sin x)\,\mathrm{d}x$

Finally in this subsection, as promised near the start, here's an explanation of why

$$\int f(u)\,\frac{\mathrm{d}u}{\mathrm{d}x}\,\mathrm{d}x = \int f(u)\,\mathrm{d}u,$$

where the variable u is a function of x, and f is a function that you can integrate. To justify this, we let F be an antiderivative of f, and apply the chain rule to $F(u)$. This gives

$$\frac{\mathrm{d}}{\mathrm{d}x}\left(F(u)\right) = f(u)\,\frac{\mathrm{d}u}{\mathrm{d}x}.$$

It follows from this equation that

$$\int f(u)\,\frac{\mathrm{d}u}{\mathrm{d}x}\,\mathrm{d}x = F(u)+c.$$

However, because F is an antiderivative of f, we also have

$$\int f(u)\,\mathrm{d}u = F(u)+c.$$

Comparing the final two equations above proves the result.

3.2 Integration by substitution in practice

Usually when you have an integrand that you can integrate by substitution, it won't initially be exactly in the form

f(something) $\times$ the derivative of the something.

You have to start by recognising that it can be rearranged into this form, then rearrange it accordingly. Here are two examples that illustrate some things that you can do.

Example 14 *Rearranging an integral before using substitution*

Find the integral

$$\int e^x \cos(e^x)\,\mathrm{d}x.$$

Solution

The derivative of e^x is e^x, so the integrand is in the required form, except that the two factors $\cos(e^x)$ and e^x are in the 'wrong' order. If you swap them, then the integrand will be exactly in the required form.

Let $u = e^x$; then $\dfrac{\mathrm{d}u}{\mathrm{d}x} = e^x$.

Imagine 'cross-multiplying' to obtain $\mathrm{d}u = e^x\,\mathrm{d}x$. Then start by swapping the two factors in the integrand.

So

$$\begin{aligned}\int e^x \cos(e^x)\,\mathrm{d}x &= \int (\cos(e^x))\; e^x\,\mathrm{d}x\\ &= \int \cos u\,\mathrm{d}u\\ &= \sin(u) + c\\ &= \sin(e^x) + c.\end{aligned}$$

Example 15 *Rearranging another integral before using substitution*

Find the integral

$$\int x^2\left(1+x^3\right)^8\,\mathrm{d}x.$$

Solution

The derivative of $1+x^3$ is $3x^2$, which is 'nearly' x^2 (it just has the wrong coefficient). So if you swap the order of the two factors x^2 and $(1+x^3)^8$, then the integrand will be nearly in the required form.

Let $u=1+x^3$; then $\dfrac{\mathrm{d}u}{\mathrm{d}x}=3x^2$.

Imagine 'cross-multiplying' to obtain $\mathrm{d}u=3x^2\,\mathrm{d}x$. Then start by swapping the two factors in the integrand.

So

$$\int x^2\left(1+x^3\right)^8\,\mathrm{d}x=\int\left(1+x^3\right)^8 x^2\,\mathrm{d}x$$

The only problem is that in place of x^2 you need $3x^2$. To achieve this, multiply by 3 inside the integral, and divide by 3 outside, to compensate. (The constant multiple rule for indefinite integrals tells you that it's okay to do this.)

$$=\tfrac{1}{3}\int\left(1+x^3\right)^8(3x^2)\,\mathrm{d}x$$

Now do the substitution.

$$\begin{aligned}&=\tfrac{1}{3}\int u^8\,\mathrm{d}u\\&=\tfrac{1}{3}\times\tfrac{1}{9}u^9+c\\&=\tfrac{1}{27}u^9+c\\&=\tfrac{1}{27}(1+x^3)^9+c.\end{aligned}$$

Here's something to notice about Example 15. You might have thought that when the integral $\int u^8\,\mathrm{d}u$ in the fourth-last line of the solution was replaced by $\frac{1}{9}u^9+c$, the solution should have proceeded like this:

$$\begin{aligned}&=\tfrac{1}{3}\int u^8\,\mathrm{d}u\\&=\tfrac{1}{3}\left(\tfrac{1}{9}u^9+c\right)\\&=\tfrac{1}{27}u^9+\tfrac{1}{3}c\\&=\tfrac{1}{27}(1+x^3)^9+\tfrac{1}{3}c.\end{aligned}$$

However, since the arbitrary constant c can take any value, the expression $\frac{1}{3}c$ can also take any value, so we may as well replace it by a single letter that can take any value. It's convenient to use the usual letter, c.

This sort of thing occurs frequently when you integrate. You never need to multiply or divide a term that is an *arbitrary* constant by a number – you just replace it by another arbitrary constant, and you usually use the standard letter, c.

Another point to remember from the working in Example 15 is the strategy of multiplying by a constant inside an integral, and dividing by the same constant outside to compensate. It's often helpful to do this when you use integration by substitution. It's okay to do it because the constant multiple rule for indefinite integrals tells you that for any function f that has an antiderivative, and any non-zero constant k,

$$\int f(x)\,\mathrm{d}x = \int \frac{1}{k}kf(x)\,\mathrm{d}x = \frac{1}{k}\int kf(x)\,\mathrm{d}x.$$

Remember that the constant k can be either positive or negative. In particular, it can be -1.

Remember too that k must be a *constant*. You can't do the same thing with an expression in x.

Here are some examples for you to try.

Activity 24 *Rearranging integrals before using substitution*

Find the following integrals.

(a) $\int x^2\cos(x^3)\,\mathrm{d}x$ (b) $\int x\sqrt{1+x^2}\,\mathrm{d}x$ (c) $\int \cos^4 x\sin x\,\mathrm{d}x$

The next example illustrates another way in which you can sometimes rearrange an integrand to get it into the right form for integration by substitution.

Example 16 *Rearranging another integral before using substitution*

Find the integral

$$\int \frac{x}{1-x^2}\,\mathrm{d}x.$$

Solution

The derivative of the denominator $1 - x^2$ is $-2x$, which is 'nearly' the numerator x (it just has the wrong coefficient). So if you split the numerator from the rest of the fraction, then the integrand will be nearly in the required form.

Let $u = 1 - x^2$; then $\dfrac{du}{dx} = -2x$.

Imagine 'cross-multiplying' to obtain $du = -2x\,dx$. Then start by splitting the numerator from the rest of the fraction in the integrand.

So

$$\int \frac{x}{1-x^2}\,dx = \int \left(\frac{1}{1-x^2}\right) x\,dx$$

The only problem is that in place of x you need $-2x$. To achieve this, multiply by -2 inside the integral, and divide by -2 outside to compensate.

$$= -\tfrac{1}{2} \int \left(\frac{1}{1-x^2}\right) (-2x)\,dx$$

Now do the substitution.

$$\begin{aligned} &= -\tfrac{1}{2} \int \frac{1}{u}\,du \\ &= -\tfrac{1}{2} \ln|u| + c \\ &= -\tfrac{1}{2} \ln|1 - x^2| + c. \end{aligned}$$

The type of integral in Example 16 occurs quite frequently, so it's worth learning to recognise it. You can use integration by substitution to find any integral of the form

$$\int \frac{\text{the derivative of the something}}{\text{something}}\,dx,$$

or

$$\int \frac{\text{the derivative of the something}}{(\text{something})^n}\,dx, \quad \text{where } n \text{ is a constant.}$$

The constant n can be any nonzero real number. Here are some integrals of this type for you to find.

Activity 25 *Rearranging more integrals before using substitution*

Find the following integrals.

(a) $\displaystyle\int \frac{x}{1+2x^2}\,dx$ (b) $\displaystyle\int \frac{e^x}{(2+e^x)^2}\,dx$ (c) $\displaystyle\int \frac{\cos x}{3-\sin x}\,dx$

Here are three more integrals for you to find by using substitution. They're a mixture of the types in Activities 24 and 25.

Activity 26 *Rearranging even more integrals before using substitution*

Find the following integrals.

(a) $\displaystyle\int e^{\cos x}\sin x\,dx$ (b) $\displaystyle\int \frac{e^x}{3-e^x}\,dx$ (c) $\displaystyle\int (1+\tfrac{1}{2}\sin x)^2\cos x\,dx$

You can use integration by substitution to find the indefinite integral of the tangent function. You're asked to do that in the next activity.

Activity 27 *Finding the indefinite integral of tan*

By writing

$$\tan x = \frac{\sin x}{\cos x},$$

and using integration by substitution, find

$$\int \tan x\,dx.$$

Remember that when you've found an indefinite integral, you can always check your answer by differentiating it. This can be particularly helpful when you found the integral by using integration by substitution.

3.3 Integrating functions of linear expressions

This subsection is about a particular type of integral that occurs frequently, and which you can find by using integration by substitution.

The integrals

$$\int \sin(5x-2)\,\mathrm{d}x, \quad \int \frac{1}{4-x}\,\mathrm{d}x, \quad \text{and} \quad \int e^{-9x}\,\mathrm{d}x$$

all have integrands of the form $f(ax+b)$, where f is a function, and a and b are constants with $a \neq 0$. In other words, each of the integrands is of the form $f(\text{a linear expression in } x)$.

You can always find an integral of this form by using integration by substitution, provided that you can integrate the function f. Here's an example.

Example 17 *Integrating a function of a linear expression*

Find the integral

$$\int \sin(5x-2)\,\mathrm{d}x.$$

Solution

The derivative of $5x-2$ is 5, so the integral will be in the required form if you rearrange it to obtain 5 inside.

Let $u = 5x-2$; then $\dfrac{\mathrm{d}u}{\mathrm{d}x} = 5$. So

$$\begin{aligned}\int \sin(5x-2)\,\mathrm{d}x &= \tfrac{1}{5}\int \sin(5x-2)\times 5\,\mathrm{d}x\\ &= \tfrac{1}{5}\int \sin(5x-2)\times 5\,\mathrm{d}x\\ &= \tfrac{1}{5}\int \sin u\,\mathrm{d}u\\ &= -\tfrac{1}{5}\cos u + c\\ &= -\tfrac{1}{5}\cos(5x-2) + c.\end{aligned}$$

Here are some similar integrals for you to find.

Activity 28 *Integrating functions of linear expressions*

Find the following integrals.

(a) $\displaystyle\int e^{6x-1}\,dx$ (b) $\displaystyle\int \sin(4x)\,dx$ (c) $\displaystyle\int e^{-9x}\,dx$

(d) $\displaystyle\int e^{-x/3}\,dx$ (e) $\displaystyle\int \cos(3-7x)\,dx$ (f) $\displaystyle\int \frac{1}{4-x}\,dx$

(g) $\displaystyle\int \frac{1}{(2x+1)^2}\,dx$ (h) $\displaystyle\int \frac{1}{(1-x)^3}\,dx$ (i) $\displaystyle\int \sec^2(6x)\,dx$

The next example uses exactly the same ideas as Example 17 and Activity 28, but is a little more complicated.

Example 18 *Integrating another function of a linear expression*

Find the integral

$$\int \frac{1}{1+9x^2}\,dx.$$

Solution

You can write the integrand as $\dfrac{1}{1+(3x)^2}$, which is of the form $f(3x)$ where $f(x)=\dfrac{1}{1+x^2}$. The integral of this function f is a standard integral. Since the derivative of $3x$ is 3, you need to rearrange the given integral to obtain 3 inside.

Let $u=3x$; then $\dfrac{du}{dx}=3$. So

$$\begin{aligned}\int \frac{1}{1+9x^2}\,dx &= \int \frac{1}{1+(3x)^2}\,dx\\ &= \tfrac{1}{3}\int \frac{1}{1+(3x)^2}\times 3\,dx\\ &= \tfrac{1}{3}\int \frac{1}{1+u^2}\,du\\ &= \tfrac{1}{3}\tan^{-1}u+c\\ &= \tfrac{1}{3}\tan^{-1}(3x)+c.\end{aligned}$$

Here are some similar integrals for you to find.

Activity 29 *Integrating more functions of linear expressions*

Find the following integrals.

(a) $\displaystyle\int \frac{1}{\sqrt{1-4x^2}}\,\mathrm{d}x$ (b) $\displaystyle\int \frac{1}{1+2x^2}\,\mathrm{d}x$ (c) $\displaystyle\int \frac{1}{4+9x^2}\,\mathrm{d}x$

Hint for part (c): start by writing

$$\frac{1}{4+9x^2} \quad \text{as} \quad \frac{1}{4}\times\frac{1}{1+\frac{9}{4}x^2}.$$

The answers to Activities 28 and 29 all follow the same general pattern, as follows. Suppose that you have an integral of the form

$$\int f(ax+b)\,\mathrm{d}x,$$

where a and b are constants with $a \neq 0$, and you know an antiderivative F of the function f. To find the integral by substitution, you put $u = ax+b$, which gives $\dfrac{\mathrm{d}u}{\mathrm{d}x} = a$. Then

$$\begin{aligned}\int f(ax+b)\,\mathrm{d}x &= \frac{1}{a}\int f(ax+b)\,a\,\mathrm{d}x\\ &= \frac{1}{a}\int f(u)\,\mathrm{d}u\\ &= \frac{1}{a}F(u)+c\\ &= \frac{1}{a}F(ax+b)+c.\end{aligned}$$

So we have the following general fact.

Indefinite integral of a function of a linear expression

If f is a function with antiderivative F, and a and b are constants with $a \neq 0$, then

$$\int f(ax+b)\,\mathrm{d}x = \frac{1}{a}F(ax+b)+c.$$

It's worth remembering this fact, and using it to deal with integrals where the integrand is a simple function of a linear expression, rather than using integration by substitution every time. For example, the fact above tells you immediately that

$$\int \cos(3x-2)\,\mathrm{d}x = \tfrac{1}{3}\sin(3x-2)+c.$$

The following special case of the fact in the box above is useful especially often.

Indefinite integral of a function of a multiple of the variable

If f is a function with antiderivative F, and a is a non-zero constant, then

$$\int f(ax)\,\mathrm{d}x = \frac{1}{a}F(ax) + c.$$

For example,

$$\int e^{-2x}\,\mathrm{d}x = -\tfrac{1}{2}e^{-2x} + c.$$

Try using the facts in the boxes above in the following activity.

Activity 30 *Integrating functions of linear expressions efficiently*

Find the following integrals.

(a) $\displaystyle\int \cos(10x)\,\mathrm{d}x$ (b) $\displaystyle\int \sin(2-5x)\,\mathrm{d}x$ (c) $\displaystyle\int e^{3t+4}\,\mathrm{d}t$

(d) $\displaystyle\int \cos(\tfrac{1}{2}\theta)\,\mathrm{d}\theta$ (e) $\displaystyle\int e^{-4p}\,\mathrm{d}p$ (f) $\displaystyle\int \frac{1}{7x-4}\,\mathrm{d}x$

(g) $\displaystyle\int \frac{1}{1-r}\,\mathrm{d}r$ (h) $\displaystyle\int \sec^2(5x-1)\,\mathrm{d}x$ (i) $\displaystyle\int \sin(\tfrac{1}{5}x)\,\mathrm{d}x$

(j) $\displaystyle\int e^{-x}\,\mathrm{d}x$ (k) $\displaystyle\int e^{x/2}\,\mathrm{d}x$ (l) $\displaystyle\int \sec^2(2-3p)\,\mathrm{d}p$

(m) $\displaystyle\int \sin\left(\frac{2-x}{3}\right)\mathrm{d}x$

The next example is of a type that you met in Unit 7, but it involves a function that you've learned how to integrate in this section.

Example 19 *Integrating in a practical situation*

The velocity of an object moving along a straight line is given by the equation

$$v = \tfrac{1}{2}\sin\left(\tfrac{3}{2}t - 2\right),$$

where t is the time in seconds since the object began moving, and v is the velocity of the object, in $\mathrm{m\,s^{-1}}$. The displacement of the object at time $t = 0$ is $3\,\mathrm{m}$.

Find an equation for the displacement s (in metres) of the object, in terms of t. Give the constant term to two decimal places.

Solution

Use the fact that displacement is an antiderivative of velocity.

The velocity v (in $\mathrm{m\,s^{-1}}$) of the object at time t (in seconds) is given by

$$v = \tfrac{1}{2}\sin\left(\tfrac{3}{2}t - 2\right).$$

Hence the displacement s (in m) of the object at time t (in seconds) is given by

$$\begin{aligned} s &= \int \tfrac{1}{2}\sin\left(\tfrac{3}{2}t - 2\right)\,\mathrm{d}t \\ &= \frac{1}{2} \times \frac{1}{3/2}\left(-\cos\left(\tfrac{3}{2}t - 2\right)\right) + c \\ &= -\tfrac{1}{3}\cos\left(\tfrac{3}{2}t - 2\right) + c, \end{aligned}$$

where c is an arbitrary constant.

Use information given in the question to find the value of c.

When $t = 0$, $s = 3$. Hence

$$\begin{aligned} 3 &= -\tfrac{1}{3}\cos\left(\tfrac{3}{2} \times 0 - 2\right) + c \\ 3 &= -\tfrac{1}{3}\cos(-2) + c \\ c &= 3 + \tfrac{1}{3}\cos(-2) \\ c &= 3 + \tfrac{1}{3}\cos 2 \\ c &= 2.86 \text{ (to 2 d.p.)}. \end{aligned}$$

So the equation for the displacement of the object in terms of time is

$$s = -\tfrac{1}{3}\cos\left(\tfrac{3}{2}t - 2\right) + 2.86.$$

Activity 31 *Integrating in a practical example*

The velocity of an object moving along a straight line is given by the equation

$$v = 2\cos\left(\tfrac{1}{2}t + 5\right),$$

where t is the time in seconds since the object began moving, and v is the velocity of the object, in $\mathrm{m\,s^{-1}}$. The displacement of the object at time $t = 0$ is $4\,\mathrm{m}$.

(a) Find an equation for the displacement s (in metres) of the object, in terms of t. Give the constant term to two decimal places.

(b) Find the displacement of the object at time $t = 10$. Give your answer in metres to two significant figures.

3.4 Integration by substitution for definite integrals

Once you've used integration by substitution to find an indefinite integral, you can find any *definite* integral with the same integrand by using the fundamental theorem of calculus in the usual way. Here's an example.

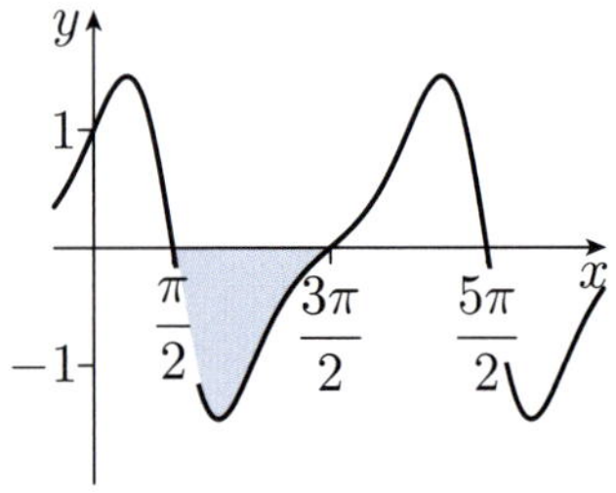

Figure 41 The definite integral in Example 20

Example 20 *Evaluating a definite integral by substitution (first method)*

Evaluate the definite integral

$$\int_{\pi/2}^{3\pi/2} e^{\sin x} \cos x \, dx.$$

(It's shown in Figure 41.) Give your answer to three significant figures.

Solution

Use integration by substitution to obtain the indefinite integral of the integrand.

Let $u = \sin x$, then $\dfrac{du}{dx} = \cos x$. So

$$\begin{aligned}\int e^{\sin x} \cos x \, dx &= \int e^u \, du \\ &= e^u + c \\ &= e^{\sin x} + c.\end{aligned}$$

Apply the fundamental theorem of calculus.

Hence

$$\begin{aligned}\int_{\pi/2}^{3\pi/2} e^{\sin x}\cos x\,\mathrm{d}x &= \left[e^{\sin x}\right]_{\pi/2}^{3\pi/2}\\ &= e^{\sin(3\pi/2)} - e^{\sin(\pi/2)}\\ &= e^{-1} - e^{1}\\ &= \frac{1}{e} - e\\ &= -2.35 \text{ (to 3 s.f.).}\end{aligned}$$

An alternative way to use integration by substitution to evaluate a definite integral is to work with the definite integral from the start, rather than first finding an indefinite integral. If you do this, then because the limits of integration are values of x, when you change the variable from x to u you must also change the limits of integration to the corresponding values of u.

This method is demonstrated in the next example. The definite integral in this example is the same as the one in Example 20.

Example 21 *Evaluating a definite integral by substitution (second method)*

Evaluate the definite integral

$$\int_{\pi/2}^{3\pi/2} e^{\sin x}\cos x\,\mathrm{d}x.$$

Give your answer to three significant figures.

Solution

Use integration by substitution.

Let $u = \sin x$, then $\dfrac{\mathrm{d}u}{\mathrm{d}x} = \cos x$.

Find the values of u that correspond to the values of x that are the limits of integration.

Putting $x = \pi/2$ into $u = \sin x$ gives $u = \sin\left(\dfrac{\pi}{2}\right) = 1$.

Putting $x = 3\pi/2$ into $u = \sin x$ gives $u = \sin\left(\dfrac{3\pi}{2}\right) = -1$.

Change the variable from x to u, remembering that you also have to change the limits of integration.

So

$$\begin{aligned}\int_{\pi/2}^{3\pi/2} e^{\sin x}\cos x\,\mathrm{d}x &= \int_{1}^{-1} e^{u}\,\mathrm{d}u \\ &= \left[e^{u}\right]_{1}^{-1} \\ &= e^{-1} - e^{1} \\ &= -2.35 \text{ (to 3 s.f.).}\end{aligned}$$

You can try using this method in the next activity.

Activity 32 *Evaluating definite integrals by substitution*

Evaluate the following definite integrals. Give your answers to three significant figures.

(a) $\displaystyle\int_0^1 \frac{x}{1+2x^2}\,\mathrm{d}x$ (b) $\displaystyle\int_{\pi/2}^{\pi} \cos^3 x \sin x\,\mathrm{d}x$ (c) $\displaystyle\int_{-1}^{0} xe^{-3x^2}\,\mathrm{d}x$

3.5 Finding more complicated integrals by substitution

You can sometimes use integration by substitution to find an integral even if the integrand isn't obviously of the form

f(something) × the derivative of the something.

In this subsection, you'll see some examples where the integrand is nearly of this form, except that it also contains a further factor, which is an expression in x. If you can express this further factor in terms of the variable u that you use for the substitution, then you may be able to simplify the integrand into a form that you can integrate. Here's an example.

Example 22 *Expressing a further factor in terms of u when integrating by substitution*

Find the integral

$$\int (x+1)(3x+1)^9 \, dx.$$

Solution

If the integral were just

$$\int (3x+1)^9 \, dx,$$

then we could use the substitution $u = 3x+1$, and find the integral by first writing it as

$$\tfrac{1}{3} \int (3x+1)^9 \times 3 \, dx.$$

However, the integrand contains a further factor, $x+1$. Try using the substitution, expressing the further factor $x+1$ in terms of u.

Let $u = 3x+1$. Then $\dfrac{du}{dx} = 3$.

Also, rearranging the equation $u = 3x+1$ gives

$$\begin{aligned} u &= 3x+1 \\ 3x &= u-1 \\ x &= \tfrac{1}{3}(u-1). \end{aligned}$$

Hence

$$\begin{aligned} x+1 &= \tfrac{1}{3}(u-1)+1 \\ &= \tfrac{1}{3}(u+2). \end{aligned}$$

So

$$\begin{aligned} \int (x+1)(3x+1)^9 \, dx &= \tfrac{1}{3} \int (x+1)(3x+1)^9 \times 3 \, dx \\ &= \tfrac{1}{3} \int \tfrac{1}{3}(u+2)u^9 \, du \\ &= \tfrac{1}{9} \int (u+2)u^9 \, du \\ &= \tfrac{1}{9} \int (u^{10} + 2u^9) \, du \\ &= \tfrac{1}{9} \left(\tfrac{1}{11}u^{11} + 2 \times \tfrac{1}{10}u^{10} \right) + c \\ &= \tfrac{1}{99}u^{11} + \tfrac{1}{45}u^{10} + c \\ &= \tfrac{1}{99}(3x+1)^{11} + \tfrac{1}{45}(3x+1)^{10} + c. \end{aligned}$$

Here are two similar integrals for you to try.

Activity 33 *Expressing further factors in terms of u*

Use integration by substitution to find the following indefinite integrals.

(a) $\displaystyle\int x\sqrt{1+x}\,\mathrm{d}x$ (b) $\displaystyle\int \frac{x-2}{(x+3)^3}\,\mathrm{d}x$

You'll learn more about integration by substitution in the module MST125 *Essential mathematics 2*, if it's in your study programme.

4 Integration by parts

In this section you'll meet another integration technique, *integration by parts*. Like integration by substitution, this technique is useful for integrating functions that have certain characteristics. It's based on the product rule for differentiation.

4.1 Basic integration by parts

Suppose that f and g are functions, and that the function G is an antiderivative of the function g. Consider the equation that you obtain when you use the product rule to differentiate the product $f(x)G(x)$:

$$\frac{\mathrm{d}}{\mathrm{d}x}(f(x)G(x)) = f(x)\left(\frac{\mathrm{d}}{\mathrm{d}x}G(x)\right) + G(x)\left(\frac{\mathrm{d}}{\mathrm{d}x}f(x)\right),$$

that is,

$$\frac{\mathrm{d}}{\mathrm{d}x}(f(x)G(x)) = f(x)g(x) + G(x)f'(x).$$

If you swap the order of $G(x)$ and $f'(x)$ in the final term, and subtract this term from both sides of the equation, then you obtain

$$\frac{\mathrm{d}}{\mathrm{d}x}(f(x)G(x)) - f'(x)G(x) = f(x)g(x).$$

Swapping the sides of this equation gives

$$f(x)g(x) = \frac{\mathrm{d}}{\mathrm{d}x}(f(x)G(x)) - f'(x)G(x).$$

If you now integrate both sides of this equation with respect to x, then you obtain

$$\int f(x)g(x)\,dx = f(x)G(x) - \int f'(x)G(x)\,dx.$$

There's no need to add an explicit arbitrary constant here, because it's included in the integral $\int f(x)g(x)\,dx$ on the left-hand side, and in the integral $\int f'(x)G(x)\,dx$ on the right-hand side.

The formula above is known as the **integration by parts formula**. It's stated again in the box below, for easy reference. If you've met it before, then you may have seen it stated in a different form. This is explained at the end of this section.

Integration by parts formula (Lagrange notation)

$$\int f(x)g(x)\,dx = f(x)G(x) - \int f'(x)G(x)\,dx.$$

Here G is an antiderivative of g.

The integration by parts formula is useful for integrating some products of the form $f(x)g(x)$. You can see that it changes the problem of integrating $f(x)g(x)$ into the problem of integrating $f'(x)G(x)$, where G is an antiderivative of g.

So it's useful for integrating products of the form $f(x)g(x)$ that have the following characteristics.

- You can find an antiderivative G of g; in other words, you can integrate the expression $g(x)$.
- The product $f'(x)G(x)$ is easier to integrate than the product $f(x)g(x)$. For example, this may be the case if $f'(x)$ is simpler than $f(x)$.

When we integrate by using the integration by parts formula, we say that we're **integrating by parts**. Here's an example.

Example 23 *Integrating by parts*

Find the integral $\int x \sin x \, dx$.

Solution

The integrand is a product of two expressions, x and $\sin x$. You can integrate the second expression, $\sin x$, and differentiating the first expression, x, makes it simpler. So try integration by parts.

Let $f(x) = x$ and $g(x) = \sin x$.

Then $f'(x) = 1$, and an antiderivative of $g(x)$ is $G(x) = -\cos x$.

So the integration by parts formula gives

$$\begin{aligned}\int x \sin x \, dx &= \int f(x)g(x) \, dx \\ &= f(x)G(x) - \int f'(x)G(x) \, dx \\ &= x \times (-\cos x) - \int 1 \times (-\cos x) \, dx \\ &= -x \cos x + \int \cos x \, dx\end{aligned}$$

Find the integral in this expression.

$$= -x \cos x + \sin x + c.$$

You can see that in Example 23 the use of the integration by parts formula changed the problem of integrating $x \sin x$ into the problem of integrating $\cos x$, which is easier. Here are two similar examples for you to try.

Activity 34 *Integrating by parts*

Find the following integrals.

(a) $\int x \cos x \, dx$ (b) $\int xe^x \, dx$

As you become more familiar with integration by parts, you'll probably find that you can carry it out more quickly and conveniently if you remember the informal version of the formula stated below. You can recite this version in your head as you apply integration by parts, in a similar way to the informal versions of the product and quotient rules for differentiation that you met in Unit 7. It's useful for applying integration by parts in fairly simple cases, like the ones that you've seen so far, which is usually all that you need to do.

Integration by parts formula (informal)

$$\begin{aligned}\text{integral of product} &= (\text{first}) \times \begin{pmatrix}\text{antiderivative}\\ \text{of second}\end{pmatrix}\\ &\quad - \text{integral of}\left(\begin{pmatrix}\text{derivative}\\ \text{of first}\end{pmatrix} \times \begin{pmatrix}\text{antiderivative}\\ \text{of second}\end{pmatrix}\right).\end{aligned}$$

In the next example, the integration in Example 23 is repeated, using the informal version of the integration by parts formula.

Example 24 *Integrating by parts, using the informal formula*

Find the integral $\int x \sin x \, dx$.

Solution

The integrand is a product of two expressions, x and $\sin x$. You can integrate the second expression, $\sin x$, and differentiating the first expression, x, makes it simpler. So try integration by parts.

Recite the informal version of the formula in your head. As you think 'first', write down the first expression, then as you think 'antiderivative of second', write down an antiderivative of the second expression, and so on.

Integrating by parts gives

$$\begin{aligned}\int x \sin x \, dx &= x \times (-\cos x) - \int 1 \times (-\cos x) \, dx\\ &= -x \cos x + \int \cos x \, dx\\ &= -x \cos x + \sin x + c.\end{aligned}$$

You can practise using the informal version of the integration by parts formula in the next activity. Alternatively, you might prefer to continue using the formal version for a little longer, and move to the informal version when you feel more confident with it. You can use the exercises on integration by parts in the practice quiz and exercise booklet for this unit for further practice.

Activity 35 *Integrating by parts*

Find the following integrals.

(a) $\int x\sin(5x)\,\mathrm{d}x$ (b) $\int xe^{2x}\,\mathrm{d}x$ (c) $\int xe^{-x}\,\mathrm{d}x$

When you use integration by parts, remember that it matters which expression in the integrand you take to be the 'first expression', and which you take to be the 'second expression'. If you're trying to use integration by parts, but you find that it leads to an integral that's more complicated than the original integral, then try swapping the two expressions in the original integrand.

4.2 Integration by parts in practice

There are several frequently occurring types of integrals that often have the right characteristics for integration by parts. In this subsection you'll meet three of these types. It's helpful for you to learn to recognise them.

Integrands of the form $x^n\,g(x)$

You can sometimes find an integral of the form

$$\int x^n\,g(x)\,\mathrm{d}x,$$

where n is a positive integer and $g(x)$ is an expression that you can integrate, by using integration by parts n times.

All the integrals in the previous subsection were of this form, with $n = 1$, so you could find them by using integration by parts once. Here's an example where you have to use integration by parts twice.

Example 25 *Integrating by parts twice*

Find the integral $\int x^2e^{4x}\,\mathrm{d}x$.

Solution

The integral is of the form described above, so try integration by parts.

Integrating by parts gives

$$\int x^2 e^{4x}\,dx = x^2 \times \tfrac{1}{4}e^{4x} - \int 2x \times \tfrac{1}{4}e^{4x}\,dx$$
$$= \tfrac{1}{4}x^2 e^{4x} - \tfrac{1}{2}\int xe^{4x}\,dx$$

The integral in this expression, $\int xe^{4x}\,dx$, looks easier to find than the original integral, but it's still not straightforward to find. However, it is itself of the form described above, so try integration by parts again.

$$= \tfrac{1}{4}x^2 e^{4x} - \tfrac{1}{2}\left(x \times \tfrac{1}{4}e^{4x} - \int 1 \times \tfrac{1}{4}e^{4x}\,dx\right)$$
$$= \tfrac{1}{4}x^2 e^{4x} - \tfrac{1}{2}\left(\tfrac{1}{4}xe^{4x} - \tfrac{1}{4}\int e^{4x}\,dx\right)$$

Multiply out the brackets. Take care with the minus sign in front of the brackets – remember that it turns the minus sign inside the brackets into a plus sign.

$$= \tfrac{1}{4}x^2 e^{4x} - \tfrac{1}{8}xe^{4x} + \tfrac{1}{8}\int e^{4x}\,dx$$

Now find the integral in this expression, and simplify the resulting expression.

$$= \tfrac{1}{4}x^2 e^{4x} - \tfrac{1}{8}xe^{4x} + \tfrac{1}{8} \times \tfrac{1}{4}e^{4x} + c$$
$$= \tfrac{1}{4}x^2 e^{4x} - \tfrac{1}{8}xe^{4x} + \tfrac{1}{32}e^{4x} + c$$
$$= \tfrac{1}{32}e^{4x}(8x^2 - 4x + 1) + c.$$

Activity 36 *Integrating by parts twice*

Find the following integrals.

(a) $\int x^2 e^x\,dx$ (b) $\int x^2 \cos(2x)\,dx$

The method introduced above for dealing with integrands of the form $x^n g(x)$ works just as well if the expression x^n is replaced by any polynomial expression in x of degree n. For example, you can find the integral $\int (3x^2 - 1)e^{4x}\,dx$ by integrating by parts twice, and you can find the integral in the next activity by integrating by parts once.

Activity 37 *Integrating an expression of the form* $(ax + b)g(x)$

Use integration by parts to find the integral $\int (5x + 2)e^{-5x}\,dx$.

Integrands of the form $x^r \ln x$

Suppose that you want to integrate an expression of the form

$$x^r \ln x,$$

where r is a constant (any real number). Your first thought, particularly if r is a positive integer, might be that this is similar to the expressions that you saw how to integrate earlier in this subsection. However, you can't apply the integration by parts formula with $\ln x$ as the second expression, because $\ln x$ isn't an expression that you've seen how to integrate.

In fact, you can find an integral of the form above by applying the integration by parts formula with $\ln x$ as the *first* expression. Here's an example.

Example 26 *Integrating an expression of the form* $x^n \ln x$

Find the integral

$$\int x^2 \ln x\,dx.$$

Solution

The integral is of the form discussed above, so swap the expressions x^2 and $\ln x$, then integrate by parts.

$$\begin{aligned}\int x^2 \ln x\,dx &= \int (\ln x)\,x^2\,dx \\ &= (\ln x) \times \tfrac{1}{3}x^3 - \int \frac{1}{x} \times \tfrac{1}{3}x^3\,dx \\ &= \tfrac{1}{3}x^3 \ln x - \tfrac{1}{3}\int x^2\,dx \\ &= \tfrac{1}{3}x^3 \ln x - \tfrac{1}{3}\left(\tfrac{1}{3}x^3\right) + c \\ &= \tfrac{1}{3}x^3 \ln x - \tfrac{1}{9}x^3 + c \\ &= \tfrac{1}{9}x^3(3\ln x - 1) + c\end{aligned}$$

Activity 38 *Integrating expressions of the form $x^r \ln x$*

Find the following integrals.

(a) $\int x^3 \ln x \, dx$ (b) $\int \ln x \, dx$

Hint for part (b): write $\ln x$ as $(\ln x) \times 1$.

You can use the ideas that you used in Activity 38 to deal with other, similar integrals that involve the natural logarithm function. Try the following activity.

Activity 39 *Integrating another expression containing the function ln*

Use integration by parts to find the integral $\int (x-2) \ln(3x) \, dx$.

Integrands of the form $e^{ax} \sin(bx)$ or $e^{ax} \cos(bx)$

You can integrate an expression of the form

$$e^{ax} \sin(bx) \quad \text{or} \quad e^{ax} \cos(bx),$$

where a and b are nonzero numbers, by integrating by parts twice. When you do this, you obtain an equation that expresses the original integral in terms of itself, and you can rearrange this equation to find the original integral. The method is demonstrated in the example below.

When you use this method, you can arrange the exponential expression and the trigonometric expression in the integrand in either order before you integrate by parts for the first time. For example, you can start with either $\int e^{ax} \sin(bx) \, dx$ or $\int \sin(bx) e^{ax} \, dx$. However, you must choose the *same order* (exponential before trigonometric, or trigonometric before exponential) when you integrate by parts for the second time. If you don't do that, then the equation that you'll obtain will reduce to $0 = 0$, which isn't helpful!

Example 27 *Finding an integral by expressing it in terms of itself*

Find the indefinite integral $\int e^x \cos(2x)\,dx$.

Solution

The integral is of the form discussed above, so use integration by parts twice. Start by swapping the expressions e^x and $\cos(2x)$, as this makes the working slightly easier (it avoids introducing fractions in the early stages).

We use integration by parts twice.

$$\int e^x \cos(2x)\,dx = \int \cos(2x)\,e^x\,dx.$$

Integrate by parts.

$$\begin{aligned}\int \cos(2x)\,e^x\,dx &= \cos(2x)e^x - \int (-2\sin(2x))e^x\,dx \\ &= \cos(2x)e^x + 2\int \sin(2x)e^x\,dx\end{aligned}$$

Integrate by parts again.

$$\begin{aligned}&= \cos(2x)e^x + 2\left(\sin(2x)e^x - \int (2\cos(2x))e^x\,dx\right) \\ &= \cos(2x)e^x + 2\sin(2x)e^x - 4\int \cos(2x)e^x\,dx\end{aligned}$$

The final term here contains the original integral. Move this term to the other side of the equation. When you do this, the right-hand side no longer contains an indefinite integral, so you have to add an arbitrary constant to that side.

So

$$\int \cos(2x)e^x\,dx + 4\int \cos(2x)e^x\,dx = \cos(2x)e^x + 2\sin(2x)e^x + c;$$

that is,

$$5\int \cos(2x)e^x\,dx = \cos(2x)e^x + 2\sin(2x)e^x + c.$$

To obtain the final answer, divide both sides by 5. (Remember that you leave the arbitrary constant as c rather than $c/5$.) Also, it's slightly tidier to write the exponential expression before the trigonometric expression in each term, and to take out a common factor.

This gives

$$\begin{aligned}\int e^x \cos(2x)\,dx &= \tfrac{1}{5}e^x \cos(2x) + \tfrac{2}{5}e^x \sin(2x) + c \\ &= \tfrac{1}{5}e^x(\cos(2x) + 2\sin(2x)) + c.\end{aligned}$$

Activity 40 *Finding integrals by expressing them in terms of themselves*

Find the following integrals.

(a) $\int e^x \sin(3x)\,dx$ (b) $\int e^{2x} \cos x\,dx$

4.3 Integration by parts for definite integrals

Once you've used integration by parts to find an indefinite integral, you can find any definite integral with the same integrand by using the fundamental theorem of calculus in the usual way. Here's an example.

Example 28 *Evaluating a definite integral by parts (first method)*

Evaluate the definite integral

$$\int_0^1 xe^{7x}\,dx.$$

Solution

Use integration by parts to obtain the indefinite integral of the integrand.

Integration by parts gives

$$\begin{aligned}\int xe^{7x}\,dx &= x \times \tfrac{1}{7}e^{7x} - \int 1 \times \tfrac{1}{7}e^{7x}\,dx \\ &= \tfrac{1}{7}xe^{7x} - \tfrac{1}{7}\int e^{7x}\,dx \\ &= \tfrac{1}{7}xe^{7x} - \tfrac{1}{7} \times \tfrac{1}{7}e^{7x} + c \\ &= \tfrac{1}{49}e^{7x}(7x - 1) + c.\end{aligned}$$

Apply the fundamental theorem of calculus.

It follows that

$$\begin{aligned}\int_0^1 xe^{7x}\,\mathrm{d}x &= \left[\tfrac{1}{49}e^{7x}(7x-1)\right]_0^1\\ &= \tfrac{1}{49}\left[e^{7x}(7x-1)\right]_0^1\\ &= \tfrac{1}{49}\left(e^{7\times 1}(7\times 1-1) - e^{7\times 0}(7\times 0-1)\right)\\ &= \tfrac{1}{49}(6e^7-(-1))\\ &= \tfrac{1}{49}(6e^7+1).\end{aligned}$$

An alternative way to use integration by parts to evaluate a definite integral is to apply the limits of integration 'as you go along'. To do this, you use the following formula, which is just the usual integration by parts formula with the limits of integration included.

Integration by parts formula for definite integrals

$$\int_a^b f(x)g(x)\,\mathrm{d}x = \left[f(x)G(x)\right]_a^b - \int_a^b f'(x)G(x)\,\mathrm{d}x.$$

Here G is an antiderivative of g.

You can remember this formula as an informal version, by applying the limits of integration a and b in the appropriate way to the informal version of the usual integration by parts formula.

The next example demonstrates how to apply the limits of integration as you go along when you integrate by parts. The definite integral in this example is the same as the one in Example 28.

Example 29 *Evaluating a definite integral by parts (second method)*

Evaluate the definite integral

$$\int_0^1 xe^{7x}\,\mathrm{d}x.$$

Solution

Integrate by parts, applying the limits of integration.

$$\int_0^1 xe^{7x}\,dx = \left[x \times \tfrac{1}{7}e^{7x}\right]_0^1 - \int_0^1 1 \times \tfrac{1}{7}e^{7x}\,dx$$
$$= \tfrac{1}{7}\left[xe^{7x}\right]_0^1 - \tfrac{1}{7}\int_0^1 e^{7x}\,dx$$

Evaluate the first term, and do the integration in the second term.

$$= \tfrac{1}{7}(1 \times e^{7\times 1} - 0 \times e^{7\times 0}) - \tfrac{1}{7}\left[\tfrac{1}{7}e^{7x}\right]_0^1$$
$$= \tfrac{1}{7}e^7 - \tfrac{1}{49}\left[e^{7x}\right]_0^1$$
$$= \tfrac{1}{7}e^7 - \tfrac{1}{49}(e^{7\times 1} - e^{7\times 0})$$
$$= \tfrac{1}{7}e^7 - \tfrac{1}{49}(e^7 - 1)$$
$$= \tfrac{1}{49}(7e^7 - (e^7 - 1))$$
$$= \tfrac{1}{49}(6e^7 + 1)$$

Activity 41 *Evaluating a definite integral by parts*

Use the integration by parts formula for definite integrals (the method of Example 29) to evaluate the definite integral

$$\int_0^1 xe^{3x}\,dx.$$

Give your answer to four decimal places.

In the next two activities you're asked to use integration by parts to find areas.

Activity 42 *Using integration by parts to find an area*

This question is about the function $f(x) = x\sin(3x)$.

(a) Explain why the graph of f lies above the x-axis for values of x in the interval $[\pi/12, \pi/6]$.

(b) Write down a definite integral that gives the area between the graph of f and the x-axis from $x = \pi/12$ to $x = \pi/6$.

(c) Use the integration by parts formula for definite integrals to find the area described in part (b). Give your answer to four decimal places.

Activity 43 *Using integration by parts to find another area*

This question is about the function $f(x) = (x-2)(e^{-x}-1)$.

(a) Explain why the graph of f lies above the x-axis for $0 < x < 2$ and below the x-axis for $x > 2$. You might find it convenient to use a table of signs as part of your explanation.

(b) Find the total area (not the total *signed area*) between the graph of f and the x-axis, from $x = 0$ to $x = 4$. Give your answer to four significant figures.

As mentioned earlier, in some texts the integration by parts formula is expressed in a different form.

Here's the formula that you've seen in this section:

$$\int f(x)g(x)\,\mathrm{d}x = f(x)G(x) - \int f'(x)G(x)\,\mathrm{d}x.$$

It involves a function G and its derivative g.

If you denote this pair of functions by g and g' instead, then you obtain the alternative version below.

Integration by parts formula (Lagrange notation, alternative version)

$$\int f(x)g'(x)\,\mathrm{d}x = f(x)g(x) - \int f'(x)g(x)\,\mathrm{d}x$$

As with all calculus formulas, the integration by parts formula can be stated in Leibniz notation.

If you put $u = f(x)$ and $v = g(x)$ in the formula in the box above, then you obtain the following form of the formula.

Integration by parts formula (Leibniz notation)

$$\int u\frac{\mathrm{d}v}{\mathrm{d}x}\,\mathrm{d}x = uv - \int v\frac{\mathrm{d}u}{\mathrm{d}x}\,\mathrm{d}x$$

5 More integration

You've now met several integration techniques, and practised using them on a variety of integrals.

In this section, you'll start by meeting one further approach, which can be helpful for integrals that contain trigonometric expressions. Then you'll have an opportunity to practise identifying which of the integration techniques that you've met are appropriate for which integrals, for a variety of different types of integrals.

Finally, you'll learn how to use a computer for integration.

5.1 Trigonometric integrals

In Section 3 you saw that you can integrate some expressions that contain trigonometric functions by using integration by substitution. Another approach that's often helpful when you want to integrate an expression that contains trigonometric functions is to start by using one of the trigonometric identities that you met in Unit 4 to write the expression in a different form. This can give you an expression that's easier to integrate.

For example, you can use the identities below to help you integrate the expressions $\sin^2 x$, $\cos^2 x$ and $\sin x \cos x$, as illustrated in the next example. The first two identities are the half-angle identities, and the third identity is one of the double-angle identities.

Three trigonometric identities

$$\sin^2 \theta = \tfrac{1}{2}(1 - \cos(2\theta))$$
$$\cos^2 \theta = \tfrac{1}{2}(1 + \cos(2\theta))$$
$$\sin(2\theta) = 2\sin\theta\cos\theta$$

Example 30 *Integrating by using a trigonometric identity*

Find the integral

$$\int \sin^2 x \, dx.$$

Solution

By the half-angle identity for sine,

$$\begin{aligned}\int \sin^2 x \, dx &= \int \tfrac{1}{2}(1 - \cos(2x)) \, dx \\ &= \tfrac{1}{2} \int (1 - \cos(2x)) \, dx \\ &= \tfrac{1}{2}(x - \tfrac{1}{2}\sin(2x)) + c \\ &= \tfrac{1}{2}x - \tfrac{1}{4}\sin(2x) + c \\ &= \tfrac{1}{4}(2x - \sin(2x)) + c.\end{aligned}$$

In the next activity you're asked to use two of the trigonometric identities stated near the beginning of this subsection to integrate another two expressions that contain trigonometric functions.

Activity 44 *Integrating by using trigonometric identities*

(a) Use a half-angle identity to find the integral $\int \cos^2 \theta \, d\theta$.

(b) Use the double-angle identity in the box above to find the integral $\int \sin x \cos x \, dx$.

The trigonometric integral in Example 30 and the one in Activity 44(a) can't easily be found without first using a trigonometric identity to write them in a different form.

However, there's an alternative way to find the integral in Activity 44(b). Since the derivative of $\sin x$ is $\cos x$, you can use integration by substitution, taking $u = \sin x$. You're asked to do that in the next activity.

Activity 45 *Integrating a trigonometric expression by using substitution*

Use integration by substitution to find the integral $\int \sin x \cos x \, dx$.

In fact there are even two different ways to find the integral in Activity 44(b), namely

$$\int \sin x \cos x \, dx,$$

by using integration by substitution. One way is as suggested above: you use the fact that the derivative of $\sin x$ is $\cos x$ and hence make the substitution $u = \sin x$. Alternatively, you can use the fact that the derivative of $\cos x$ is $-\sin x$, write the integral as

$$-\int (\cos x)(-\sin x) \, dx,$$

and make the substitution $u = \cos x$. The working for each of these methods is given in the solution to Activity 45.

So you've now seen three different ways to integrate the expression $\sin x \cos x$: one way by using a trigonometric identity, and two ways by using substitution. These three different ways lead to three answers that look different, as you can see in the solutions to Activities 44(b) and 45. This illustrates that there can be more than one way to integrate an expression, and different ways of integrating it can lead to answers that look different, though of course these answers will be equivalent.

If an answer that you obtain for an integral is different from the answer given in a book or by a computer, but you think that you haven't made a mistake, then you may be able to check that the two answers are equivalent by using algebraic manipulation. You may need to use trigonometric identities – there are examples of this in the solution to Activity 45.

A rougher, but sometimes quicker, check is to plot the graphs of the two different answers on a computer (taking the arbitrary constants to be zero, for example), and check that they seem to be vertical translations of each other. This shows that the two different answers seem to be equivalent once the arbitrary constants are taken into account. Again, there's an example of this in the solution to Activity 45.

In this subsection you've seen just a few examples of how you can use trigonometric identities and integration by substitution to help you integrate expressions that contain trigonometric functions. There are many more possibilities, some of which are covered in the module *Essential mathematics 2* (MST125).

5.2 Choosing a method of integration

When you have an integral that you want to find, it may not be obvious which method to try. Also, sometimes you might need to combine two or more techniques – for example you might be able to use algebraic manipulation, the constant multiple rule and the sum rule to express an integral in terms of simpler integrals, then use integration by substitution or integration by parts to find the simpler integrals.

Here are some things to think about when you have an integral that you want to find.

Choosing a method for finding an integral

- Is it a standard integral? Consult the table in the *Handbook*.
- Can you rearrange the integral to express it as a sum of constant multiples of simpler integrals?
- Is the integrand of the form $f(ax)$ or $f(ax+b)$, where a and b are constants and f is a function that you can integrate? If so, use the rule for integrating a function of a linear expression, or use integration by substitution, with $u = ax$ or $u = ax + b$ as appropriate.
- Can the integrand be written in the form

  $$f(\text{something}) \times \text{the derivative of the something},$$

  where f is a function that you can integrate? If so, use integration by substitution. Start by setting the 'something' equal to u.
- Is the integrand of the form

  $$f(x)g(x),$$

  where f is a function that becomes simpler when differentiated, and g is a function that you can integrate? (For example, $f(x)$ might be x or x^2 or $\ln x$.) If so, try integration by parts.
- If the integrand contains trigonometric functions, can you use trigonometric identities to rewrite it in a form that's easier to integrate?

When you're finding an indefinite integral, try to always remember to add the arbitrary constant '$+c$'. It's easy to forget it!

In the next activity you can practise identifying which method of integration to use.

Activity 46 *Choosing a method of integration*

For each integral below, suggest a method of integration to use. Where you suggest substitution, also suggest a suitable equation for u in terms of x.

You are not asked to actually find the integrals, but solutions are included so you can do so if you wish.

(a) $\displaystyle\int x^3\cos(9x^4)\,dx$ (b) $\displaystyle\int x\cos(5x)\,dx$ (c) $\displaystyle\int (x+\cos(5x))\,dx$

(d) $\displaystyle\int x(x^2+3)\,dx$ (e) $\displaystyle\int \frac{1}{1+3x^2}\,dx$ (f) $\displaystyle\int \frac{x^2}{(7-x^3)^7}\,dx$

(g) $\displaystyle\int xe^{-x/3}\,dx$ (h) $\displaystyle\int \sin\left(\tfrac{1}{2}x\right)\cos\left(\tfrac{1}{2}x\right)\,dx$ (i) $\displaystyle\int (x-1)^4\,dx$

(j) $\displaystyle\int e^x(1-2e^x)\,dx$ (k) $\displaystyle\int x(1+\sin(x^2))\,dx$ (l) $\displaystyle\int x(3x^2-2)^8\,dx$

You'll learn further useful strategies for integration in the module *Essential mathematics 2* (MST125), if it's in your study programme. In particular, you'll learn more about integration by substitution and integration by using trigonometric identities, and you'll study *partial fractions*, which are useful when you want to integrate rational functions.

5.3 Integration using a computer

There are many situations in which it's helpful to use a computer for integration, such as when you want to integrate an expression that's difficult to integrate by hand. In the first activity of this subsection you can learn how to use the module computer algebra system to find indefinite integrals and evaluate definite integrals.

Activity 47 *Using a computer for integration*

Work through Section 9 of the *Computer algebra guide*.

You can use the skills that you learned in Activity 47 in the next activity.

Activity 48 *Using a computer to work with a complicated function*

This activity is about the function

$$f(x) = \frac{x^2 - 10x + 15}{1 + \sqrt{x+1}}.$$

Do parts (b) to (d) below by using the module computer algebra system. Give your answers to parts (c) and (d) to three significant figures.

(a) What is the domain of f?

(b) Plot the graph of f.

(c) Find the two x-intercepts of f.

(d) Find the area between the graph of f and the x-axis, between the two x-intercepts.

The next two activities are the final ones of the unit. In them, you're asked to use your knowledge of integration to solve practical problems. The definite integrals involved in the problems can't easily be evaluated by using the fundamental theorem of calculus, as it's difficult to find antiderivatives of $\sqrt{x^3+1}$ and $\sin(t^2/150)$. However, you can use a computer to find approximate values.

Activity 49 *Using integration to solve a practical problem*

Consider the plastic edging shown below. The shape of its cross-section is given by the equation

$$y = \tfrac{1}{3}\sqrt{x^3+1} \quad (-1 \le x \le 1.5),$$

where x and y are measured in centimetres.

Calculate the volume of plastic that is required to make one metre of the edging. Give your answer in cubic centimetres to three significant figures.

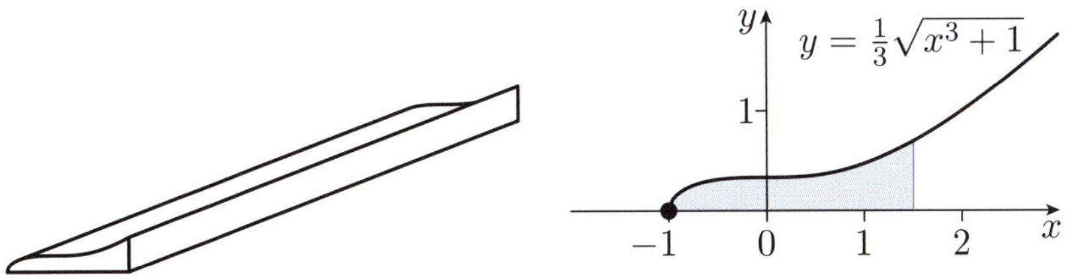

Activity 50 *Using integration to solve another practical problem*

Suppose that an object moves in a straight line, and its velocity v (in $\mathrm{m\,s^{-1}}$) at time t (in seconds) is given by the equation

$$v = \sin\left(\frac{t^2}{150}\right) \quad (0 \le t \le 20),$$

as shown in Figure 42.

If the displacement of the object from a particular reference point on the line at time $t = 0$ is 6 m, as shown below, what is the displacement of the object from this reference point at time $t = 10$? Give your answer to the nearest centimetre.

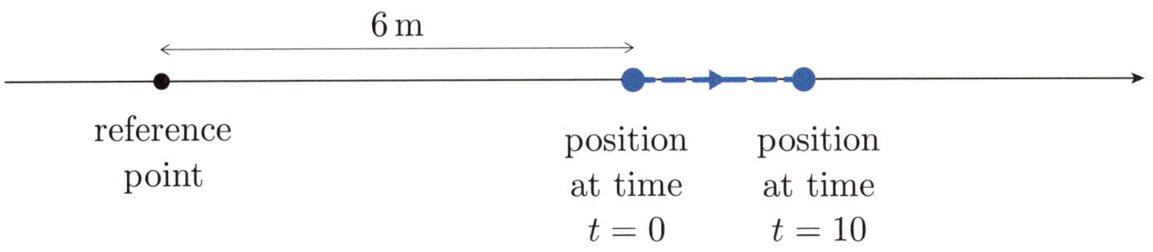

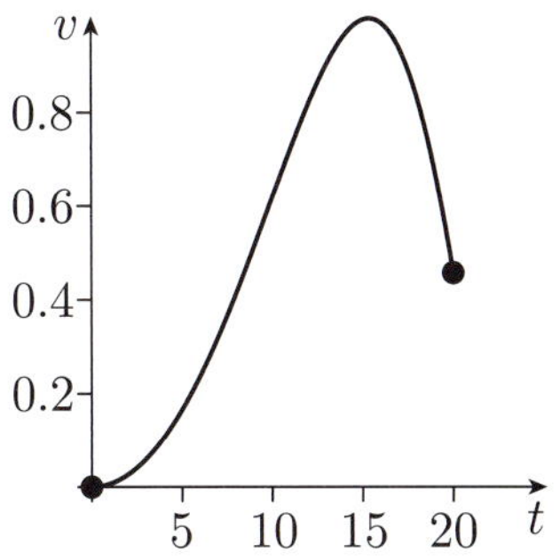

Figure 42 The graph of $v = \sin(x^2/150)$ $(0 \le t \le 20)$

You've now come to the end of the introduction to basic calculus in this module. You can use the skills that you've acquired to solve many different types of problems in mathematics, engineering and science, for example. These skills also equip you to study many further topics in these areas. You'll make a start on that in Unit 11, in which you'll use calculus in your study of *Taylor polynomials*. If your study programme includes further mathematics modules, then in some of these you may study more advanced calculus, with even more powerful applications.

Learning outcomes

After studying this unit, you should be able to:

- understand what is meant by a definite integral
- understand how to use subintervals to find an approximate value for a definite integral
- use integral notation
- understand the fundamental theorem of calculus
- evaluate some definite integrals by using the fundamental theorem of calculus
- integrate by substitution
- integrate by parts
- appreciate how to use trigonometric identities and integration by substitution to help you integrate some trigonometric expressions
- identify a suitable method of integration for some types of integrals
- use a computer for integration.

Solutions to activities

Solution to Activity 1

If we divide the interval $[-18, 18]$ into six subintervals of equal width, then each subinterval has width $36/6 = 6$.

The left endpoints of the six subintervals are

$$-18, \quad -18+6, \quad -18+2\times 6,$$
$$-18+3\times 6, \quad -18+4\times 6, \quad -18+5\times 6$$

that is

$$-18, \quad -12, \quad -6, \quad 0, \quad 6, \quad 12.$$

So the heights of the rectangles are:

$$f(-18) = 9 - \tfrac{1}{36}\times(-18)^2 = 0$$
$$f(-12) = 9 - \tfrac{1}{36}\times(-12)^2 = 5$$
$$f(-6) = 9 - \tfrac{1}{36}\times(-6)^2 = 8$$
$$f(0) = 9 - \tfrac{1}{36}\times 0^2 = 9$$
$$f(6) = 9 - \tfrac{1}{36}\times 6^2 = 8$$
$$f(12) = 9 - \tfrac{1}{36}\times 12^2 = 5.$$

So we obtain the following approximate value for the area:

$$(0\times 6) + (5\times 6) + (8\times 6)$$
$$+ (9\times 6) + (8\times 6) + (5\times 6) = 210.$$

Solution to Activity 2

(a) The signed area from $x = 1$ to $x = 5$ is -6.3.

(b) The signed area from $x = 5$ to $x = 7$ is 1.3.

(c) The signed area from $x = 1$ to $x = 7$ is $-6.3 + 1.3 = -5$.

(d) The signed area from $x = 1$ to $x = 9$ is $-6.3 + 1.3 - 5.9 = -10.9$.

Solution to Activity 3

If we divide the interval $[-3, 3]$ into six subintervals of equal width, then each subinterval has width $6/6 = 1$.

The left endpoints of the six subintervals are

$$-3, -2, -1, 0, 1, 2.$$

So the heights of the rectangles are:

$$f(-3) = 3 - (-3)^2 = -6$$
$$f(-2) = 3 - (-2)^2 = -1$$
$$f(-1) = 3 - (-1)^2 = 2$$
$$f(0) = 3 - 0^2 = 3$$
$$f(1) = 3 - 1^2 = 2$$
$$f(2) = 3 - 2^2 = -1.$$

So we obtain the following approximate value for the signed area:

$$(-6\times 1) + (-1\times 1) + (2\times 1)$$
$$+ (3\times 1) + (2\times 1) + (-1\times 1) = -1.$$

Solution to Activity 4

The exact value of the signed area between the graph of the function $f(x) = 3 - x^2$ and the x-axis from $x = -3$ to $x = 3$ is 0.

Solution to Activity 5

(a) The signed area from $x = 2$ to $x = 6$ is 4.0.

(b) The signed area from $x = 2$ to $x = 6$ is 4.0, so the signed area from $x = 6$ to $x = 2$ is -4.0.

(c) The signed area from $x = 2$ to $x = 10$ is $4.0 - 5.4 = -1.4$.

(d) From part (c), the signed area from $x = 2$ to $x = 10$ is -1.4, so the signed area from $x = 10$ to $x = 2$ is 1.4.

(e) The signed area from $x = 8$ to $x = 8$ is 0.

(f) The signed area from $x = -3$ to $x = 2$ is $4.1 - 2.4 = 1.7$, so the signed area from $x = 2$ to $x = -3$ is -1.7.

(g) The signed area from $x = -1$ to $x = 10$ is $-2.4 + 4.0 - 5.4 = -3.8$, so the signed area from $x = 10$ to $x = -1$ is 3.8.

Solution to Activity 6

(a) $\int_{-5}^{-3} f(x)\,dx = -5.$

(b) $\int_{-3}^{2} f(x)\,dx = 6.$

(c) $\int_{2}^{7} f(x)\,dx = -11.$

(d) $\int_{-5}^{2} f(x)\,dx = -5 + 6 = 1.$

(e) $\int_{-3}^{7} f(x)\,dx = 6 - 11 = -5.$

(f) $\int_{-5}^{7} f(x)\,dx = -5 + 6 - 11 = -10.$

(g) $\int_{5}^{5} f(x)\,dx = 0.$

(h) $\int_{2}^{-3} f(x)\,dx = -\int_{-3}^{2} f(x)\,dx = -6.$

(i) $\int_{7}^{2} f(x)\,dx = -\int_{2}^{7} f(x)\,dx = -(-11) = 11.$

(j) $\int_{-3}^{-5} f(x)\,dx = -\int_{-5}^{-3} f(x)\,dx = -(-5) = 5.$

(k)
$$\begin{aligned}\int_{7}^{-3} f(x)\,dx &= -\int_{-3}^{7} f(x)\,dx \\ &= -(-5) = 5,\end{aligned}$$
by part (e).

(l)
$$\begin{aligned}\int_{7}^{-5} f(x)\,dx &= -\int_{-5}^{7} f(x)\,dx \\ &= -(-10) = 10,\end{aligned}$$
by part (f).

Solution to Activity 7

$$\int_2^9 f(x)\,dx = \int_2^7 f(x)\,dx + \int_7^9 f(x)\,dx$$
so
$$-15 = -11 + \int_7^9 f(x)\,dx$$
and hence
$$\int_7^9 f(x)\,dx = -15 - (-11) = -4.$$

Solution to Activity 8

(a) $\left[\frac{1}{2}x^2\right]_3^5 = \frac{1}{2} \times 5^2 - \frac{1}{2} \times 3^2 = \frac{1}{2}(25 - 9) = 8.$

(b) $\left[\cos x\right]_0^{2\pi} = \cos(2\pi) - \cos 0 = 1 - 1 = 0.$

(c) $\left[e^x\right]_{-1}^{1} = e^1 - e^{-1} = e - (1/e) = 2.35$ (to 3 s.f.).

Solution to Activity 9

(a) An antiderivative of x^2 is $\frac{1}{3}x^3$, so
$$\begin{aligned}\int_{-1}^{1} x^2\,dx &= \left[\tfrac{1}{3}x^3\right]_{-1}^{1} \\ &= \tfrac{1}{3} \times 1^3 - \tfrac{1}{3} \times (-1)^3 \\ &= \tfrac{2}{3}.\end{aligned}$$

(b) An antiderivative of e^t is e^t, so
$$\begin{aligned}\int_0^2 e^t\,dt &= \left[e^t\right]_0^2 \\ &= e^2 - e^0 \\ &= e^2 - 1 = 6.39 \text{ (to 3 s.f.)}.\end{aligned}$$

(c) An antiderivative of $1/u$ (for $u > 0$) is $\ln u$, so
$$\begin{aligned}\int_1^4 \frac{1}{u}\,du &= \left[\ln u\right]_1^4 \\ &= \ln 4 - \ln 1 \\ &= \ln 4 = 1.39 \text{ (to 3 s.f.)}.\end{aligned}$$

(d) An antiderivative of $\sec^2\theta$ is $\tan\theta$, so
$$\begin{aligned}\int_{-\pi/4}^{\pi/4} \sec^2\theta\,d\theta &= \left[\tan\theta\right]_{-\pi/4}^{\pi/4} \\ &= \tan\frac{\pi}{4} - \tan\left(-\frac{\pi}{4}\right) \\ &= 1 - (-1) \\ &= 2.\end{aligned}$$

(e) An antiderivative of $1/(1 + u^2)$ is $\tan^{-1} u$, so
$$\begin{aligned}\int_{-1}^{1} \frac{1}{1+u^2}\,du &= \left[\tan^{-1} u\right]_{-1}^{1} \\ &= \tan^{-1}(1) - \tan^{-1}(-1) \\ &= \frac{\pi}{4} - \left(-\frac{\pi}{4}\right) \\ &= \pi/2 \\ &= 1.57 \text{ (to 3 s.f.)}.\end{aligned}$$

Solution to Activity 10

(a) $$\int_1^4 3\sqrt{x}\,\mathrm{d}x = \int_1^4 3x^{1/2}\,\mathrm{d}x$$
$$= \left[3 \times \frac{1}{3/2}x^{3/2}\right]_1^4$$
$$= \left[2x^{3/2}\right]_1^4$$
$$= 2 \times 4^{3/2} - 2 \times 1^{3/2}$$
$$= 2 \times 8 - 2$$
$$= 14.$$

(b) $$\int_{-1}^0 x(1+x)\,\mathrm{d}x$$
$$= \int_{-1}^0 (x + x^2)\,\mathrm{d}x$$
$$= \left[\tfrac{1}{2}x^2 + \tfrac{1}{3}x^3\right]_{-1}^0$$
$$= (\tfrac{1}{2} \times 0^2 + \tfrac{1}{3} \times 0^3) - (\tfrac{1}{2}(-1)^2 + \tfrac{1}{3}(-1)^3)$$
$$= -(\tfrac{1}{2} - \tfrac{1}{3})$$
$$= -\tfrac{1}{6}.$$

(c) $$\int_2^4 \frac{r+5}{r}\,\mathrm{d}r = \int_2^4 \left(1 + \frac{5}{r}\right)\mathrm{d}r$$
$$= \left[r + 5\ln r\right]_2^4$$
$$= (4 + 5\ln 4) - (2 + 5\ln 2)$$
$$= 2 + 5(\ln 4 - \ln 2)$$
$$= 2 + 5\ln(4/2)$$
$$= 2 + 5\ln 2.$$

Solution to Activity 11

(a) $$\int_3^5 x^{3/2}\,\mathrm{d}x = \left[\frac{1}{5/2}x^{5/2}\right]_3^5$$
$$= \left[\tfrac{2}{5}x^{5/2}\right]_3^5$$
$$= \tfrac{2}{5}\left[x^{5/2}\right]_3^5$$
$$= \tfrac{2}{5}(5^{5/2} - 3^{5/2})$$
$$= 16.1 \text{ (to 3 s.f.)}.$$

(b) $$\int_{-1/2}^1 (u - 2u^2)\,\mathrm{d}u$$
$$= \left[\tfrac{1}{2}u^2 - 2 \times \tfrac{1}{3}u^3\right]_{-1/2}^1$$
$$= \left[\tfrac{1}{2}u^2 - \tfrac{2}{3}u^3\right]_{-1/2}^1$$
$$= \left[\tfrac{1}{2}u^2\right]_{-1/2}^1 - \left[\tfrac{2}{3}u^3\right]_{-1/2}^1$$
$$= \tfrac{1}{2}\left[u^2\right]_{-1/2}^1 - \tfrac{2}{3}\left[u^3\right]_{-1/2}^1$$
$$= \tfrac{1}{2}\left(1^2 - (-\tfrac{1}{2})^2\right) - \tfrac{2}{3}\left(1^3 - (-\tfrac{1}{2})^3\right)$$
$$= \tfrac{1}{2}(1 - \tfrac{1}{4}) - \tfrac{2}{3}(1 + \tfrac{1}{8})$$
$$= \tfrac{1}{2} \times \tfrac{3}{4} - \tfrac{2}{3} \times \tfrac{9}{8}$$
$$= \tfrac{3}{8} - \tfrac{3}{4}$$
$$= -\tfrac{3}{8}.$$

Solution to Activity 12

(a) $$\int_{\sqrt{3}}^{\sqrt{7}} \tfrac{1}{2}x^3\,\mathrm{d}x = \tfrac{1}{2}\int_{\sqrt{3}}^{\sqrt{7}} x^3\,\mathrm{d}x$$
$$= \tfrac{1}{2}\left[\tfrac{1}{4}x^4\right]_{\sqrt{3}}^{\sqrt{7}}$$
$$= \tfrac{1}{8}\left[x^4\right]_{\sqrt{3}}^{\sqrt{7}}$$
$$= \tfrac{1}{8}\left((\sqrt{7})^4 - (\sqrt{3})^4\right)$$
$$= \tfrac{1}{8}(7^2 - 3^2)$$
$$= \tfrac{1}{8}(49 - 9)$$
$$= 5.$$

(b) $$\int_{-\pi/4}^{\pi/4} (\sin x + \cos x)\,\mathrm{d}x$$
$$= \int_{-\pi/4}^{\pi/4} \sin x\,\mathrm{d}x + \int_{-\pi/4}^{\pi/4} \cos x\,\mathrm{d}x$$
$$= \left[-\cos x\right]_{-\pi/4}^{\pi/4} + \left[\sin x\right]_{-\pi/4}^{\pi/4}$$
$$= -\left[\cos x\right]_{-\pi/4}^{\pi/4} + \left[\sin x\right]_{-\pi/4}^{\pi/4}$$
$$= -\left(\cos\left(\frac{\pi}{4}\right) - \cos\left(-\frac{\pi}{4}\right)\right) + \sin\left(\frac{\pi}{4}\right) - \sin\left(-\frac{\pi}{4}\right)$$
$$= -\left(\frac{1}{\sqrt{2}} - \frac{1}{\sqrt{2}}\right) + \frac{1}{\sqrt{2}} - \left(-\frac{1}{\sqrt{2}}\right)$$
$$= \frac{2}{\sqrt{2}}$$
$$= \sqrt{2}.$$

Solution to Activity 13

The required area is given by

$$\begin{aligned}
&\int_{-18}^{18}\left(9-\tfrac{1}{36}x^2\right)dx\\
&=\left[9x-\tfrac{1}{36}\times\tfrac{1}{3}x^3\right]_{-18}^{18}\\
&=\left[9x-\tfrac{1}{108}x^3\right]_{-18}^{18}\\
&=\left[9x\right]_{-18}^{18}-\left[\tfrac{1}{108}x^3\right]_{-18}^{18}\\
&=9\left[x\right]_{-18}^{18}-\tfrac{1}{108}\left[x^3\right]_{-18}^{18}\\
&=9(18-(-18))-\tfrac{1}{108}(18^3-(-18)^3)\\
&=9\times 36-\tfrac{1}{108}(18^3-(-18^3))\\
&=9\times 36-\tfrac{1}{108}(2\times 18^3)\\
&=324-108\\
&=216.
\end{aligned}$$

So the area of the cross-section of the roof is $216\,\text{m}^2$.

Solution to Activity 14

(a) The x-intercepts are given by

$$\begin{aligned}
\tfrac{1}{32}x^2-200&=0\\
\tfrac{1}{32}x^2&=200\\
x^2&=6400\\
x&=\pm\sqrt{6400}\\
x&=\pm 80.
\end{aligned}$$

So the x-intercepts are -80 and 80.

(b) The shaded area is

$$\begin{aligned}
&\int_{-80}^{80}\left(\tfrac{1}{32}x^2-200\right)dx\\
&=\left[\tfrac{1}{32}\times\tfrac{1}{3}x^3-200x\right]_{-80}^{80}\\
&=\left[\tfrac{1}{96}x^3-200x\right]_{-80}^{80}\\
&=\tfrac{1}{96}\left[x^3\right]_{-80}^{80}-200\left[x\right]_{-80}^{80}\\
&=\tfrac{1}{96}(80^3-(-80)^3)-200(80-(-80))\\
&=\tfrac{1}{96}(2\times 80^3)-200\times 160\\
&=\frac{80^3}{48}-32\,000\\
&=-\frac{64\,000}{3}\\
&=-21\,333\tfrac{1}{3}.
\end{aligned}$$

Hence the shaded area is $21\,333\,\text{cm}^2$, to the nearest cm^2.

Solution to Activity 15

(a) (i) From $t=0$ to $t=1$ the displacement of the object changes by $2.3\,\text{cm}$.

(ii) From $t=3$ to $t=4$ the displacement of the object changes by $-1.8\,\text{cm}$. That is, it changes by $1.8\,\text{cm}$ in the negative direction.

(iii) From $t=0$ to $t=3$ the displacement of the object changes by

$$(2.3+5.9+2.7)\,\text{cm}=10.9\,\text{cm}.$$

(iv) From $t=3$ to $t=6$ the displacement of the object changes by

$$(-1.8-2.0-0.3)\,\text{cm}=-4.1\,\text{cm}.$$

That is, it changes by $4.1\,\text{cm}$ in the negative direction.

(b) (i) The displacement of the object from its starting position at time $t=2$ is

$$(2.3+5.9)\,\text{cm}=8.2\,\text{cm}.$$

(ii) By (a) parts (iii) and (iv), the displacement of the object from its starting position at time $t=6$ is

$$(10.9-4.1)\,\text{cm}=6.8\,\text{cm}.$$

(c) By parts (a)(iii) and (a)(iv), the object travels $10.9\,\text{cm}$ in the positive direction, then $4.1\,\text{cm}$ in the negative direction, so the total distance that it travels is

$$(10.9+4.1)\,\text{cm}=15\,\text{cm}.$$

Solution to Activity 16

(a) The change in the displacement of the object from time $t=0$ to time $t=1$ is given by

$$\begin{aligned}
\int_0^1\left(3-\tfrac{3}{2}t\right)dt&=\left[3t-\tfrac{3}{2}\times\tfrac{1}{2}t^2\right]_0^1\\
&=\left[3t-\tfrac{3}{4}t^2\right]_0^1\\
&=3\left[t\right]_0^1-\tfrac{3}{4}\left[t^2\right]_0^1\\
&=3(1-0)-\tfrac{3}{4}(1^2-0^2)\\
&=3-\tfrac{3}{4}\\
&=\tfrac{9}{4}=2.25.
\end{aligned}$$

That is, the object travels $2.25\,\text{m}$ in that time (in the positive direction).

(b) The change in the displacement of the object from time $t = 4$ to time $t = 5$ is given by

$$\begin{aligned}\int_4^5 (3 - \tfrac{3}{2}t)\mathrm{d}t &= \left[3t - \tfrac{3}{2} \times \tfrac{1}{2}t^2\right]_4^5 \\ &= \left[3t - \tfrac{3}{4}t^2\right]_4^5 \\ &= 3\left[t\right]_4^5 - \tfrac{3}{4}\left[t^2\right]_4^5 \\ &= 3(5-4) - \tfrac{3}{4}(5^2 - 4^2) \\ &= 3 - \tfrac{3}{4} \times 9 \\ &= -\tfrac{15}{4} \\ &= -3.75.\end{aligned}$$

That is, the object travels 3.75 m in that time (in the negative direction).

Solution to Activity 17

The change in the displacement of the car during those four seconds is given by

$$\begin{aligned}\int_0^4 (28 - 7t)\,\mathrm{d}t &= \left[28t - \tfrac{7}{2}t^2\right]_0^4 \\ &= 28\left[t\right]_0^4 - \tfrac{7}{2}\left[t^2\right]_0^4 \\ &= 28(4-0) - \tfrac{7}{2}(4^2 - 0) \\ &= 112 - 56 \\ &= 56.\end{aligned}$$

That is, the car travels 56 m during those four seconds.

(In fact, since the required answer is the area of the shaded triangle in Figure 39, you can calculate it in the following more straightforward way.

The change in the displacement of the car during the four seconds is

$$(\tfrac{1}{2} \times 4 \times 28)\,\mathrm{m} = 56\,\mathrm{m}.)$$

Solution to Activity 18

The change in the displacement of the car is the area shaded in Figure 40, which is given by

$$\begin{aligned}\tfrac{1}{2} \times 5 \times 15 - \tfrac{1}{2} \times 4 \times 12 &= \tfrac{1}{2}(75 - 48) \\ &= \tfrac{1}{2} \times 27 = 13.5.\end{aligned}$$

So the car travels 13.5 m during the fifth second.

(There are various ways to work out the area shaded in Figure 40. It is worked out above by subtracting the area of one triangle from the area of another, but alternatively you could add the areas of a rectangle and a triangle, or use the formula for the area of a trapezium.)

Solution to Activity 19

(a) $\displaystyle\int e^x\,\mathrm{d}x = e^x + c.$

(b) $$\begin{aligned}\int (x+3)(x-2)\,\mathrm{d}x &= \int (x^2 + x - 6)\,\mathrm{d}x \\ &= \tfrac{1}{3}x^3 + \tfrac{1}{2}x^2 - 6x + c.\end{aligned}$$

(c) $$\begin{aligned}\int \frac{1}{x^3}\,\mathrm{d}x &= \int x^{-3}\,\mathrm{d}x \\ &= \frac{1}{-2}x^{-2} + c \\ &= -\frac{1}{2x^2} + c.\end{aligned}$$

(d) $\displaystyle\int (\sin\theta - \cos\theta)\,\mathrm{d}\theta = -\cos\theta - \sin\theta + c.$

(e) $\displaystyle\int \frac{1}{1+x^2}\,\mathrm{d}x = \tan^{-1}x + c.$

(f) $\displaystyle\int \frac{-3}{1+u^2}\,\mathrm{d}u = -3\tan^{-1}u + c.$

(g) $\displaystyle\int \frac{1}{2(1+r^2)}\,\mathrm{d}r = \tfrac{1}{2}\tan^{-1}r + c.$

(h) $$\begin{aligned}\int \frac{(1+x^2)}{2}\,\mathrm{d}x &= \tfrac{1}{2}(x + \tfrac{1}{3}x^3) + c \\ &= \tfrac{1}{6}(3x + x^3) + c.\end{aligned}$$

Solution to Activity 20

(a) $$\begin{aligned}\int \left(\frac{3}{x} - \frac{1}{1+x^2}\right)\mathrm{d}x &= \int \frac{3}{x}\,\mathrm{d}x - \int \frac{1}{1+x^2}\,\mathrm{d}x \\ &= 3\int \frac{1}{x}\,\mathrm{d}x - \int \frac{1}{1+x^2}\,\mathrm{d}x \\ &= 3\ln|x| - \tan^{-1}x + c.\end{aligned}$$

(b) $$\begin{aligned}&\int (\operatorname{cosec} x)(\operatorname{cosec} x + \cot x)\,\mathrm{d}x \\ &= \int (\operatorname{cosec}^2 x + \operatorname{cosec} x \cot x)\,\mathrm{d}x \\ &= \int \operatorname{cosec}^2 x\,\mathrm{d}x + \int \operatorname{cosec} x \cot x\,\mathrm{d}x \\ &= -\cot x - \operatorname{cosec} x + c.\end{aligned}$$

Solution to Activity 21

(a) Let $u = \cos x$; then $\dfrac{du}{dx} = -\sin x$. So

$$\begin{aligned}\int e^{\cos x}(-\sin x)\,dx &= \int e^u\,du\\ &= e^u + c\\ &= e^{\cos x} + c.\end{aligned}$$

(b) Let $u = x^3$; then $\dfrac{du}{dx} = 3x^2$. So

$$\begin{aligned}\int \left(\sin(x^3)\right)(3x^2)\,dx &= \int \sin(u)\,du\\ &= -\cos u + c\\ &= -\cos(x^3) + c.\end{aligned}$$

(c) Let $u = \sin x$; then $\dfrac{du}{dx} = \cos x$. So

$$\begin{aligned}\int \left(\frac{1}{\sin x}\right)\cos x\,dx &= \int \frac{1}{u}\,du\\ &= \ln|u| + c\\ &= \ln|\sin x| + c.\end{aligned}$$

(d) Let $u = \cos x$; then $\dfrac{du}{dx} = -\sin x$. So

$$\begin{aligned}\int \left(\frac{1}{\cos^2 x}\right)(-\sin x)\,dx &= \int \frac{1}{u^2}\,du\\ &= \int u^{-2}\,du\\ &= \frac{1}{-1}u^{-1} + c\\ &= -\frac{1}{u} + c\\ &= -\frac{1}{\cos x} + c\\ &= -\sec x + c.\end{aligned}$$

(A quicker way to find this indefinite integral, without using substitution, is to write the integrand as $-\sec x \tan x$ and use the table of standard indefinite integrals.)

(e) Let $u = \sin x$; then $\dfrac{du}{dx} = \cos x$. So

$$\begin{aligned}\int \sin^4 x \cos x\,dx &= \int u^4\,du\\ &= \tfrac{1}{5}u^5 + c\\ &= \tfrac{1}{5}\sin^5 x + c.\end{aligned}$$

(f) Let $u = \sin x$; then $\dfrac{du}{dx} = \cos x$. So

$$\begin{aligned}\int \left(\frac{1}{1+\sin^2 x}\right)\cos x\,dx &= \int \left(\frac{1}{1+u^2}\right)du\\ &= \tan^{-1} u + c\\ &= \tan^{-1}(\sin x) + c.\end{aligned}$$

Solution to Activity 22

(a) Let $u = 4 + \cos x$; then $\dfrac{du}{dx} = -\sin x$. So

$$\begin{aligned}\int (4+\cos x)^7(-\sin x)\,dx &= \int u^7\,du\\ &= \tfrac{1}{8}u^8 + c\\ &= \tfrac{1}{8}(4+\cos x)^8 + c.\end{aligned}$$

(b) Let $u = 1 + x^2$; then $\dfrac{du}{dx} = 2x$. So

$$\begin{aligned}\int \sqrt{1+x^2}\,(2x)\,dx &= \int \sqrt{u}\,du\\ &= \int u^{1/2}\,du\\ &= \frac{1}{3/2}u^{3/2} + c\\ &= \tfrac{2}{3}(1+x^2)^{3/2} + c.\end{aligned}$$

(c) Let $u = x^5 - 8$; then $\dfrac{du}{dx} = 5x^4$. So

$$\begin{aligned}\int (x^5-8)^{10}(5x^4)\,dx &= \int u^{10}\,du\\ &= \tfrac{1}{11}u^{11} + c\\ &= \tfrac{1}{11}(x^5-8)^{11} + c.\end{aligned}$$

(d) Let $u = e^x + 5$; then $\dfrac{du}{dx} = e^x$. So

$$\begin{aligned}\int \left(\frac{1}{e^x+5}\right)e^x\,dx &= \int \frac{1}{u}\,du\\ &= \ln|u| + c\\ &= \ln|e^x+5| + c\\ &= \ln(e^x+5) + c.\end{aligned}$$

(e) Let $u = 5 + 2x^3$; then $\dfrac{du}{dx} = 6x^2$. So

$$\begin{aligned}\int \left(\sin(5+2x^3)\right)(6x^2)\,dx &= \int \sin u\,du\\ &= -\cos u + c\\ &= -\cos(5+2x^3) + c.\end{aligned}$$

Solution to Activity 23

(a) Let $u = e^x$; then $\dfrac{du}{dx} = e^x$. So

$$\begin{aligned}\int (\cos(e^x))\, e^x \, dx &= \int \cos u \, du \\ &= \sin u + c \\ &= \sin(e^x) + c.\end{aligned}$$

(b) Let $u = 1 + \sin x$; then $\dfrac{du}{dx} = \cos x$. So

$$\begin{aligned}\int e^{1+\sin x}(\cos x)\, dx &= \int e^u \, du \\ &= e^u + c \\ &= e^{1+\sin x} + c.\end{aligned}$$

(c) Let $u = x^{10} + 6$; then $\dfrac{du}{dx} = 10x^9$. So

$$\begin{aligned}\int \left(\frac{1}{x^{10}+6}\right)(10x^9)\, dx &= \int \frac{1}{u}\, du \\ &= \ln|u| + c \\ &= \ln|x^{10}+6| + c \\ &= \ln(x^{10}+6) + c.\end{aligned}$$

(d) Let $u = \cos x$; then $\dfrac{du}{dx} = -\sin x$. So

$$\begin{aligned}\int (\cos^3 x)(-\sin x)\, dx &= \int u^3 \, du \\ &= \tfrac{1}{4}u^4 + c \\ &= \tfrac{1}{4}\cos^4 x + c.\end{aligned}$$

Solution to Activity 24

(a) Let $u = x^3$; then $\dfrac{du}{dx} = 3x^2$. So

$$\begin{aligned}\int x^2 \cos(x^3)\, dx &= \int (\cos(x^3))\, x^2 \, dx \\ &= \tfrac{1}{3}\int (\cos(x^3))(3x^2)\, dx \\ &= \tfrac{1}{3}\int \cos u \, du \\ &= \tfrac{1}{3}\sin u + c \\ &= \tfrac{1}{3}\sin(x^3) + c.\end{aligned}$$

(b) Let $u = 1 + x^2$; then $\dfrac{du}{dx} = 2x$. So

$$\begin{aligned}\int x\sqrt{1+x^2}\, dx &= \int \left(\sqrt{1+x^2}\right) x \, dx \\ &= \tfrac{1}{2}\int \left(\sqrt{1+x^2}\right) \times 2x \, dx \\ &= \tfrac{1}{2}\int \sqrt{u}\, du \\ &= \tfrac{1}{2}\int u^{1/2}\, du \\ &= \tfrac{1}{2} \times \tfrac{2}{3}u^{3/2} + c \\ &= \tfrac{1}{3}u^{3/2} + c \\ &= \tfrac{1}{3}(1+x^2)^{3/2} + c.\end{aligned}$$

(c) Let $u = \cos x$; then $\dfrac{du}{dx} = -\sin x$. So

$$\begin{aligned}\int \cos^4 x \sin x \, dx &= -\int (\cos^4 x)(-\sin x)\, dx \\ &= -\int u^4 \, du \\ &= -\tfrac{1}{5}u^5 + c \\ &= -\tfrac{1}{5}\cos^5 x + c.\end{aligned}$$

Solution to Activity 25

(a) Let $u = 1 + 2x^2$; then $\dfrac{du}{dx} = 4x$. So

$$\begin{aligned}\int \frac{x}{1+2x^2}\, dx &= \int \left(\frac{1}{1+2x^2}\right) x \, dx \\ &= \tfrac{1}{4}\int \left(\frac{1}{1+2x^2}\right)(4x)\, dx \\ &= \tfrac{1}{4}\int \frac{1}{u}\, du \\ &= \tfrac{1}{4}\ln|u| + c \\ &= \tfrac{1}{4}\ln|1+2x^2| + c \\ &= \tfrac{1}{4}\ln(1+2x^2) + c\end{aligned}$$

(since $1 + 2x^2$ is always positive).

(b) Let $u = 2 + e^x$; then $\dfrac{du}{dx} = e^x$. So

$$\begin{aligned}\int \frac{e^x}{(2+e^x)^2}\,dx &= \int \left(\frac{1}{(2+e^x)^2}\right) e^x\,dx\\ &= \int \frac{1}{u^2}\,du\\ &= \int u^{-2}\,du\\ &= -u^{-1} + c\\ &= -\frac{1}{u} + c\\ &= -\frac{1}{2+e^x} + c.\end{aligned}$$

(c) Let $u = 3 - \sin x$; then $\dfrac{du}{dx} = -\cos x$. So

$$\begin{aligned}\int \frac{\cos x}{3-\sin x}\,dx &= -\int \left(\frac{1}{3-\sin x}\right)(-\cos x)\,dx\\ &= -\int \frac{1}{u}\,du\\ &= -\ln|u| + c\\ &= -\ln|3-\sin x| + c\\ &= -\ln(3-\sin x) + c\end{aligned}$$

(since $3 - \sin x$ is always positive).

Solution to Activity 26

(a) Let $u = \cos x$; then $\dfrac{du}{dx} = -\sin x$. So

$$\begin{aligned}\int e^{\cos x}\sin x\,dx &= -\int e^{\cos x}(-\sin x)\,dx\\ &= -\int e^u\,du\\ &= -e^u + c\\ &= -e^{\cos x} + c.\end{aligned}$$

(b) Let $u = 3 - e^x$; then $\dfrac{du}{dx} = -e^x$. So

$$\begin{aligned}\int \frac{e^x}{3-e^x}\,dx &= -\int \left(\frac{1}{3-e^x}\right)(-e^x)\,dx\\ &= -\int \frac{1}{u}\,du\\ &= -\ln|u| + c\\ &= -\ln|3-e^x| + c.\end{aligned}$$

(c) Let $u = 1 + \frac{1}{2}\sin x$; then $\dfrac{du}{dx} = \frac{1}{2}\cos x$. So

$$\begin{aligned}&\int (1+\tfrac{1}{2}\sin x)^2\cos x\,dx\\ &= 2\int (1+\tfrac{1}{2}\sin x)^2(\tfrac{1}{2}\cos x)\,dx\\ &= 2\int u^2\,du\\ &= 2\times\tfrac{1}{3}u^3 + c\\ &= \tfrac{2}{3}u^3 + c\\ &= \tfrac{2}{3}(1+\tfrac{1}{2}\sin x)^3 + c.\end{aligned}$$

Solution to Activity 27

Let $u = \cos x$; then $\dfrac{du}{dx} = -\sin x$. So

$$\begin{aligned}\int \tan x\,dx &= \int \frac{\sin x}{\cos x}\,dx\\ &= \int \left(\frac{1}{\cos x}\right)(\sin x)\,dx\\ &= -\int \left(\frac{1}{\cos x}\right)(-\sin x)\,dx\\ &= -\int \frac{1}{u}\,du\\ &= -\ln|u| + c\\ &= -\ln|\cos x| + c.\end{aligned}$$

Solution to Activity 28

(a) Let $u = 6x - 1$; then $\dfrac{du}{dx} = 6$. So

$$\begin{aligned}\int e^{6x-1}\,dx &= \tfrac{1}{6}\int e^{6x-1}\times 6\,dx\\ &= \tfrac{1}{6}\int e^u\,du\\ &= \tfrac{1}{6}e^u + c\\ &= \tfrac{1}{6}e^{6x-1} + c.\end{aligned}$$

(b) Let $u = 4x$; then $\dfrac{du}{dx} = 4$. So

$$\begin{aligned}\int \sin(4x)\,dx &= \tfrac{1}{4}\int (\sin(4x))\times 4\,dx\\ &= \tfrac{1}{4}\int \sin u\,du\\ &= \tfrac{1}{4}(-\cos u) + c\\ &= -\tfrac{1}{4}\cos u + c\\ &= -\tfrac{1}{4}\cos(4x) + c.\end{aligned}$$

(c) Let $u = -9x$; then $\dfrac{\mathrm{d}u}{\mathrm{d}x} = -9$. So

$$\begin{aligned}\int e^{-9x}\,\mathrm{d}x &= -\tfrac{1}{9}\int e^{-9x} \times (-9)\,\mathrm{d}x\\ &= -\tfrac{1}{9}\int e^{u}\,\mathrm{d}u\\ &= -\tfrac{1}{9}e^{u} + c\\ &= -\tfrac{1}{9}e^{-9x} + c.\end{aligned}$$

(d) Let $u = -x/3$; then $\dfrac{\mathrm{d}u}{\mathrm{d}x} = -\tfrac{1}{3}$. So

$$\begin{aligned}\int e^{-x/3}\,\mathrm{d}x &= -3\int e^{-x/3} \times (-\tfrac{1}{3})\,\mathrm{d}x\\ &= -3\int e^{u}\,\mathrm{d}u\\ &= -3e^{u} + c\\ &= -3e^{-x/3} + c.\end{aligned}$$

(e) Let $u = 3 - 7x$; then $\dfrac{\mathrm{d}u}{\mathrm{d}x} = -7$. So

$$\begin{aligned}\int \cos(3-7x)\,\mathrm{d}x &= -\tfrac{1}{7}\int (\cos(3-7x)) \times (-7)\,\mathrm{d}x\\ &= -\tfrac{1}{7}\int \cos u\,\mathrm{d}u\\ &= -\tfrac{1}{7}\sin u + c\\ &= -\tfrac{1}{7}\sin(3-7x) + c.\end{aligned}$$

(f) Let $u = 4 - x$; then $\dfrac{\mathrm{d}u}{\mathrm{d}x} = -1$. So

$$\begin{aligned}\int \frac{1}{4-x}\,\mathrm{d}x &= -\int \left(\frac{1}{4-x}\right) \times (-1)\,\mathrm{d}x\\ &= -\int \frac{1}{u}\,\mathrm{d}u\\ &= -\ln|u| + c\\ &= -\ln|4-x| + c.\end{aligned}$$

(g) Let $u = 2x + 1$; then $\dfrac{\mathrm{d}u}{\mathrm{d}x} = 2$. So

$$\begin{aligned}\int \frac{1}{(2x+1)^2}\,\mathrm{d}x &= \tfrac{1}{2}\int \left(\frac{1}{(2x+1)^2}\right) \times 2\,\mathrm{d}x\\ &= \tfrac{1}{2}\int \frac{1}{u^2}\,\mathrm{d}u\\ &= \tfrac{1}{2}\int u^{-2}\,\mathrm{d}u\\ &= -\tfrac{1}{2}u^{-1} + c\\ &= -\frac{1}{2} \times \frac{1}{u} + c\\ &= -\frac{1}{2(2x+1)} + c.\end{aligned}$$

(h) Let $u = 1 - x$; then $\dfrac{\mathrm{d}u}{\mathrm{d}x} = -1$. So

$$\begin{aligned}\int \frac{1}{(1-x)^3}\,\mathrm{d}x &= -\int \left(\frac{1}{(1-x)^3}\right) \times (-1)\,\mathrm{d}x\\ &= -\int \frac{1}{u^3}\,\mathrm{d}u\\ &= -\int u^{-3}\,\mathrm{d}u\\ &= -(-\tfrac{1}{2}u^{-2}) + c\\ &= \frac{1}{2u^2} + c\\ &= \frac{1}{2(1-x)^2} + c.\end{aligned}$$

(i) Let $u = 6x$; then $\dfrac{\mathrm{d}u}{\mathrm{d}x} = 6$. So

$$\begin{aligned}\int \sec^2(6x)\,\mathrm{d}x &= \tfrac{1}{6}\int (\sec^2(6x)) \times 6\,\mathrm{d}x\\ &= \tfrac{1}{6}\int \sec^2 u\,\mathrm{d}u\\ &= \tfrac{1}{6}\tan u + c\\ &= \tfrac{1}{6}\tan(6x) + c.\end{aligned}$$

Solution to Activity 29

(a) Let $u = 2x$; then $\dfrac{du}{dx} = 2$. So

$$\begin{aligned}\int \frac{1}{\sqrt{1-4x^2}}\,dx &= \int \frac{1}{\sqrt{1-(2x)^2}}\,dx\\ &= \tfrac{1}{2}\int \frac{1}{\sqrt{1-(2x)^2}} \times 2\,dx\\ &= \tfrac{1}{2}\int \frac{1}{\sqrt{1-u^2}}\,du\\ &= \tfrac{1}{2}\sin^{-1} u + c\\ &= \tfrac{1}{2}\sin^{-1}(2x) + c.\end{aligned}$$

(b) Let $u = \sqrt{2}\,x$; then $\dfrac{du}{dx} = \sqrt{2}$. So

$$\begin{aligned}\int \frac{1}{1+2x^2}\,dx &= \frac{1}{\sqrt{2}}\int \frac{1}{1+\left(\sqrt{2}\,x\right)^2} \times \sqrt{2}\,dx\\ &= \frac{1}{\sqrt{2}}\int \frac{1}{1+u^2}\,du\\ &= \frac{1}{\sqrt{2}}\tan^{-1} u + c\\ &= \frac{1}{\sqrt{2}}\tan^{-1}\left(\sqrt{2}\,x\right) + c.\end{aligned}$$

(c) Let $u = \frac{3}{2}x$; then $\dfrac{du}{dx} = \frac{3}{2}$. So

$$\begin{aligned}\int \frac{1}{4+9x^2}\,dx &= \tfrac{1}{4}\int \frac{1}{1+\frac{9}{4}x^2}\,dx\\ &= \tfrac{1}{4}\int \frac{1}{1+\left(\frac{3}{2}x\right)^2}\,dx\\ &= \tfrac{1}{4}\times\tfrac{2}{3}\int \frac{1}{1+\left(\frac{3}{2}x\right)^2} \times \tfrac{3}{2}\,dx\\ &= \tfrac{1}{6}\int \frac{1}{1+\left(\frac{3}{2}x\right)^2} \times \tfrac{3}{2}\,dx\\ &= \tfrac{1}{6}\int \frac{1}{1+u^2}\,du\\ &= \tfrac{1}{6}\tan^{-1} u + c\\ &= \tfrac{1}{6}\tan^{-1}\left(\tfrac{3}{2}x\right) + c.\end{aligned}$$

Solution to Activity 30

(a) $\displaystyle\int \cos(10x)\,dx = \tfrac{1}{10}\sin(10x) + c.$

(b) $$\begin{aligned}\int \sin(2-5x)\,dx &= -\tfrac{1}{5}(-\cos(2-5x)) + c\\ &= \tfrac{1}{5}\cos(2-5x) + c.\end{aligned}$$

(c) $\displaystyle\int e^{3t+4}\,dt = \tfrac{1}{3}e^{3t+4} + c.$

(d) $\displaystyle\int \cos(\tfrac{1}{2}\theta)\,d\theta = 2\sin(\tfrac{1}{2}\theta) + c.$

(e) $\displaystyle\int e^{-4p}\,dp = -\tfrac{1}{4}e^{-4p} + c.$

(f) $\displaystyle\int \frac{1}{7x-4}\,dx = \tfrac{1}{7}\ln|7x-4| + c.$

(g) $\displaystyle\int \frac{1}{1-r}\,dr = -\ln|1-r| + c.$

(h) $\displaystyle\int \sec^2(5x-1)\,dx = \tfrac{1}{5}\tan(5x-1) + c.$

(i) $\displaystyle\int \sin(\tfrac{1}{5}x)\,dx = -5\cos(\tfrac{1}{5}x) + c.$

(j) $\displaystyle\int e^{-x}\,dx = -e^{-x} + c.$

(k) $\displaystyle\int e^{x/2}\,dx = 2e^{x/2} + c.$

(l) $$\begin{aligned}\int \sec^2(2-3p)\,dp &= -\tfrac{1}{3}\tan(2-3p) + c\\ &= \tfrac{1}{3}\tan(3p-2) + c.\end{aligned}$$

(The final expression was obtained by using the trigonometric identity $\tan(-\theta) = -\tan\theta$.)

(m) $$\begin{aligned}\int \sin\left(\frac{2-x}{3}\right)dx &= \int \sin\left(\tfrac{2}{3}-\tfrac{1}{3}x\right)dx\\ &= -3\left(-\cos\left(\tfrac{2}{3}-\tfrac{1}{3}x\right)\right) + c\\ &= 3\cos\left(\frac{2-x}{3}\right) + c.\end{aligned}$$

Solution to Activity 31

(a) The velocity v (in $\mathrm{m\,s^{-1}}$) of the object at time t (in seconds) is given by

$$v = 2\cos\left(\tfrac{1}{2}t+5\right).$$

Hence the displacement s (in m) of the object at time t (in seconds) is given by

$$\begin{aligned}s &= \int 2\cos\left(\tfrac{1}{2}t+5\right)dt\\ &= 2\times 2\sin\left(\tfrac{1}{2}t+5\right) + c\\ &= 4\sin\left(\tfrac{1}{2}t+5\right) + c,\end{aligned}$$

where c is an arbitrary constant.

When $t = 0$, $s = 4$. Hence

$$\begin{aligned} 4 &= 4\sin\left(\tfrac{1}{2}\times 0 + 5\right) + c \\ 4 &= 4\sin(5) + c \\ c &= 4 - 4\sin(5) \\ c &= 7.835\ldots \\ c &= 7.84 \text{ (to 2 d.p.)}. \end{aligned}$$

So the equation for the displacement of the object in terms of time is

$$s = 4\sin\left(\tfrac{1}{2}t + 5\right) + 7.84.$$

(b) When $t = 10$,

$$\begin{aligned} s &= 4\sin\left(\tfrac{1}{2}\times 10 + 5\right) + 7.835\ldots \\ &= 4\sin(10) + 7.835\ldots \\ &= 5.7 \text{ (to 2 s.f.)}. \end{aligned}$$

So the displacement of the object at time 10 seconds is 5.7 m (to 2 s.f.).

Solution to Activity 32

(a) Let $u = 1 + 2x^2$; then $\dfrac{\mathrm{d}u}{\mathrm{d}x} = 4x$.

Putting $x = 0$ gives $u = 1$ and putting $x = 1$ gives $u = 3$. So

$$\begin{aligned} \int_0^1 \frac{x}{1+2x^2}\,\mathrm{d}x &= \tfrac{1}{4}\int_0^1 \left(\frac{1}{1+2x^2}\right)(4x)\,\mathrm{d}x \\ &= \tfrac{1}{4}\int_1^3 \frac{1}{u}\,\mathrm{d}u \\ &= \tfrac{1}{4}\Big[\ln|u|\Big]_1^3 \\ &= \tfrac{1}{4}(\ln 3 - \ln 1) \\ &= \tfrac{1}{4}\ln 3 \\ &= 0.275 \text{ (to 3 s.f.)}. \end{aligned}$$

(b) Let $u = \cos x$; then $\dfrac{\mathrm{d}u}{\mathrm{d}x} = -\sin x$. Putting $x = \pi/2$ gives $u = 0$ and putting $x = \pi$ gives $u = -1$. So

$$\begin{aligned} \int_{\pi/2}^{\pi} \cos^3 x \sin x\,\mathrm{d}x &= -\int_{\pi/2}^{\pi} \cos^3 x(-\sin x)\,\mathrm{d}x \\ &= -\int_0^{-1} u^3\,\mathrm{d}u \\ &= -\left[\tfrac{1}{4}u^4\right]_0^{-1} \\ &= -\tfrac{1}{4}\left[u^4\right]_0^{-1} \\ &= -\tfrac{1}{4}\left((-1)^4 - 0^4\right) \\ &= -\tfrac{1}{4} \\ &= -0.25. \end{aligned}$$

(Notice that in the second line of the manipulation above we obtain an integral, $\int_0^{-1} u^3$, in which the lower limit of integration is greater than the upper limit of integration. An alternative way to proceed from this line is to use the fact that, in general, $\int_b^a f(x)\,\mathrm{d}x = -\int_a^b f(x)\,\mathrm{d}x$. Then the manipulation continues like this:

$$\begin{aligned} \int_0^{-1} u^3\,\mathrm{d}u &= \int_{-1}^{0} u^3\,\mathrm{d}u \\ &= \left[\tfrac{1}{4}u^4\right]_{-1}^{0} \\ &= \tfrac{1}{4}\left[u^4\right]_{-1}^{0} \\ &= \tfrac{1}{4}\left(0^4 - (-1)^4\right) \\ &= -\tfrac{1}{4} \\ &= -0.25.) \end{aligned}$$

(c) Let $u = -3x^2$; then $\dfrac{\mathrm{d}u}{\mathrm{d}x} = -6x$.

Putting $x = -1$ gives $u = -3$ and putting $x = 0$ gives $u = 0$. So

$$\begin{aligned} \int_{-1}^{0} xe^{-3x^2}\,\mathrm{d}x &= -\tfrac{1}{6}\int_{-1}^{0} \left(e^{-3x^2}\right)(-6x)\,\mathrm{d}x \\ &= -\tfrac{1}{6}\int_{-3}^{0} e^u\,du \\ &= -\tfrac{1}{6}\left[e^u\right]_{-3}^{0} \\ &= -\tfrac{1}{6}\left(e^0 - e^{-3}\right) \\ &= \tfrac{1}{6}\left(e^{-3} - 1\right) \\ &= -0.158 \text{ (to 3 s.f.)}. \end{aligned}$$

Solution to Activity 33

(a) Let $u = 1 + x$; then $\dfrac{du}{dx} = 1$. Also, $x = u - 1$. So

$$\begin{aligned}\int x\sqrt{1+x}\,dx &= \int (u-1)\sqrt{u}\,du\\ &= \int \left(u^{3/2} - u^{1/2}\right)du\\ &= \frac{u^{5/2}}{5/2} - \frac{u^{3/2}}{3/2} + c\\ &= \tfrac{2}{5}(1+x)^{5/2} - \tfrac{2}{3}(1+x)^{3/2} + c.\end{aligned}$$

(b) Let $u = x + 3$; then $\dfrac{du}{dx} = 1$. Also, $x = u - 3$, so $x - 2 = u - 5$. Hence

$$\begin{aligned}\int \frac{x-2}{(x+3)^3}\,dx &= \int \frac{u-5}{u^3}\,du\\ &= \int (u^{-2} - 5u^{-3})\,du\\ &= -u^{-1} + \tfrac{5}{2}u^{-2} + c\\ &= -\frac{1}{u} + \frac{5}{2u^2} + c\\ &= -\frac{1}{x+3} + \frac{5}{2(x+3)^2} + c.\end{aligned}$$

Solution to Activity 34

(a) Let $f(x) = x$ and $g(x) = \cos x$. Then $f'(x) = 1$ and an antiderivative of $g(x)$ is $G(x) = \sin x$. Hence

$$\begin{aligned}\int x\cos x\,dx &= \int f(x)g(x)\,dx\\ &= f(x)G(x) - \int f'(x)G(x)\,dx\\ &= x\sin x - \int \sin x\,dx\\ &= x\sin x - (-\cos x) + c\\ &= x\sin x + \cos x + c.\end{aligned}$$

(b) Let $f(x) = x$ and $g(x) = e^x$. Then $f'(x) = 1$ and an antiderivative of $g(x)$ is $G(x) = e^x$. Hence

$$\begin{aligned}\int xe^x\,dx &= \int f(x)g(x)\,dx\\ &= f(x)G(x) - \int f'(x)G(x)\,dx\\ &= xe^x - \int e^x\,dx\\ &= xe^x - e^x + c\\ &= e^x(x-1) + c.\end{aligned}$$

Solution to Activity 35

(a)
$$\begin{aligned}\int x\sin(5x)\,dx &= x\left(-\tfrac{1}{5}\cos(5x)\right)\\ &\quad - \int 1\times\left(-\tfrac{1}{5}\cos(5x)\right)dx\\ &= -\tfrac{1}{5}x\cos(5x) + \tfrac{1}{5}\int \cos(5x)\,dx\\ &= -\tfrac{1}{5}x\cos(5x) + \tfrac{1}{5}\times\tfrac{1}{5}\sin(5x) + c\\ &= -\tfrac{1}{5}x\cos(5x) + \tfrac{1}{25}\sin(5x) + c\\ &= \tfrac{1}{25}\sin(5x) - \tfrac{1}{5}x\cos(5x) + c.\end{aligned}$$

(b)
$$\begin{aligned}\int xe^{2x}\,dx &= x\left(\tfrac{1}{2}e^{2x}\right) - \int 1\times\left(\tfrac{1}{2}e^{2x}\right)dx\\ &= \tfrac{1}{2}xe^{2x} - \tfrac{1}{2}\int e^{2x}\,dx\\ &= \tfrac{1}{2}xe^{2x} - \tfrac{1}{2}\times\tfrac{1}{2}e^{2x} + c\\ &= \tfrac{1}{2}xe^{2x} - \tfrac{1}{4}e^{2x} + c\\ &= \tfrac{1}{4}e^{2x}(2x-1) + c.\end{aligned}$$

(c)
$$\begin{aligned}\int xe^{-x}\,dx &= x\left(-e^{-x}\right) - \int 1\times\left(-e^{-x}\right)dx\\ &= -xe^{-x} + \int e^{-x}\,dx\\ &= -xe^{-x} + (-e^{-x}) + c\\ &= -xe^{-x} - e^{-x} + c\\ &= -e^{-x}(x+1) + c.\end{aligned}$$

Solution to Activity 36

(a)
$$\begin{aligned}\int x^2e^x\,\mathrm{d}x &= x^2e^x - \int 2xe^x\,\mathrm{d}x\\ &= x^2e^x - 2\int xe^x\,\mathrm{d}x\\ &= x^2e^x - 2\left(xe^x - \int 1\times e^x\,\mathrm{d}x\right)\\ &= x^2e^x - 2xe^x + 2\int e^x\,\mathrm{d}x\\ &= x^2e^x - 2xe^x + 2e^x + c\\ &= e^x(x^2 - 2x + 2) + c.\end{aligned}$$

(b)
$$\begin{aligned}&\int x^2\cos(2x)\,\mathrm{d}x\\ &= x^2(\tfrac{1}{2}\sin(2x)) - \int 2x(\tfrac{1}{2}\sin(2x))\,\mathrm{d}x\\ &= \tfrac{1}{2}x^2\sin(2x) - \int x\sin(2x)\,\mathrm{d}x\\ &= \tfrac{1}{2}x^2\sin(2x)\\ &\quad - \left(x(-\tfrac{1}{2}\cos(2x)) - \int 1\times(-\tfrac{1}{2}\cos(2x))\,\mathrm{d}x\right)\\ &= \tfrac{1}{2}x^2\sin(2x) + \tfrac{1}{2}x\cos(2x) - \tfrac{1}{2}\int\cos(2x)\,\mathrm{d}x\\ &= \tfrac{1}{2}x^2\sin(2x) + \tfrac{1}{2}x\cos(2x) - \tfrac{1}{2}\times\tfrac{1}{2}\sin(2x) + c\\ &= \tfrac{1}{4}(2x^2-1)\sin(2x) + \tfrac{1}{2}x\cos(2x) + c.\end{aligned}$$

Solution to Activity 37

$$\begin{aligned}&\int (5x+2)e^{-5x}\,\mathrm{d}x\\ &= (5x+2)\left(\frac{1}{-5}\right)e^{-5x} - \int 5\left(\frac{1}{-5}\right)e^{-5x}\,\mathrm{d}x\\ &= -\tfrac{1}{5}(5x+2)e^{-5x} + \int e^{-5x}\,\mathrm{d}x\\ &= -\tfrac{1}{5}(5x+2)e^{-5x} - \tfrac{1}{-5}e^{-5x}\,\mathrm{d}x + c\\ &= -\tfrac{1}{5}e^{-5x}(5x+2+1) + c\\ &= -\tfrac{1}{5}(5x+3)e^{-5x} + c.\end{aligned}$$

(An alternative way to find this integral is to start by writing it as

$$5\int xe^{-5x}\,\mathrm{d}x + 2\int e^{-5x}\,\mathrm{d}x.$$

The first integral in this expression has an integrand of the form $xg(x)$, where $g(x)$ is an expression that you can integrate, so you can find this integral by using integration by parts. The second integral is straightforward to find.)

Solution to Activity 38

(a)
$$\begin{aligned}\int x^3\ln x\,\mathrm{d}x &= \int(\ln x)x^3\,\mathrm{d}x\\ &= (\ln x)\times\tfrac{1}{4}x^4 - \int\frac{1}{x}\times\tfrac{1}{4}x^4\,\mathrm{d}x\\ &= \tfrac{1}{4}x^4\ln x - \tfrac{1}{4}\int x^3\,\mathrm{d}x\\ &= \tfrac{1}{4}x^4\ln x - \tfrac{1}{4}\times\tfrac{1}{4}x^4 + c\\ &= \tfrac{1}{16}x^4(4\ln x - 1) + c.\end{aligned}$$

(b)
$$\begin{aligned}\int\ln x\,\mathrm{d}x &= \int(\ln x)\times 1\,\mathrm{d}x\\ &= (\ln x)x - \int\frac{1}{x}\times x\,\mathrm{d}x\\ &= x\ln x - \int 1\,\mathrm{d}x\\ &= x\ln x - x + c\\ &= x(\ln x - 1) + c.\end{aligned}$$

Solution to Activity 39

$$\begin{aligned}&\int(x-2)\ln(3x)\,\mathrm{d}x\\ &= \int(\ln(3x))(x-2)\,\mathrm{d}x\\ &= \ln(3x)\times(\tfrac{1}{2}x^2 - 2x) - \int\frac{1}{x}(\tfrac{1}{2}x^2-2x)\,\mathrm{d}x\\ &= \tfrac{1}{2}x(x-4)\ln(3x) - \int(\tfrac{1}{2}x - 2)\,\mathrm{d}x\\ &= \tfrac{1}{2}x(x-4)\ln(3x) - \tfrac{1}{2}\int(x-4)\,\mathrm{d}x\\ &= \tfrac{1}{2}x(x-4)\ln(3x) - \tfrac{1}{2}(\tfrac{1}{2}x^2 - 4x) + c\\ &= \tfrac{1}{2}x(x-4)\ln(3x) - \tfrac{1}{4}x(x-8) + c.\end{aligned}$$

Solution to Activity 40

(a) We write

$$\int e^x \sin(3x)\,dx = \int \sin(3x)\,e^x\,dx.$$

Integrating by parts twice gives

$$\begin{aligned}\int \sin(3x)\,e^x\,dx &= \sin(3x)e^x - \int (3\cos(3x))e^x\,dx \\ &= \sin(3x)e^x - 3\int \cos(3x)e^x\,dx \\ &= \sin(3x)e^x \\ &\quad - 3\left(\cos(3x)e^x - \int (-3\sin(3x))e^x\,dx\right) \\ &= \sin(3x)e^x - 3\cos(3x)e^x - 9\int \sin(3x)e^x\,dx.\end{aligned}$$

So

$$10\int \sin(3x)\,e^x\,dx = \sin(3x)e^x - 3\cos(3x)e^x + c$$

and hence

$$\int \sin(3x)\,e^x\,dx = \tfrac{1}{10}\sin(3x)e^x - \tfrac{3}{10}\cos(3x)e^x + c.$$

That is,

$$\begin{aligned}\int e^x \sin(3x)\,dx &= \tfrac{1}{10}e^x\sin(3x) - \tfrac{3}{10}e^x\cos(3x) + c \\ &= \tfrac{1}{10}e^x(\sin(3x) - 3\cos(3x)) + c.\end{aligned}$$

(b) Integrating by parts twice gives

$$\begin{aligned}\int e^{2x}\cos x\,dx &= e^{2x}\sin x - \int (2e^{2x})\sin x\,dx \\ &= e^{2x}\sin x - 2\int e^{2x}\sin x\,dx \\ &= e^{2x}\sin x \\ &\quad - 2\left(e^{2x}(-\cos x) - \int (2e^{2x})(-\cos x)\,dx\right) \\ &= e^{2x}\sin x + 2e^{2x}\cos x - 4\int e^{2x}\cos x\,dx.\end{aligned}$$

So

$$5\int e^{2x}\cos x\,dx = e^{2x}\sin x + 2e^{2x}\cos x + c$$

and hence

$$\int e^{2x}\cos x\,dx = \tfrac{1}{5}e^{2x}\sin x + \tfrac{2}{5}e^{2x}\cos x + c.$$

That is,

$$\int e^{2x}\cos x\,dx = \tfrac{1}{5}e^{2x}(\sin x + 2\cos x) + c.$$

Solution to Activity 41

$$\begin{aligned}\int_0^1 xe^{3x}\,dx &= \left[x \times \tfrac{1}{3}e^{3x}\right]_0^1 - \int_0^1 1 \times \tfrac{1}{3}e^{3x}\,dx \\ &= \tfrac{1}{3}\left[xe^{3x}\right]_0^1 - \tfrac{1}{3}\int_0^1 e^{3x}\,dx \\ &= \tfrac{1}{3}(e^3 - 0) - \tfrac{1}{3}\left[\tfrac{1}{3}e^{3x}\right]_0^1 \\ &= \tfrac{1}{3}e^3 - \tfrac{1}{9}\left[e^{3x}\right]_0^1 \\ &= \tfrac{1}{3}e^3 - \tfrac{1}{9}(e^3 - 1) \\ &= \tfrac{1}{9}(3e^3 - e^3 + 1) \\ &= \tfrac{1}{9}(2e^3 + 1) \\ &= 4.5746 \text{ (to 4 d.p.)}.\end{aligned}$$

Solution to Activity 42

(a) If $x \in [\pi/12, \pi/6]$, then $3x \in [\pi/4, \pi/2]$.

Now

$$\sin\theta > 0 \text{ when } \theta \in [\pi/4, \pi/2],$$

so

$$\sin(3x) > 0 \text{ when } x \in [\pi/12, \pi/6].$$

Also

$$x > 0 \text{ when } x \in [\pi/12, \pi/6].$$

Hence

$$x\sin(3x) > 0 \text{ when } x \in [\pi/12, \pi/6].$$

That is, the graph of $f(x) = x\sin(3x)$ lies above the x-axis for values of x in the interval $[\pi/12, \pi/6]$.

(b) It follows from part (a) that the area between this graph and the x-axis from $x = \pi/12$ to $x = \pi/6$ is

$$\int_{\pi/12}^{\pi/6} x\sin(3x)\,dx.$$

(c) Integrating by parts gives

$$\int_{\pi/12}^{\pi/6} x\sin(3x)\,dx$$
$$= \left[x(-\tfrac{1}{3}\cos(3x))\right]_{\pi/12}^{\pi/6} - \int_{\pi/12}^{\pi/6}(-\tfrac{1}{3}\cos(3x))\,dx$$
$$= -\tfrac{1}{3}\left[x\cos(3x)\right]_{\pi/12}^{\pi/6} + \tfrac{1}{3}\int_{\pi/12}^{\pi/6}\cos(3x)\,dx$$
$$= -\tfrac{1}{3}\left(\frac{\pi}{6}\cos\left(\frac{\pi}{2}\right) - \frac{\pi}{12}\cos\left(\frac{\pi}{4}\right)\right) + \tfrac{1}{3}\left[\tfrac{1}{3}\sin(3x)\right]_{\pi/12}^{\pi/6}$$
$$= -\tfrac{1}{3}\left(\frac{\pi}{6}\times 0 - \frac{\pi}{12}\times\frac{1}{\sqrt{2}}\right) + \tfrac{1}{9}\left[\sin(3x)\right]_{\pi/12}^{\pi/6}$$
$$= \frac{\pi}{36\sqrt{2}} + \tfrac{1}{9}\left(\sin\left(\frac{\pi}{2}\right) - \sin\left(\frac{\pi}{4}\right)\right)$$
$$= \frac{\pi}{36\sqrt{2}} + \tfrac{1}{9}\left(1 - \frac{1}{\sqrt{2}}\right)$$
$$= 0.0943 \text{ (to 4 d.p.)}.$$

(The area found in this solution is shown below.)

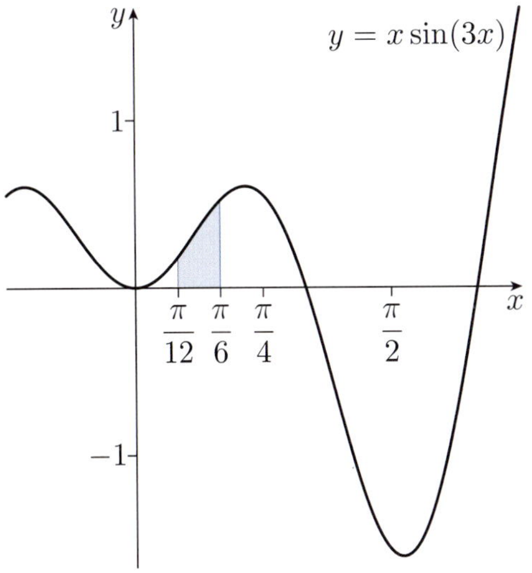

Solution to Activity 43

(a) To help us construct a table of signs for the function $f(x) = (x-2)(e^{-x}-1)$, first we consider the two factors of the expression $(x-2)(e^{-x}-1)$ separately.

The function $y = x - 2$ is increasing on its whole domain, and the x-intercept of its graph is $x = 2$.

The function $y = e^{-x}$ is decreasing on its whole domain, so the function $y = e^{-x} - 1$ is also decreasing on its whole domain. The x-intercept of its graph is given by $e^{-x} - 1 = 0$, which gives

$$e^{-x} = 1$$
$$-x = \ln 1$$
$$-x = 0$$
$$x = 0.$$

So its x-intercept is $x = 0$.

So we have the following table of signs, for $x > 0$.

x	$(0,2)$	2	$(2,\infty)$
$x-2$	$-$	0	$+$
$e^{-x}-1$	$-$	$-$	$-$
$(x-2)(e^{-x}-1)$	$+$	0	$-$

The table shows that the graph of the function $f(x) = (x-2)(e^{-x}-1)$ lies above the x-axis for $0 < x < 2$ and below the x-axis for $x > 2$.

(b) Because the graph of f lies above the x-axis for $0 < x < 2$ and below the x-axis for $x > 2$, to find the total area between the graph of f and the x-axis from $x = 0$ to $x = 4$ we have to work out the signed area from $x = 0$ to $x = 2$ and the signed area from $x = 2$ to $x = 4$ separately.

The signed area between the graph of f and the x-axis from $x = 0$ to $x = 2$ is given by

$$\int_0^2 (x-2)(e^{-x}-1)\,dx$$
$$= \left[(x-2)(-e^{-x}-x)\right]_0^2 - \int_0^2(-e^{-x}-x)\,dx$$
$$= 0 - (-2)(-e^{-0}) + \int_0^2(e^{-x}+x)\,dx$$
$$= 2(-1) + \left[-e^{-x} + \tfrac{1}{2}x^2\right]_0^2$$
$$= -2 + (-e^{-2}+2) - (-e^{-0}+0)$$
$$= -2 - e^{-2} + 2 + 1$$
$$= 1 - e^{-2}.$$

Since the graph of f lies above the x-axis for $0 < x < 2$, it follows that the area between the graph of f and the x-axis from $x = 0$ to $x = 2$ is $1 - e^{-2}$.

Similarly, the signed area between the graph of f and the x-axis from $x = 2$ to $x = 4$ is given by

$$\begin{aligned}&\int_2^4 (x-2)(e^{-x}-1)\,dx\\ &= \Big[(x-2)(-e^{-x}-x)\Big]_2^4 - \int_2^4 (-e^{-x}-x)\,dx\\ &= 2(-e^{-4}-4) - 0 + \int_2^4 (e^{-x}+x)\,dx\\ &= -2e^{-4} - 8 + \Big[-e^{-x} + \tfrac{1}{2}x^2\Big]_2^4\\ &= -2e^{-4} - 8 + (-e^{-4}+8) - (-e^{-2}+2)\\ &= -3e^{-4} + e^{-2} - 2.\end{aligned}$$

Since the graph of f lies below the x-axis for $2 < x < 4$, it follows that the area between the graph of f and the x-axis from $x = 2$ to $x = 4$ is

$$-(-3e^{-4} + e^{-2} - 2) = 3e^{-4} - e^{-2} + 2.$$

So the total area between the graph of f and the x-axis from $x = 0$ to $x = 4$ is

$$\begin{aligned}&(1-e^{-2}) + (3e^{-4} - e^{-2} + 2)\\ &= 3 - 2e^{-2} + 3e^{-4}\\ &= 2.784 \text{ (to 4 s.f.).}\end{aligned}$$

(The area found in this solution is shown below.)

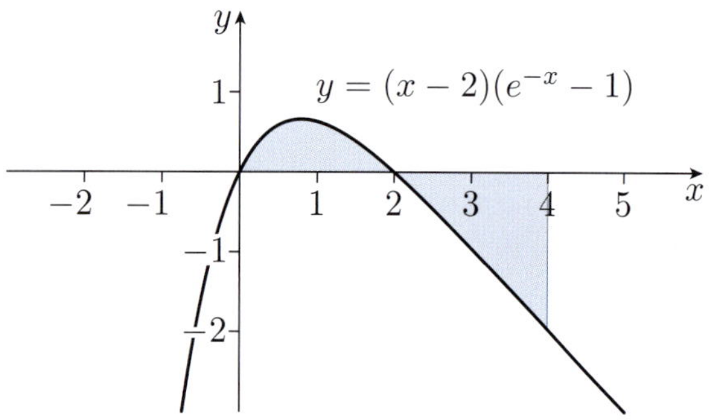

Solution to Activity 44

(a)
$$\begin{aligned}\int \cos^2\theta\,d\theta &= \int \tfrac{1}{2}(1+\cos(2\theta))\,d\theta\\ &= \tfrac{1}{2}\int (1+\cos(2\theta))\,d\theta\\ &= \tfrac{1}{2}(\theta + \tfrac{1}{2}\sin(2\theta)) + c\\ &= \tfrac{1}{4}(2\theta + \sin(2\theta)) + c.\end{aligned}$$

(b)
$$\begin{aligned}\int \sin x\,\cos x\,dx &= \int \tfrac{1}{2}\sin(2x)\,dx\\ &= \tfrac{1}{2}\times\tfrac{1}{2}(-\cos(2x)) + c\\ &= -\tfrac{1}{4}\cos(2x) + c.\end{aligned}$$

Solution to Activity 45

Let $u = \sin x$; then $\dfrac{du}{dx} = \cos x$. So

$$\begin{aligned}\int \sin x\,\cos x\,dx &= \int u\,du\\ &= \tfrac{1}{2}u^2 + c\\ &= \tfrac{1}{2}\sin^2 x + c.\end{aligned}$$

(There's an alternative approach to finding this integral, as follows. Let $u = \cos x$; then $\dfrac{du}{dx} = -\sin x$. So

$$\begin{aligned}\int \sin x\,\cos x\,dx &= \int \cos x \sin x\,dx\\ &= -\int (\cos x)(-\sin x)\,dx\\ &= -\int u\,du\\ &= -\tfrac{1}{2}u^2 + c\\ &= -\tfrac{1}{2}\cos^2 x + c.\end{aligned}$$

This approach gives a different answer, but of course it's equivalent to the first answer. You can confirm this by using the identity $\sin^2\theta + \cos^2\theta = 1$, as follows:

$$\begin{aligned}-\tfrac{1}{2}\cos^2 x + c &= -\tfrac{1}{2}(1-\sin^2 x) + c\\ &= \tfrac{1}{2}\sin^2 x - \tfrac{1}{2} + c\\ &= \tfrac{1}{2}\sin^2 x + d,\end{aligned}$$

where $d = -\frac{1}{2} + c$ is an arbitrary constant.

Notice also that the solution to Activity 44(b) gives a third different answer:

$$\int \sin x\,\cos x\,dx = -\tfrac{1}{4}\cos(2x) + c.$$

You can confirm that this answer is equivalent to the first answer above by using the identity $\cos(2\theta) = 1 - 2\sin^2\theta$, as follows:

$$\begin{aligned}-\tfrac{1}{4}\cos(2x) + c &= -\tfrac{1}{4}(1 - 2\sin^2 x) + c\\ &= \tfrac{1}{2}\sin^2 x - \tfrac{1}{4} + c\\ &= \tfrac{1}{2}\sin^2 x + d,\end{aligned}$$

where $d = -\frac{1}{4} + c$ is an arbitrary constant.)

(The diagram below shows the graphs of the three different antiderivatives of $f(x) = \sin x \cos x$ that were found in the solutions to Activity 44(b) and Activity 45. The arbitrary constant has been taken to be zero in each case. You can see that, as you'd expect, the three graphs appear to be vertical translations of each other.)

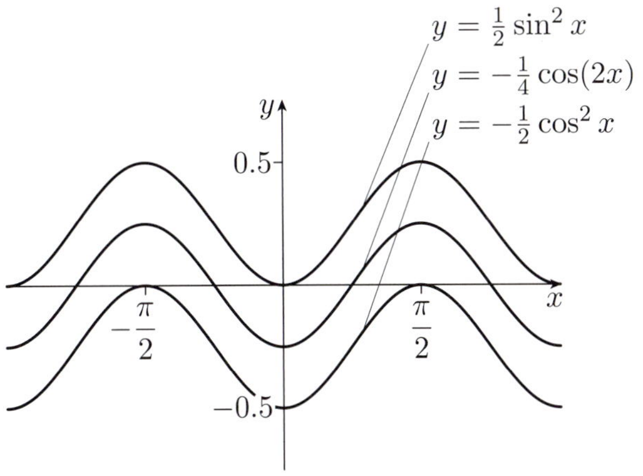

Solution to Activity 46

(a) Since the derivative of $9x^4$ is $36x^3$, which is 'nearly' x^3, you can use the substitution $u = 9x^4$.

Let $u = 9x^4$; then $\dfrac{du}{dx} = 36x^3$. So

$$\begin{aligned}\int x^3\cos(9x^4)\,dx &= \tfrac{1}{36}\int \cos(9x^4)\times 36x^3\,dx\\ &= \tfrac{1}{36}\int \cos u\,du\\ &= \tfrac{1}{36}\sin u + c\\ &= \tfrac{1}{36}\sin(9x^4) + c.\end{aligned}$$

(b) The integrand is of the form $xg(x)$, where g is a function that you can integrate, which suggests that you should try integration by parts. To integrate the 'second' expression, you can use the fact that it is a simple function of a linear expression.

$$\begin{aligned}\int x\cos(5x)\,dx &= \tfrac{1}{5}x\sin(5x) - \tfrac{1}{5}\int \sin(5x)\,dx\\ &= \tfrac{1}{5}x\sin(5x) - \tfrac{1}{5}\left(-\tfrac{1}{5}\cos(5x)\right) + c\\ &= \tfrac{1}{5}x\sin(5x) + \tfrac{1}{25}\cos(5x) + c.\end{aligned}$$

(c) The integrand is the sum of two expressions that are straightforward to integrate. The second expression is a simple function of a linear expression.

$$\int (x + \cos(5x))\,dx = \tfrac{1}{2}x^2 + \tfrac{1}{5}\sin(5x) + c.$$

(d) You could use integration by substitution (taking $u = x^2 + 3$) or integration by parts, but the simplest method is to start by multiplying out the brackets.

$$\begin{aligned}\int x(x^2+3)\,dx &= \int (x^3 + 3x)\,dx\\ &= \tfrac{1}{4}x^4 + \tfrac{3}{2}x^2 + c\\ &= \tfrac{1}{4}x^2(x^2+6) + c.\end{aligned}$$

(e) The integrand can be written as $\dfrac{1}{1+\left(\sqrt{3}\,x\right)^2}$, so it's a function of the linear expression $\sqrt{3}\,x$. So you can use the substitution $u = \sqrt{3}\,x$, together with the indefinite integral of $\dfrac{1}{1+x^2}$, which is a standard integral.

Let $u = \sqrt{3}\,x$; then $du/dx = \sqrt{3}$. So

$$\begin{aligned}\int \frac{1}{1+3x^2}\,dx &= \frac{1}{\sqrt{3}}\int \frac{1}{1+\left(\sqrt{3}\,x\right)^2}\times\sqrt{3}\,dx\\ &= \frac{1}{\sqrt{3}}\int \frac{1}{1+u^2}\,du\\ &= \frac{1}{\sqrt{3}}\tan^{-1}u + c\\ &= \frac{1}{\sqrt{3}}\tan^{-1}\left(\sqrt{3}\,x\right) + c.\end{aligned}$$

(f) Since the derivative of $7 - x^3$ is $-3x^2$, which is 'nearly' the numerator x^2, you can use the substitution $u = 7 - x^3$.

Let $u = 7 - x^3$; then $\dfrac{du}{dx} = -3x^2$. So

$$\begin{aligned}\int \frac{x^2}{(7-x^3)^7}\,dx &= -\tfrac{1}{3}\int \frac{1}{(7-x^3)^7}(-3x^2)\,dx\\ &= -\tfrac{1}{3}\int \frac{1}{u^7}\,du\\ &= -\tfrac{1}{3}\int u^{-7}\,du\\ &= -\frac{1}{3}\times\frac{1}{-6}u^{-6} + c\\ &= \frac{1}{18u^6} + c\\ &= \frac{1}{18(7-x^3)^6} + c.\end{aligned}$$

(g) The integrand is of the form $xg(x)$, where g is a function that you can integrate, which suggests that you should try integration by parts. To integrate the 'second' expression, you can use the fact that it's a simple function of a linear expression.

$$\int xe^{-x/3}\,dx$$
$$= x\left(-3e^{-x/3}\right) - \int 1 \times \left(-3e^{-x/3}\right)\,dx$$
$$= -3xe^{-x/3} + 3\int e^{-x/3}\,dx$$
$$= -3xe^{-x/3} - 9e^{-x/3} + c$$
$$= -3e^{-x/3}(x+3) + c.$$

(h) You can use the trigonometric identity $\sin(2\theta) = 2\sin\theta\cos\theta$.

$$\int \sin\left(\tfrac{1}{2}x\right)\cos\left(\tfrac{1}{2}x\right)\,dx = \int \tfrac{1}{2}\sin x\,dx$$
$$= \tfrac{1}{2}\int \sin x\,dx$$
$$= -\tfrac{1}{2}\cos x + c.$$

(i) You can use the fact that the integrand is a simple function of a linear expression, or you can use the substitution $u = x - 1$.

$$\int (x-1)^4\,dx = \frac{(x-1)^5}{5} + c.$$

(j) You could use integration by substitution (taking $u = 1 - 2e^x$), but the simplest method is to start by multiplying out the brackets.

$$\int e^x(1-2e^x)\,dx = \int (e^x - 2(e^x)^2)\,dx$$
$$= \int (e^x - 2e^{2x})\,dx$$
$$= e^x - 2 \times \tfrac{1}{2}e^{2x} + c$$
$$= e^x - e^{2x} + c$$
$$= e^x - (e^x)^2 + c$$
$$= e^x(1-e^x) + c.$$

(Here's how the integration goes if you use integration by substitution.

Let $u = 1 - 2e^x$; then $\dfrac{du}{dx} = -2e^x$. So

$$\int e^x(1-2e^x)\,dx = -\tfrac{1}{2}\int (1-2e^x)(-2e^x)\,dx$$
$$= -\tfrac{1}{2}\int u\,du$$
$$= -\tfrac{1}{2} \times \tfrac{1}{2}u^2 + c$$
$$= -\tfrac{1}{4}u^2 + c$$
$$= -\tfrac{1}{4}(1-2e^x)^2 + c$$
$$= -\tfrac{1}{4}(1 - 4e^x + 4(e^x)^2) + c$$
$$= -\tfrac{1}{4}(1 - 4e^x + 4e^{2x}) + c$$
$$= e^x - e^{2x} - \tfrac{1}{4} + c$$
$$= e^x - (e^x)^2 - \tfrac{1}{4} + c$$
$$= e^x(1-e^x) - \tfrac{1}{4} + c$$
$$= e^x(1-e^x) + d,$$

where $d = -\frac{1}{4} + c$ is an arbitrary constant.)

(k) You can start by multiplying out the brackets and applying the sum rule. Then you have two integrals to find. One of them is straightforward, and you can find the other by using integration by substitution, taking $u = x^2$.

$$\int x(1+\sin(x^2))\,dx$$
$$= \int (x + x\sin(x^2))\,dx$$
$$= \int x\,dx + \int x\sin(x^2)\,dx$$
$$= \tfrac{1}{2}x^2 + \int x\sin(x^2)\,dx$$

Let $u = x^2$; then $du/dx = 2x$. The expression above becomes

$$\tfrac{1}{2}x^2 + \tfrac{1}{2}\int (\sin(x^2)) \times 2x\,dx$$
$$= \tfrac{1}{2}x^2 + \tfrac{1}{2}\int \sin u\,du$$
$$= \tfrac{1}{2}x^2 + \tfrac{1}{2}(-\cos u) + c$$
$$= \tfrac{1}{2}x^2 - \tfrac{1}{2}\cos(x^2) + c$$
$$= \tfrac{1}{2}(x^2 - \cos(x^2)) + c.$$

(l) Since the derivative of $3x^2 - 2$ is $6x$, which is 'nearly' x, you can use the substitution $u = 3x^2 - 2$.

Let $u = 3x^2 - 2$; then $\dfrac{du}{dx} = 6x$. So

$$\begin{aligned}\int x(3x^2-2)^8\,dx &= \tfrac{1}{6}\int (3x^2-2)^8 \times 6x\,dx\\ &= \tfrac{1}{6}\int u^8\,du\\ &= \tfrac{1}{6} \times \tfrac{1}{9}u^9\\ &= \tfrac{1}{54}(3x^2-2)^9 + c.\end{aligned}$$

Solution to Activity 48

(a) The domain of f is the interval $[-1, \infty)$ (because the expression $\sqrt{x+1}$ is defined only when x is in this interval).

(b) The graph of f is shown below.

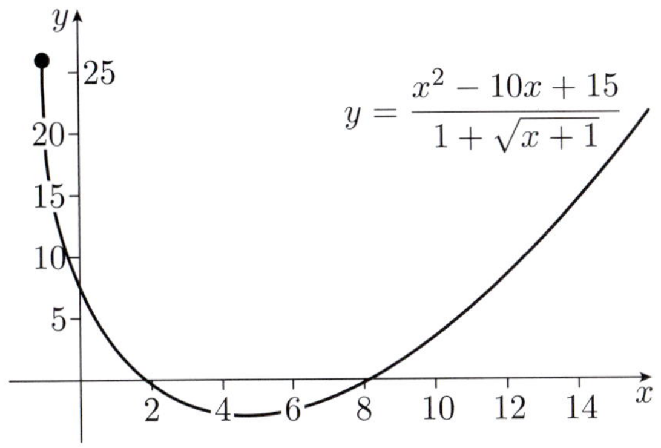

(c) The x-intercepts of f are 1.84 and 8.16 (to 3 s.f.).

(The exact values are $5 \pm \sqrt{10}$.)

(d) The area between the graph of f and the x-axis, between the two x-intercepts, is

$$-\int_{1.837...}^{8.162...} \frac{x^2 - 10x + 15}{1 + \sqrt{x+1}}\,dx.$$

A computer gives the value of this expression as 12.4 (to 3 s.f.).

(Details of how to use the CAS for this activity are given in the *Computer algebra guide*, in the section 'Computer methods for CAS activities in Books A–D'.)

Solution to Activity 49

The area of the cross-section, in square centimetres, is

$$\int_{-1}^{1.5} \tfrac{1}{3}\sqrt{x^3+1}\,dx.$$

A computer gives the value of this definite integral as 0.939 (to 3 s.f.). So the volume of plastic required to make one metre of the edging is approximately $93.9\,\text{cm}^3$.

(Details of how to use the CAS to find the indefinite integral are given in the *Computer algebra guide*, in the section 'Computer methods for CAS activities in Books A–D'.)

Solution to Activity 50

The change in the displacement of the object from time $t = 0$ to time $t = 10$ is

$$\int_0^{10} \sin\left(\frac{t^2}{150}\right) dt.$$

A computer gives the value of this definite integral as 2.15 (to 2 d.p.).

So the displacement of the object at time $t = 10$ is $(6 + 2.15)\,\text{m} = 8.15\,\text{m}$, to the nearest centimetre.

(Details of how to use the CAS to find the indefinite integral are given in the *Computer algebra guide*, in the section 'Computer methods for CAS activities in Books A–D'.)

Acknowledgements

Grateful acknowledgement is made to the following source:

Page 193: Mark Hobbs, for the idea for the cartoon

Unit 9

Matrices

Introduction

This unit introduces mathematical objects called *matrices* (the singular form is *matrix*). At their simplest, matrices are arrays of numbers used to hold information in an orderly way. In particular, they provide a useful way of storing data that can be organised in rows and columns. Everyday examples of such data include transport timetables and tables holding survey results.

In this unit you will discover that arrays of numbers can be thought of as 'multi-dimensional numbers'. For example, you can add, subtract and multiply them, just as you can add, subtract and multiply numbers. You've met a similar idea before: in Unit 5, you saw that column vectors are numerical entities made up of more than one number, and they can be added, subtracted and multiplied by scalars, for example.

A magic square (an example of an array of numbers) on the facade of the Sagrada Familia basilica in Barcelona

One advantage of arranging large amounts of numerical data in matrix form is precisely that a block of data can be treated as a single entity. Many problems that involve large amounts of numerical data are first reduced to matrix problems before being solved with a computer. So matrices are an important topic in mathematics, and find widespread use in mathematical modelling, as well as in computer graphics (where they are used to move and rotate objects), medical imaging, economic models, electrical networks, data encryption and many other applications.

As a simple practical example, consider a publisher that supplies its books to a number of bookshops. The publisher's sales department might collect the weekly order quantities for three of its bestselling books in a grid such as the one in Table 1.

Table 1 A publisher's order quantities for three books in a certain week

	Bestseller 1	Bestseller 2	Bestseller 3
Bookshop 1	10	25	12
Bookshop 2	5	10	5
Bookshop 3	7	5	0
Bookshop 4	10	8	10

The information in Table 1 can be stored more succinctly by omitting the row and column headings and enclosing the data in brackets, if it is understood that each row corresponds to a bookshop and each column corresponds to a book:

$$\begin{pmatrix} 10 & 25 & 12 \\ 5 & 10 & 5 \\ 7 & 5 & 0 \\ 10 & 8 & 10 \end{pmatrix}.$$

This is an example of a matrix. Information about order quantities can be collected every week for the life of the three books and stored in this form.

For example, the information for two consecutive weeks can be recorded in two separate matrices:

$$\begin{pmatrix} 10 & 25 & 12 \\ 5 & 10 & 5 \\ 7 & 5 & 0 \\ 10 & 8 & 10 \end{pmatrix}, \quad \begin{pmatrix} 15 & 30 & 20 \\ 10 & 20 & 10 \\ 20 & 8 & 5 \\ 15 & 0 & 10 \end{pmatrix}.$$

The order quantities for a fortnight can now be calculated by adding corresponding numbers in the two arrays, which gives

$$\begin{pmatrix} 10+15 & 25+30 & 12+20 \\ 5+10 & 10+20 & 5+10 \\ 7+20 & 5+8 & 0+5 \\ 10+15 & 8+0 & 10+10 \end{pmatrix}, \quad \text{that is,} \quad \begin{pmatrix} 25 & 55 & 32 \\ 15 & 30 & 15 \\ 27 & 13 & 5 \\ 25 & 8 & 20 \end{pmatrix}.$$

This calculation amounts to *adding* the two matrices, in a sense that will be made precise in Section 1.

You might wonder what the advantage is in carrying out calculations in this way. After all, the total number of orders for each bookshop can be worked out in a straightforward way without resorting to matrices. In a real-life situation, however, a large publisher will supply hundreds of bookshops with hundreds of books. Calculating order quantities and total charges requires careful accounting, and it is precisely in this sort of situation that matrices become a useful way to store data (especially electronically) and carry out calculations with them.

Matrices were first introduced in Western mathematics in their current form by the British mathematician Arthur Cayley (1821–1895). However, they appeared as early as 300–200 BC in the Chinese text *The nine chapters on the mathematical art*, where matrix methods are used to solve simultaneous equations. You'll meet a modern version of these methods in Section 4.

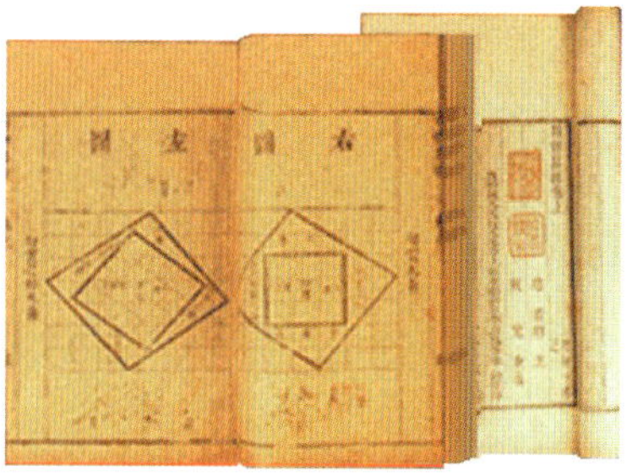

Pages from *The nine chapters on the mathematical art*

There are other useful operations that can be carried out when data are stored in matrix form, such as the multiplication of matrices by single numbers, and the multiplication of two matrices. You'll learn about these operations in Section 1, and see an application of matrix multiplication in Section 2. In Section 3 you'll learn about *matrix inverses*, and in Section 4 you'll meet a connection between matrices and systems of linear equations.

1 Matrices and matrix operations

In this section you'll meet matrices and standard matrix notation, and learn about matrix addition, multiplication of matrices by single numbers (known as *scalar multiplication* of matrices), matrix multiplication and matrix powers.

1.1 Matrix notation and terminology

A **matrix** (pronounced 'may-tricks') is a rectangular array of numbers, usually enclosed in brackets. Here are some examples:

$$\begin{pmatrix} 1 & 2 & -1 \\ 5 & 3 & 0 \end{pmatrix}, \quad \begin{pmatrix} 1 & 2 \\ 3 & 1 \end{pmatrix}, \quad \begin{pmatrix} \frac{1}{3} & 4 \\ 7 & -6 \\ 22 & 0 \end{pmatrix}.$$

First rule of consulting: never, *never* call this a "table". We can bill four times as much when we call it a "matrix".

Some texts use square brackets for matrices, like this:

$$\begin{bmatrix} 1 & 2 \\ 3 & 1 \end{bmatrix}.$$

In this module we'll always use round brackets.

We often use capital letters to denote matrices. For example, we might write

$$\mathbf{A} = \begin{pmatrix} 1 & 2 & -1 \\ 5 & 3 & 0 \end{pmatrix}, \quad \mathbf{B} = \begin{pmatrix} 1 & 2 \\ 3 & 1 \end{pmatrix}, \quad \mathbf{C} = \begin{pmatrix} \frac{1}{3} & 4 \\ 7 & -6 \\ 22 & 0 \end{pmatrix}.$$

In the text in this module, the capital letters that denote matrices are in bold font, but you don't need to do anything special when you handwrite them. In particular, you don't need to underline them, unlike letters that denote vectors.

The numbers in a matrix are called its **elements** (or its **entries**, in some texts). For example, matrices **A** and **C** above each have six elements, and matrix **B** has four elements. In this module, the elements of matrices are always real numbers.

A horizontal line of numbers in a matrix is called a **row**, and a vertical line of numbers is called a **column**. For instance, in the matrix **A** above the first row is $\begin{pmatrix} 1 & 2 & -1 \end{pmatrix}$ and the second row is $\begin{pmatrix} 5 & 3 & 0 \end{pmatrix}$. Likewise, the first column is $\begin{pmatrix} 1 \\ 5 \end{pmatrix}$, the second column is $\begin{pmatrix} 2 \\ 3 \end{pmatrix}$ and the third column is $\begin{pmatrix} -1 \\ 0 \end{pmatrix}$.

The matrix **A** has two rows and three columns: we say that it is a 2×3 matrix. In general, a matrix with m rows and n columns is described as an $m \times n$ matrix, or a matrix of **size** $m \times n$. The first number in this notation is always the number of rows, and the second number is always the number of columns. The size $m \times n$ of a matrix is read as 'm by n'.

A matrix with the same number of rows as columns is called a **square matrix**. Thus, for example, matrix **B** above is square.

The column notation for vectors that you met in Unit 5 is a particular instance of matrix notation: two-dimensional column vectors are 2×1 matrices, and three-dimensional column vectors are 3×1 matrices. In general, the word **vector** is used to mean any matrix with a single column, even if it has more than 3 elements and hence has no geometric interpretation in the plane or three-dimensional space. For instance, the following matrices are vectors:

$$\begin{pmatrix} 1 \\ 3 \\ 5 \\ 7 \end{pmatrix}, \quad \begin{pmatrix} \sqrt{2} \\ -3 \\ \frac{1}{2} \\ 0 \\ 7 \end{pmatrix} \quad \text{and} \quad \begin{pmatrix} w \\ x \\ y \\ z \end{pmatrix}, \quad \text{where } w,\ x,\ y,\ z \text{ are variables.}$$

The elements of a vector are often called its **components**, as in Unit 5. A vector with n components is called an n-**dimensional** vector.

There is a useful notation for the elements of a matrix in terms of their row and column positions. The element in row i and column j of a matrix $\mathbf{A}$ is denoted by a_{ij}. Similarly, the element in row i and column j of a matrix $\mathbf{B}$ is denoted by b_{ij}, and so on.

For instance, if

$$\mathbf{A} = \begin{pmatrix} \frac{1}{3} & 4 \\ 7 & -6 \end{pmatrix},$$

then $a_{11} = \frac{1}{3}$, $a_{12} = 4$, $a_{21} = 7$ and $a_{22} = -6$.

This notation is used for matrices of any size. Thus, for instance, a general 3×4 matrix can be denoted by

$$\mathbf{A} = \begin{pmatrix} a_{11} & a_{12} & a_{13} & a_{14} \\ a_{21} & a_{22} & a_{23} & a_{24} \\ a_{31} & a_{32} & a_{33} & a_{34} \end{pmatrix}.$$

Activity 1 *Using matrix notation*

Write down the elements a_{33}, a_{34}, a_{43}, a_{14}, a_{44} and a_{42} of the matrix

$$\mathbf{A} = \begin{pmatrix} 2 & 8 & 0 & -1 \\ 4 & 7 & 5 & -2 \\ 9 & -1 & 3 & -4 \\ 6 & -5 & -7 & -9 \end{pmatrix}.$$

What is the size of $\mathbf{A}$?

Two matrices are equal if they have the same numbers of rows and columns, and all corresponding elements are equal. For example,

$$\begin{pmatrix} 1 & 2 \\ 3 & 4 \end{pmatrix} = \begin{pmatrix} 1 & 2 \\ 3 & 4 \end{pmatrix},$$

but

$$\begin{pmatrix}1 & 2\\ 3 & 4\end{pmatrix} \neq \begin{pmatrix}1 & 2\\ 3 & 4\\ 5 & 6\end{pmatrix} \quad \text{and} \quad \begin{pmatrix}1 & 2\\ 3 & 4\end{pmatrix} \neq \begin{pmatrix}1 & 3\\ 2 & 4\end{pmatrix}.$$

You can carry out operations on matrices in a similar way to numbers. For example, you can add, subtract and multiply them, as you'll see in this section. There's also a way to manipulate matrices that in certain circumstances plays the role of division, which you'll learn about in Section 3. While matrix addition is similar to addition of vectors, which you met in Unit 5, matrix multiplication is more surprising. We'll start by looking at matrix addition.

Arthur Cayley gave the rules for matrix operations in a paper presented to the Royal Society in 1858. The paper also contained the conditions under which a matrix has an inverse (you'll meet matrix inverses in Section 3). However, it was Cayley's friend and fellow mathematician James Joseph Sylvester who coined the term 'matrix', in 1850, and he also contributed to the theory of matrices.

Before being appointed to a mathematics professorship at the University of Cambridge, Cayley worked for 14 years as a lawyer in London. At the time, Sylvester was employed as an actuary. The two worked together at the courts of Lincoln's Inn, which gave them the opportunity to discuss mathematics.

Sylvester had a more varied mathematical career than Cayley, including professorships at University College London, Johns Hopkins University in the USA, and the University of Oxford. It is said that he tutored Florence Nightingale in mathematics, though firm evidence for this has yet to be found.

Arthur Cayley (1821–1895)

James Joseph Sylvester (1814–1897)

1.2 Matrix addition and subtraction

You can add two matrices only if they have the same size. The sum of two matrices is obtained by adding their corresponding elements. For example:

$$\begin{pmatrix}-2 & 4\\ 1 & 5\\ 0 & -3\end{pmatrix} + \begin{pmatrix}3 & 1\\ -2 & 10\\ -6 & 4\end{pmatrix} = \begin{pmatrix}-2+3 & 4+1\\ 1+(-2) & 5+10\\ 0+(-6) & -3+4\end{pmatrix} = \begin{pmatrix}1 & 5\\ -1 & 15\\ -6 & 1\end{pmatrix}.$$

The rule for adding two matrices can be expressed formally as follows.

Matrix addition

If $\mathbf{A}$ and $\mathbf{B}$ are $m \times n$ matrices, then $\mathbf{A} + \mathbf{B}$ is the $m \times n$ matrix whose element in row i and column j is $a_{ij} + b_{ij}$.

Addition of two matrices of different sizes is not defined.

For instance, you cannot add a 3×3 matrix and a 3×5 matrix.

The rule for adding matrices extends to the sum of any number of matrices of the same size: you just add corresponding elements.

Activity 2 *Adding matrices*

For each of the pairs of matrices below, work out the matrix sum $\mathbf{A} + \mathbf{B}$.

(a) $\mathbf{A} = \begin{pmatrix} 2 & 1 \\ 3 & 0 \end{pmatrix}, \quad \mathbf{B} = \begin{pmatrix} 4 & 6 \\ 2 & -1 \end{pmatrix}$

(b) $\mathbf{A} = \begin{pmatrix} 2 & 0 & -4 \\ -1 & 6 & 0 \end{pmatrix}, \quad \mathbf{B} = \begin{pmatrix} 1 & 0 & 1 \\ 0 & 1 & 1 \end{pmatrix}$

(c) $\mathbf{A} = \begin{pmatrix} a & b \\ c & d \end{pmatrix}, \quad \mathbf{B} = \begin{pmatrix} 3 & -1 \\ -1 & 2 \end{pmatrix}$

In Unit 5, you saw that vector addition shares many properties with the addition of numbers. This happens for matrix addition, too. For example, you can see from the definition of matrix addition that

$$\mathbf{A} + \mathbf{B} = \mathbf{B} + \mathbf{A},$$

for all matrices $\mathbf{A}$ and $\mathbf{B}$ of the same size. As you saw with vectors, this property is expressed by saying that matrix addition is **commutative**.

You can also see from the definition of matrix addition that

$$(\mathbf{A} + \mathbf{B}) + \mathbf{C} = \mathbf{A} + (\mathbf{B} + \mathbf{C}),$$

for all matrices $\mathbf{A}$, $\mathbf{B}$ and $\mathbf{C}$ of the same size. In other words, when you add three matrices $\mathbf{A}$, $\mathbf{B}$ and $\mathbf{C}$ of the same size, it doesn't matter whether you add $\mathbf{A}$ and $\mathbf{B}$ and then add $\mathbf{C}$ to the result, or add $\mathbf{B}$ and $\mathbf{C}$ and then add $\mathbf{A}$ to the result, as you get the same overall result either way. This property is expressed by saying that matrix addition is **associative**. It tells you that you can write the expression

$$\mathbf{A} + \mathbf{B} + \mathbf{C}$$

without ambiguity: it doesn't matter whether it means $(\mathbf{A} + \mathbf{B}) + \mathbf{C}$ or $\mathbf{A} + (\mathbf{B} + \mathbf{C})$, as both are equal.

There are other respects in which matrix operations behave like number operations. You will discover more similarities between matrices and numbers as you meet other matrix operations.

You can also subtract matrices of the same size by subtracting their corresponding elements. For example,

$$\begin{pmatrix} 4 & 5 & 3 \\ 7 & -6 & 0 \end{pmatrix} - \begin{pmatrix} 1 & 3 & 4 \\ 3 & 9 & 1 \end{pmatrix} = \begin{pmatrix} 4-1 & 5-3 & 3-4 \\ 7-3 & -6-9 & 0-1 \end{pmatrix}$$
$$= \begin{pmatrix} 3 & 2 & -1 \\ 4 & -15 & -1 \end{pmatrix}.$$

Here is the formal definition of matrix subtraction.

Matrix subtraction

If $\mathbf{A}$ and $\mathbf{B}$ are $m \times n$ matrices, then $\mathbf{A} - \mathbf{B}$ is the $m \times n$ matrix whose element in row i and column j is $a_{ij} - b_{ij}$.

Subtraction of two matrices of different sizes is not defined.

Like the subtraction of numbers, matrix subtraction is neither commutative nor associative. For example, it is *not* true that $\mathbf{A} - \mathbf{B} = \mathbf{B} - \mathbf{A}$ for all matrices $\mathbf{A}$ and $\mathbf{B}$ of the same size, as illustrated by parts (a) and (b) of the next activity.

Activity 3 *Subtracting matrices*

Let $\mathbf{A} = \begin{pmatrix} 2 & -2 \\ 1 & 0 \end{pmatrix}$, $\mathbf{B} = \begin{pmatrix} \frac{1}{2} & 4 \\ -1 & 2 \end{pmatrix}$ and $\mathbf{C} = \begin{pmatrix} 0 & -x \\ x & 0 \end{pmatrix}$. Work out the following matrices.

(a) $\mathbf{A} - \mathbf{B}$ (b) $\mathbf{B} - \mathbf{A}$ (c) $\mathbf{B} - \mathbf{C}$

Since vectors are single-column matrices, they are added and subtracted by adding or subtracting corresponding components, as in the examples below.

$$\begin{pmatrix} 1 \\ -3 \\ 4 \\ \frac{1}{2} \\ 0 \end{pmatrix} + \begin{pmatrix} 7 \\ -2 \\ 1 \\ \frac{3}{2} \\ -1 \end{pmatrix} = \begin{pmatrix} 1+7 \\ -3+(-2) \\ 4+1 \\ \frac{1}{2}+\frac{3}{2} \\ 0+(-1) \end{pmatrix} = \begin{pmatrix} 8 \\ -5 \\ 5 \\ 2 \\ -1 \end{pmatrix}$$

$$\begin{pmatrix} 10 \\ 3 \\ 1 \\ -1 \end{pmatrix} - \begin{pmatrix} -2 \\ 3 \\ 2 \\ 1 \end{pmatrix} = \begin{pmatrix} 10-(-2) \\ 3-3 \\ 1-2 \\ -1-1 \end{pmatrix} = \begin{pmatrix} 12 \\ 0 \\ -1 \\ -2 \end{pmatrix}$$

In other words, the ideas of addition and subtraction of column vectors that you met in Unit 5 extend in the obvious way to vectors with more than three components.

A matrix each of whose elements is zero is called a **zero matrix**, and is usually denoted by $\mathbf{0}$. For example, the 2×2 zero matrix is

$$\mathbf{0} = \begin{pmatrix} 0 & 0 \\ 0 & 0 \end{pmatrix}.$$

Also, as you'd expect, if $\mathbf{A}$ is any matrix, then the matrix formed by changing each element of $\mathbf{A}$ to its negative is called the **negative** of $\mathbf{A}$, and denoted by $-\mathbf{A}$. For example,

$$-\begin{pmatrix} 0 & 1 \\ -2 & -4 \end{pmatrix} = \begin{pmatrix} 0 & -1 \\ 2 & 4 \end{pmatrix}.$$

It follows from the definitions of matrix addition and subtraction that for any matrices $\mathbf{A}$ and $\mathbf{B}$ of the same size,

$$\mathbf{A} - \mathbf{B} = \mathbf{A} + (-\mathbf{B}),$$

as you'd expect.

1.3 Scalar multiplication of matrices

Any matrix can be added to itself, and a matrix sum $\mathbf{A} + \mathbf{A}$ can be written as $2\mathbf{A}$. Each element in $2\mathbf{A}$ is twice the corresponding element in $\mathbf{A}$, as in the following example:

$$\begin{pmatrix} 1 & 0 & 5 \\ -1 & -2 & 3 \\ 3 & \frac{1}{2} & 1 \end{pmatrix} + \begin{pmatrix} 1 & 0 & 5 \\ -1 & -2 & 3 \\ 3 & \frac{1}{2} & 1 \end{pmatrix} = \begin{pmatrix} 2 & 0 & 10 \\ -2 & -4 & 6 \\ 6 & 1 & 2 \end{pmatrix}.$$

Similarly, adding $\mathbf{A}$ to itself n times, for any natural number n, gives a matrix in which each element is the corresponding element in $\mathbf{A}$ multiplied by n.

This idea can be generalised to any real number k, so $k\mathbf{A}$ is the matrix obtained by multiplying each element of $\mathbf{A}$ by k. For example,

$$\frac{1}{2}\begin{pmatrix} 4 & 3 \\ 0 & -1 \end{pmatrix} = \begin{pmatrix} \frac{1}{2} \times 4 & \frac{1}{2} \times 3 \\ \frac{1}{2} \times 0 & \frac{1}{2} \times (-1) \end{pmatrix} = \begin{pmatrix} 2 & \frac{3}{2} \\ 0 & -\frac{1}{2} \end{pmatrix}.$$

The matrix $k\mathbf{A}$ is called the **scalar multiple** of the matrix $\mathbf{A}$ by the real number k, and the operation of forming $k\mathbf{A}$ is called **scalar multiplication**.

Scalar multiplication

If $\mathbf{A}$ is an $m \times n$ matrix and k is any real number, then $k\mathbf{A}$ is the matrix whose element in row i and column j is ka_{ij}.

The number k in a scalar multiple $k\mathbf{A}$ of a matrix $\mathbf{A}$ is sometimes referred to as the **scalar** k in this context, to emphasise that it's a number rather than a matrix.

Activity 4 *Multiplying matrices by scalars*

(a) Calculate the following products of scalars and matrices, simplifying your answers.

(i) $3\begin{pmatrix} 4 & 3 \\ 2 & 1 \end{pmatrix}$ (ii) $\frac{1}{2}\begin{pmatrix} 4 & 6 \\ 2 & -1 \end{pmatrix}$ (iii) $\frac{2}{3}\begin{pmatrix} 6 & x \\ 3 & y \end{pmatrix}$

(b) Simplify each of the following matrices by writing it as a scalar multiple of a matrix with integer elements.

(i) $\begin{pmatrix} 4.5 & 3 \\ 2 & 1.5 \end{pmatrix}$ (ii) $\begin{pmatrix} \frac{3}{2} \\ \frac{2}{3} \end{pmatrix}$ (iii) $\begin{pmatrix} 2x & 3x \\ 0 & 5x \end{pmatrix}$

Notice that it follows from the definition of scalar multiplication that, for any matrix $\mathbf{A}$,

$$(-1)\mathbf{A} = -\mathbf{A},$$

as you'd expect.

In the next activity you can practise the matrix operations that you've learned so far.

Activity 5 *Combining matrix addition, subtraction and scalar multiplication*

Let $\mathbf{A} = \begin{pmatrix} \frac{1}{2} & 1 & 2 \\ 4 & -2 & 0 \end{pmatrix}$, $\mathbf{B} = \begin{pmatrix} \frac{1}{2} & -1 \\ -2 & 2 \end{pmatrix}$ and $\mathbf{C} = \begin{pmatrix} \frac{3}{2} & 0 & -1 \\ -1 & 3 & 5 \end{pmatrix}$.

Where possible, evaluate $\mathbf{A} - \mathbf{C}$, $\mathbf{B} + 2\mathbf{A}$ and $\mathbf{C} - 2\mathbf{A}$.

Here's a summary of some useful properties of matrix addition and scalar multiplication of matrices. As you'd expect (since vectors are single-column matrices), these properties are the same as the properties of vector algebra that you met in Unit 5, but with vectors replaced by matrices.

Properties of matrix addition and scalar multiplication

The following properties hold for all matrices $\mathbf{A}$, $\mathbf{B}$ and $\mathbf{C}$ for which the sums mentioned are defined, and for all scalars m and n.

1. $\mathbf{A} + \mathbf{B} = \mathbf{B} + \mathbf{A}$
2. $(\mathbf{A} + \mathbf{B}) + \mathbf{C} = \mathbf{A} + (\mathbf{B} + \mathbf{C})$
3. $\mathbf{A} + \mathbf{0} = \mathbf{A}$
4. $\mathbf{A} + (-\mathbf{A}) = \mathbf{0}$
5. $m(\mathbf{A} + \mathbf{B}) = m\mathbf{A} + m\mathbf{B}$
6. $(m + n)\mathbf{A} = m\mathbf{A} + n\mathbf{A}$
7. $m(n\mathbf{A}) = (mn)\mathbf{A}$
8. $1\mathbf{A} = \mathbf{A}$

(In properties 3 and 4, $\mathbf{0}$ is the zero matrix of the same size as $\mathbf{A}$.)

1.4 Matrix multiplication

In Subsection 1.3 you learned how to multiply a number and a matrix. Under certain conditions, it is also possible to multiply two matrices. The way that this is done is less obvious than matrix addition or scalar multiplication.

Recall the example in the introduction to this unit, where a matrix was used to store the numbers of copies of three bestselling books that four bookshops ordered from a publisher in a certain week:

$$\begin{array}{r} \\ \text{Bookshop 1} \\ \text{Bookshop 2} \\ \text{Bookshop 3} \\ \text{Bookshop 4} \end{array} \begin{array}{c} \begin{array}{ccc} \text{Bestseller 1} & \text{Bestseller 2} & \text{Bestseller 3} \end{array} \\ \begin{pmatrix} 10 & 25 & 12 \\ 5 & 10 & 5 \\ 7 & 5 & 0 \\ 10 & 8 & 10 \end{pmatrix} \end{array}.$$

Now suppose that the publisher charges the bookshops £4 per copy for Bestseller 1, £7 per copy for Bestseller 2 and £9 per copy for Bestseller 3. According to the data in the matrix, Bookshop 1 ordered 10 copies of Bestseller 1, 25 copies of Bestseller 2 and 12 copies of Bestseller 3. Therefore the total amount, in £, that Bookshop 1 owes the publisher for that week's order is

$$10 \times 4 + 25 \times 7 + 12 \times 9 = 323.$$

The expression on the left of this equation is obtained by multiplying each element in the first row of the matrix by the price of the corresponding book, and then adding the results.

If we write the prices (in £) for the three bestsellers as a three-dimensional vector, that is, as a 3×1 matrix, then we can picture this procedure as shown in Figure 1.

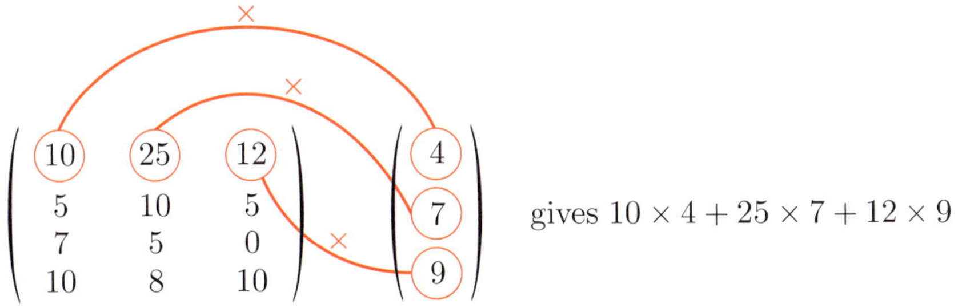

Figure 1 The procedure for calculating the amount owed by Bookshop 1

If we want to calculate the amount owed by Bookshop 2, then we can carry out the same procedure with the second row of the matrix, and likewise for Bookshops 3 and 4 with the third and fourth rows, respectively. We can write the four amounts that we obtain as a four-dimensional vector. The procedure described above combines the two matrices

$$\begin{pmatrix} 10 & 25 & 12 \\ 5 & 10 & 5 \\ 7 & 5 & 0 \\ 10 & 8 & 10 \end{pmatrix} \quad \text{and} \quad \begin{pmatrix} 4 \\ 7 \\ 9 \end{pmatrix}$$

to give

$$\begin{pmatrix} 10 \times 4 + 25 \times 7 + 12 \times 9 \\ 5 \times 4 + 10 \times 7 + 5 \times 9 \\ 7 \times 4 + 5 \times 7 + 0 \times 9 \\ 10 \times 4 + 8 \times 7 + 10 \times 9 \end{pmatrix} = \begin{pmatrix} 323 \\ 135 \\ 63 \\ 186 \end{pmatrix}.$$

This calculation shows that the amounts owed by the four bookshops are £323, £135, £63 and £186, respectively.

This procedure for combining matrices comes up so often in applications that it is used to define matrix multiplication. We say that

$$\text{the } \textbf{product} \text{ of } \begin{pmatrix} 10 & 25 & 12 \\ 5 & 10 & 5 \\ 7 & 5 & 0 \\ 10 & 8 & 10 \end{pmatrix} \quad \text{and} \quad \begin{pmatrix} 4 \\ 7 \\ 9 \end{pmatrix} \quad \text{is} \quad \begin{pmatrix} 323 \\ 135 \\ 63 \\ 186 \end{pmatrix}.$$

Now suppose that as well as calculating the amount owed by each of the four bookshops, the publisher also wants to calculate the profit made from sales to each individual bookshop. Suppose that the profits made on each copy of Bestsellers 1, 2 and 3 are £1, £2 and £2, respectively. You can use exactly the same procedure as above to calculate the profit made from the sales to each bookshop. You write the profits (in £) made on the three books as a three-dimensional vector, and form

$$\text{the product of} \quad \begin{pmatrix} 10 & 25 & 12 \\ 5 & 10 & 5 \\ 7 & 5 & 0 \\ 10 & 8 & 10 \end{pmatrix} \quad \text{and} \quad \begin{pmatrix} 1 \\ 2 \\ 2 \end{pmatrix},$$

which is

$$\begin{pmatrix} 10 \times 1 + 25 \times 2 + 12 \times 2 \\ 5 \times 1 + 10 \times 2 + 5 \times 2 \\ 7 \times 1 + 5 \times 2 + 0 \times 2 \\ 10 \times 1 + 8 \times 2 + 10 \times 2 \end{pmatrix} = \begin{pmatrix} 84 \\ 35 \\ 17 \\ 46 \end{pmatrix}.$$

So the profits made from sales to the four bookshops are £84, £35, £17 and £46, respectively.

In fact you can view the two matrix multiplications above – the one for the amounts owed, and the one for the profits – as a *single* matrix multiplication. To do this, you put the two vectors containing the prices and the profits together as a single matrix with two columns, and you put the two vectors that contain the results of the calculations together as a single matrix with two columns, and say that

$$\text{the product of} \quad \begin{pmatrix} 10 & 25 & 12 \\ 5 & 10 & 5 \\ 7 & 5 & 0 \\ 10 & 8 & 10 \end{pmatrix} \quad \text{and} \quad \begin{pmatrix} 4 & 1 \\ 7 & 2 \\ 9 & 2 \end{pmatrix} \quad \text{is} \quad \begin{pmatrix} 323 & 84 \\ 135 & 35 \\ 63 & 17 \\ 186 & 46 \end{pmatrix}. \qquad (1)$$

This calculation is an example of how matrices are multiplied in general. In Section 2 you'll meet another context that shows that it makes sense to multiply matrices in this way. Before that, you'll see a more detailed explanation of the procedure for multiplying matrices, and have a chance to practise it yourself.

Let's start by considering which pairs of matrices can be multiplied together. Remember that the element in the first row and first column of the product matrix in calculation (1) above was obtained by multiplying each element in the first *row* of the first matrix by the corresponding element in the first *column* of the second matrix, and adding the results:

$$10 \times 4 + 25 \times 7 + 12 \times 9 = 323.$$

It was possible to do this because *the number of columns in the first matrix is the same as the number of rows in the second matrix*, so each element in the first row of the first matrix *has* a corresponding element in the first column of the second matrix. You can multiply two matrices together only if this condition holds.

Thus, you can multiply together a 4×3 matrix and a 3×2 matrix. On the other hand, it is not possible to multiply together a 3×4 matrix and a 2×3 matrix. When it is not possible to multiply two matrices, we sometimes say that their product is **undefined**.

A convenient way to determine whether two matrices, of sizes $m \times n$ and $p \times q$, say, can be multiplied together is to write their sizes next to each other as follows:

$$m \times n \quad p \times q.$$

The matrices can be multiplied only if the two numbers in the middle are equal; that is, if $n = p$. If these numbers are equal, then the size of the product matrix is given by the remaining numbers. So the product of an $m \times n$ matrix and an $n \times q$ matrix is a matrix of size $m \times q$.

For example, a 4×3 matrix and a 3×2 matrix can be multiplied together, and the product has size 4×2. Here's a useful way to picture this fact:

$$4 \times \boxed{3} \quad \boxed{3} \times 2 \quad \text{gives} \quad 4 \times 2.$$

Now let's look at the procedure for multiplying two matrices step by step. It's demonstrated in the next example.

Example 1 *Multiplying matrices*

Let $\mathbf{A} = \begin{pmatrix} -2 & 1 & 3 \\ 2 & 4 & -2 \\ 0 & -1 & 0 \end{pmatrix}$ and $\mathbf{B} = \begin{pmatrix} 1 & 0 \\ \frac{1}{2} & 2 \\ -1 & 4 \end{pmatrix}$.

Check that the product matrix $\mathbf{AB}$ can be formed, determine the size of $\mathbf{AB}$, and calculate it.

Solution

Write down the sizes of $\mathbf{A}$ and $\mathbf{B}$.

$\mathbf{A}$ has size 3×3 and $\mathbf{B}$ has size 3×2.

$$3 \times \boxed{3} \quad \boxed{3} \times 2$$

The numbers in the middle are equal, so the product $\mathbf{AB}$ can be formed. It has size 3×2.

To obtain the element in the *first row* and *first column* of **AB**, multiply each element in the *first row* of **A** by the corresponding element in the *first column* of **B**, and add the results.

$$\begin{pmatrix} -2 & 1 & 3 \\ 2 & 4 & -2 \\ 0 & -1 & 0 \end{pmatrix}\begin{pmatrix} 1 & 0 \\ \frac{1}{2} & 2 \\ -1 & 4 \end{pmatrix} = \begin{pmatrix} (-2)\times 1 + 1\times\frac{1}{2} + 3\times(-1) & \\ & \\ & \end{pmatrix}$$

$$= \begin{pmatrix} -\frac{9}{2} & \\ & \\ & \end{pmatrix}$$

To obtain the element in the *first row* and *second column* of **AB**, apply the same procedure to the *first row* of **A** and the *second column* of **B**.

$$\begin{pmatrix} -2 & 1 & 3 \\ 2 & 4 & -2 \\ 0 & -1 & 0 \end{pmatrix}\begin{pmatrix} 1 & 0 \\ \frac{1}{2} & 2 \\ -1 & 4 \end{pmatrix} = \begin{pmatrix} -\frac{9}{2} & (-2)\times 0 + 1\times 2 + 3\times 4 \\ & \\ & \end{pmatrix}$$

$$= \begin{pmatrix} -\frac{9}{2} & 14 \\ & \\ & \end{pmatrix}$$

To obtain the element in the *second row* and *first column* of **AB**, apply the same procedure to the *second row* of **A** and the *first column* of **B**.

$$\begin{pmatrix} -2 & 1 & 3 \\ 2 & 4 & -2 \\ 0 & -1 & 0 \end{pmatrix}\begin{pmatrix} 1 & 0 \\ \frac{1}{2} & 2 \\ -1 & 4 \end{pmatrix} = \begin{pmatrix} -\frac{9}{2} & 14 \\ 2\times 1 + 4\times\frac{1}{2} + (-2)\times(-1) & \\ & \end{pmatrix}$$

$$= \begin{pmatrix} -\frac{9}{2} & 14 \\ 6 & \\ & \end{pmatrix}$$

In general, to obtain the element in the ith row and jth column of **AB**, multiply each element in the ith row of **A** by the corresponding element in the jth row of **B**, and add the results.

$$\begin{pmatrix} -2 & 1 & 3 \\ 2 & 4 & -2 \\ 0 & -1 & 0 \end{pmatrix}\begin{pmatrix} 1 & 0 \\ \frac{1}{2} & 2 \\ -1 & 4 \end{pmatrix}$$

$$= \begin{pmatrix} -\frac{9}{2} & 14 \\ 6 & 2\times 0 + 4\times 2 + (-2)\times 4 \\ 0\times 1 + (-1)\times\frac{1}{2} + 0\times(-1) & 0\times 0 + (-1)\times 2 + 0\times 4 \end{pmatrix}$$

$$= \begin{pmatrix} -\frac{9}{2} & 14 \\ 6 & 0 \\ -\frac{1}{2} & -2 \end{pmatrix}$$

Here's a summary of the procedure for matrix multiplication.

Matrix multiplication

Let **A** and **B** be matrices. Then the product matrix **AB** can be formed only if the number of columns of **A** is equal to the number of rows of **B**.

If **A** has size $m \times n$ and **B** has size $n \times p$, then the product **AB** has size $m \times p$.

The element in row i and column j of the product matrix **AB** is obtained by multiplying each element in the ith row of **A** by the corresponding element in the jth column of **B** and adding the results.

In element notation, if c_{ij} denotes the element in the ith row and jth column of **AB**, then

$$c_{ij} = a_{i1}b_{1j} + a_{i2}b_{2j} + \cdots + a_{in}b_{nj}.$$

Figure 2 illustrates how row i of a matrix **A** and column j of a matrix **B** are combined to give the element in row i and column j of the product matrix **AB**.

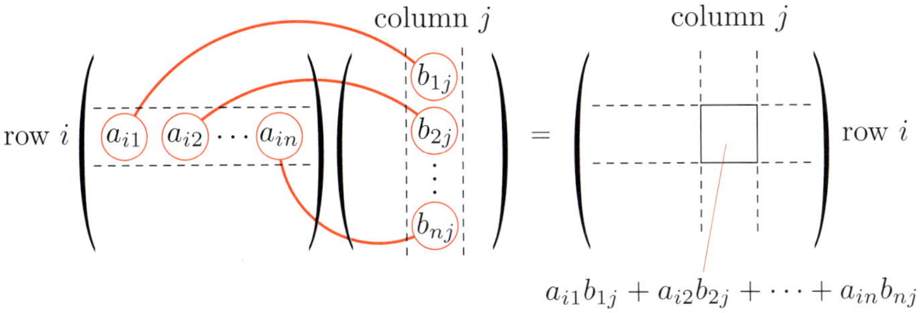

Figure 2 The element in row i and column j of a product matrix **AB** expressed in terms of elements from **A** and **B**

Matrix multiplication may seem quite complicated at first, but it is a very useful technique, and with practice you should become proficient at it.

Activity 6 *Multiplying matrices*

In each of parts (a)–(e) below, calculate the matrix product if it exists.

(a) $\begin{pmatrix} 2 & 9 \\ 7 & 1 \\ -6 & 8 \end{pmatrix} \begin{pmatrix} -2 & 3 & 1 \\ 4 & 0 & 5 \end{pmatrix}$ (b) $\begin{pmatrix} 1 & 2 & 3 \end{pmatrix} \begin{pmatrix} 9 \\ 8 \\ 7 \end{pmatrix}$

(c) $\begin{pmatrix} 2 & 9 \\ 7 & 1 \\ -6 & 8 \end{pmatrix} \begin{pmatrix} 0 & 10 \\ 4 & 1 \\ -2 & 7 \end{pmatrix}$ (d) $\begin{pmatrix} 9 \\ 8 \\ 7 \end{pmatrix} \begin{pmatrix} 1 & 2 & 3 \end{pmatrix}$

(e) $\begin{pmatrix} 1 & 2 & 3 \\ 7 & 6 & 1 \end{pmatrix} \begin{pmatrix} 1 & 0 & 1 \\ 1 & 0 & -1 \\ 2 & 1 & 3 \end{pmatrix}$

Matrix multiplication shares many of the properties of multiplication of numbers, as you'll see later in this subsection and later in the unit. However, there's an important difference between the properties for matrices and those for numbers, as you can find out in the next activity.

Activity 7 *Investigating matrix multiplication*

(a) Determine whether the products **AB** and **BA** are defined for the following pairs of matrices.

(i) $\mathbf{A} = \begin{pmatrix} -1 & 4 \\ 2 & 1 \end{pmatrix}$ and $\mathbf{B} = \begin{pmatrix} 3 \\ 5 \end{pmatrix}$

(ii) $\mathbf{A} = \begin{pmatrix} 1 & 1 \\ -1 & 0 \end{pmatrix}$ and $\mathbf{B} = \begin{pmatrix} 0 & 1 \\ 2 & 3 \end{pmatrix}$

(b) For each of parts (a)(i) and (a)(ii), if the products **AB** and **BA** are both defined, evaluate them. What do you notice?

Activity 7 illustrates the important fact that *matrix multiplication is not commutative.* First, there are matrices **A** and **B** for which the product **AB** exists but the product **BA** is not defined. Second, even in cases where both the products **AB** and **BA** are defined, these two products can be *different matrices.*

In the next activity, you can investigate whether another property that holds for multiplication of numbers also holds for multiplication of matrices.

Activity 8 *Investigating matrix multiplication further*

Let $\mathbf{A} = \begin{pmatrix} -1 & 0 \end{pmatrix}$, $\mathbf{B} = \begin{pmatrix} 1 & -1 \\ 2 & -1 \end{pmatrix}$ and $\mathbf{C} = \begin{pmatrix} 1 & 3 & 0 \\ -3 & 4 & 1 \end{pmatrix}$.

(a) What are the sizes of the products $\mathbf{AB}$ and $\mathbf{BC}$? Calculate these matrices.

(b) Do the products $(\mathbf{AB})\mathbf{C}$ and $\mathbf{A}(\mathbf{BC})$ exist? If so, calculate them. What do you notice?

In Activity 8 you should have found that, for the particular three matrices $\mathbf{A}, \mathbf{B}$ and $\mathbf{C}$ in the activity, $(\mathbf{AB})\mathbf{C} = \mathbf{A}(\mathbf{BC})$. This does not of course tell you that this property holds for matrices in general. However, this property does hold in general: whenever $\mathbf{A}, \mathbf{B}$ and $\mathbf{C}$ are matrices such that the products $\mathbf{AB}$ and $\mathbf{BC}$ are defined, we have

$$(\mathbf{AB})\mathbf{C} = \mathbf{A}(\mathbf{BC}).$$

In other words, matrix multiplication is *associative*. In this respect matrix multiplication behaves like multiplication of numbers.

In the next activity you can investigate how matrix multiplication interacts with multiplication by a scalar.

Activity 9 *Combining matrix multiplication and multiplication by a scalar*

Let $\mathbf{A} = \begin{pmatrix} 1 & 2 \\ 3 & 4 \end{pmatrix}$ and $\mathbf{B} = \begin{pmatrix} 1 & 0 \\ 2 & 1 \end{pmatrix}$. Show that

$$\mathbf{A}(2\mathbf{B}) = 2(\mathbf{AB}).$$

In Activity 9 you verified a particular instance of a general property of matrices: if the matrix product $\mathbf{AB}$ exists, then, for any scalar k,

$$\mathbf{A}(k\mathbf{B}) = (k\mathbf{A})\mathbf{B} = k(\mathbf{AB}).$$

In the next activity you're asked to investigate how matrix multiplication interacts with matrix addition.

Activity 10 *Combining matrix addition and multiplication*

Let $\mathbf{A} = \begin{pmatrix} 1 & -1 \\ 3 & 4 \end{pmatrix}$, $\mathbf{B} = \begin{pmatrix} 1 \\ 2 \end{pmatrix}$ and $\mathbf{C} = \begin{pmatrix} -1 \\ 1 \end{pmatrix}$.

By calculating $\mathbf{A}(\mathbf{B}+\mathbf{C})$ and $\mathbf{AB}+\mathbf{AC}$, show that

$$\mathbf{A}(\mathbf{B}+\mathbf{C}) = \mathbf{AB}+\mathbf{AC}.$$

In Activity 10 you verified a particular instance of another general property of matrices: if the matrix product $\mathbf{A}(\mathbf{B}+\mathbf{C})$ can be formed, then

$$\mathbf{A}(\mathbf{B}+\mathbf{C}) = \mathbf{AB}+\mathbf{AC}.$$

It's also true that if the matrix product $(\mathbf{A}+\mathbf{B})\mathbf{C}$ can be formed, then

$$(\mathbf{A}+\mathbf{B})\mathbf{C} = \mathbf{AC}+\mathbf{BC}.$$

In other words, matrix multiplication is *distributive* over matrix addition.

So, for example, if you encounter a matrix expression of the form

$$\mathbf{AB}+\mathbf{AC},$$

then you can factorise it to give

$$\mathbf{AB}+\mathbf{AC} = \mathbf{A}(\mathbf{B}+\mathbf{C}).$$

Here's a summary of the properties of matrix multiplication that you've seen so far.

Some properties of matrix multiplication

The following properties hold for all matrices $\mathbf{A}$, $\mathbf{B}$ and $\mathbf{C}$ for which the products and sums mentioned are defined.

- $(\mathbf{AB})\mathbf{C} = \mathbf{A}(\mathbf{BC})$
- $k(\mathbf{AB}) = (k\mathbf{A})\mathbf{B} = \mathbf{A}(k\mathbf{B})$, for any scalar k
- $\mathbf{A}(\mathbf{B}+\mathbf{C}) = \mathbf{AB}+\mathbf{AC}$
- $(\mathbf{A}+\mathbf{B})\mathbf{C} = \mathbf{AC}+\mathbf{BC}$

Remember also the following important fact.

Matrix multiplication is not commutative

- There are matrices $\mathbf{A}$, $\mathbf{B}$ such that the product $\mathbf{AB}$ exists but the product $\mathbf{BA}$ does not.
- Even when both products are defined, it can happen that $\mathbf{AB} \neq \mathbf{BA}$.

Matrix powers

Just as the square a^2 of a real number a is defined to be the product $a \times a$, so we define the **square** $\mathbf{A}^2$ of a matrix $\mathbf{A}$ to be the matrix product $\mathbf{AA}$, if it exists.

For example, if

$$\mathbf{A} = \begin{pmatrix} 2 & 1 \\ 3 & 0 \end{pmatrix},$$

then

$$\mathbf{A}^2 = \mathbf{AA} = \begin{pmatrix} 2 & 1 \\ 3 & 0 \end{pmatrix}\begin{pmatrix} 2 & 1 \\ 3 & 0 \end{pmatrix} = \begin{pmatrix} 7 & 2 \\ 6 & 3 \end{pmatrix}.$$

In the next activity you're asked to investigate which matrices can be squared.

Activity 11 *Investigating which matrices can be squared*

(a) Let $\mathbf{M} = \begin{pmatrix} 1 & 0 & 3 \\ 0 & 2 & 4 \end{pmatrix}$. Why is it not possible to form the product $\mathbf{MM}$?

(b) Let $\mathbf{N} = \begin{pmatrix} 2 & -1 \\ 0 & 2 \end{pmatrix}$. Is the product $\mathbf{NN}$ defined? If, so, evaluate it.

(c) Describe which matrices can be squared.

In Activity 11 you saw that if $\mathbf{A}$ is a *square* matrix (a matrix with the same number of rows as columns), then its square $\mathbf{A}^2$ exists. The matrix $\mathbf{A}^2$ has the same size as the original matrix $\mathbf{A}$.

Be careful not to confuse the two different concepts *square matrix* and *square of a matrix*.

If $\mathbf{A}$ is a square matrix, then since $\mathbf{A}^2$ is a square matrix of the same size as $\mathbf{A}$, we can also form the products $(\mathbf{A}^2)\mathbf{A}$ and $\mathbf{A}(\mathbf{A}^2)$. Since matrix multiplication is associative, we know that

$$\mathbf{A}^2\mathbf{A} = (\mathbf{AA})\mathbf{A} = \mathbf{A}(\mathbf{AA}) = \mathbf{AA}^2.$$

Hence either of these products can be written unambiguously as $\mathbf{AAA}$. This matrix is called the **cube** of $\mathbf{A}$ and is denoted by $\mathbf{A}^3$. For example, for the matrix

$$\mathbf{A} = \begin{pmatrix} 2 & 1 \\ 3 & 0 \end{pmatrix}$$

discussed at the beginning of this subsection, we have

$$\mathbf{A}^3 = \mathbf{A}^2\mathbf{A} = \begin{pmatrix} 7 & 2 \\ 6 & 3 \end{pmatrix}\begin{pmatrix} 2 & 1 \\ 3 & 0 \end{pmatrix} = \begin{pmatrix} 20 & 7 \\ 21 & 6 \end{pmatrix}.$$

In general, for any square matrix $\mathbf{A}$, the nth **power** of $\mathbf{A}$ is

$$\mathbf{A}^n = \underbrace{\mathbf{A}\mathbf{A}\cdots\mathbf{A}}_{n \text{ times}}.$$

Activity 12 *Calculating matrix powers*

(a) Calculate the squares of the following matrices:

$$\begin{pmatrix} -1 & 4 \\ 2 & 0 \end{pmatrix}, \quad \begin{pmatrix} 1 & 0 & 1 \\ 0 & 2 & 0 \\ 0 & 0 & 3 \end{pmatrix}.$$

(b) Let $\mathbf{A} = \begin{pmatrix} 2 & 1 \\ 3 & 0 \end{pmatrix}$. Calculate $\mathbf{A}^6$.

Hint: the cube $\mathbf{A}^3$ of this matrix $\mathbf{A}$ was calculated before the activity.

Matrix multiplication with a computer

Matrix multiplication can be time-consuming to carry out when the matrices involved are large. In the next activity you can find out how to use a computer to carry out matrix operations, including multiplication. This work will increase your familiarity with the way matrices behave, and help you distinguish between circumstances when it is worth using a computer to perform matrix calculations and when it is more efficient to do the calculations by hand.

Activity 13 *Working with matrices on a computer*

Work through Subsection 10.1 of the *Computer algebra guide*.

2 Matrices and networks

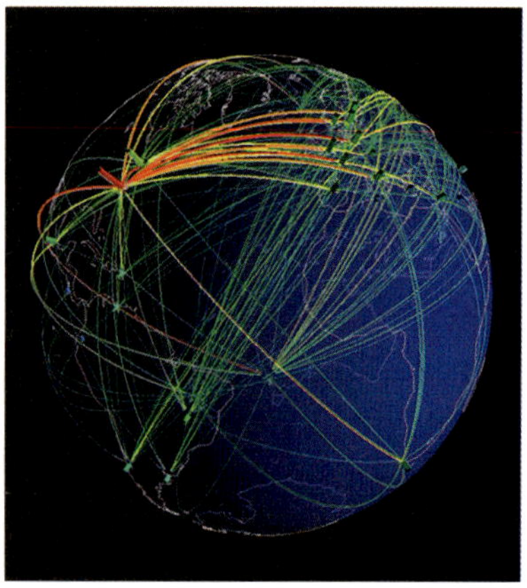

A network of world internet traffic

Matrices can be used to represent and analyse various flows in *networks*. A **network** is a collection of objects connected by links – either physical or abstract – such as cities connected by roads, telephone exchange points connected by telephone lines, or people connected by acquaintance. A foremost example is the internet, which is a network of communicating computers linked by a range of technologies. In this section you'll look at flows in networks of a particular type, and learn how to use matrices to work with them.

2.1 Networks

Figure 3 shows a simple example of a network of water pipes. Water can be poured into the system at position A or B, and it then flows down the pipes to come out at positions U, V and W. The pipes have different capacities: the three pipes that come out of position A are labelled with numbers that indicate the *proportions* of water that flow through these pipes when water is poured in at A, and the three pipes that come out of position B are labelled in the same way. For example, if a litre of water is poured in at position A, then 0.4 litres flows down the first pipe from A, 0.2 litres flows down the second pipe from A, and 0.4 litres flows down the third pipe from A.

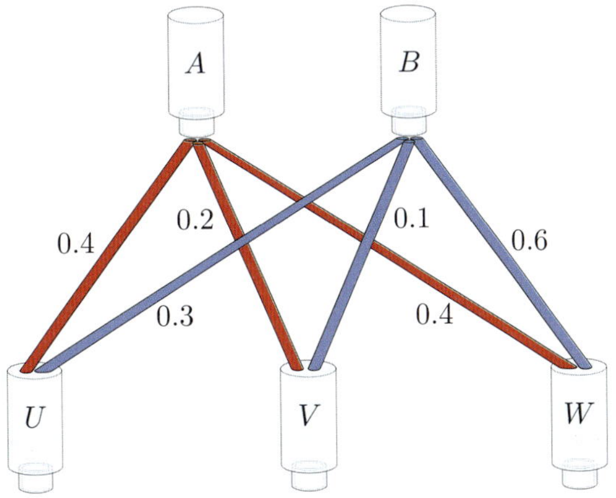

Figure 3 A network of water pipes

A network like this can be represented more simply as a **network diagram**, which is a mathematical representation of a physical network. The network diagram in Figure 4 represents the network in Figure 3.

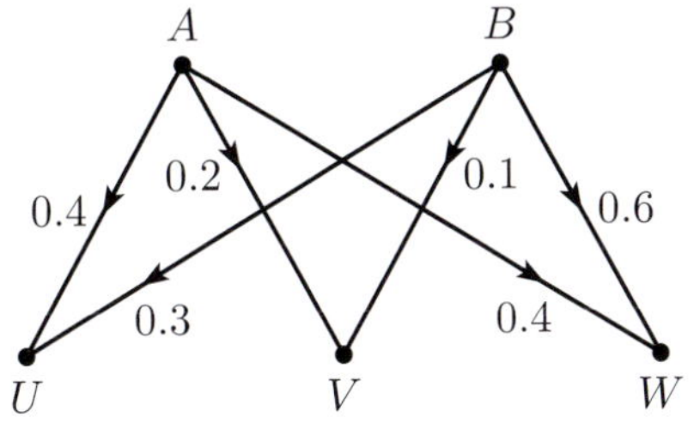

Figure 4 A network diagram representing the pipe system in Figure 3

The dots in a network diagram are called **nodes**, and the lines are called **arcs**. The arrows on the arcs in Figure 4 are included because this network diagram represents a network in which flow along an arc takes place in only one direction. When no confusion arises, we use the word 'network' to mean either 'network diagram' or 'physical network'. We'll use network diagrams to represent networks from now on in this section.

Example 2 *Calculating network outputs*

The network diagram in Figure 4 is repeated below.

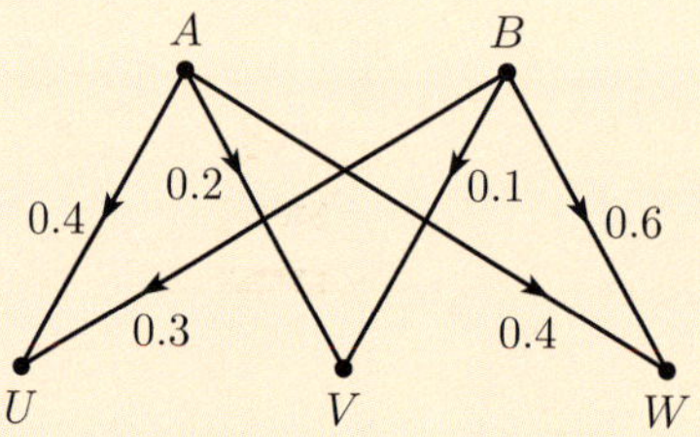

How much water is output at each of nodes U, V and W if the only input is 1 litre of water at node B?

Solution

The label on the arc joining node B to node U indicates the proportion of the water input at node B that reaches node U.

The label on the arc from node B to node U is 0.3. So 0.3 litres of water is output at node U when 1 litre is input at node B.

Read the labels on the other arcs from node B.

Similarly, 0.1 litres of water is output at node V, and 0.6 litres of water is output at node W.

Now consider what happens if water is input at *both* node A and node B in the network in Example 2. The amount of water reaching an output node is the sum of the water reaching the node from node A and from node B. For example, if 1 litre of water is input at node A and another 1 litre is input at node B, then node U will collect 0.4 litres of water from node A and 0.3 litres from node B. So, in total, 0.7 litres of water will be output at node U. Similarly, $0.2 + 0.1 = 0.3$ litres of water will be output at node V, and $0.4 + 0.6 = 1$ litre of water will be output at node W.

In general, if x litres of water is input at node A and y litres of water is input at node B, then

$0.4x + 0.3y$ litres of water is output at node U,

$0.2x + 0.1y$ litres of water is output at node V,

$0.4x + 0.6y$ litres of water is output at node W.

In the next activity you can practise calculating network outputs yourself.

Activity 14 *Calculating network outputs*

The following network diagram represents a network of water pipes.

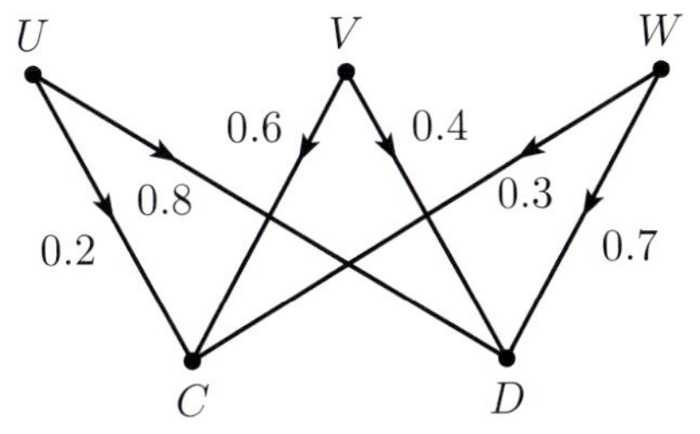

(a) How much water is output at nodes C and D if 1 litre of water is input at node U, 2 litres of water is input at node V, and there is no input at node W?

(b) How much water is output at nodes C and D if x litres of water are input at node U, y litres of water is input at node V, and z litres of water is input at node W?

Network diagrams like those that you've seen in this subsection can be used to model many other kinds of flows, such as the movement of money in and out of companies. For example, consider three corporate donors, D, E and F, say, that every month make donations to two charities, J and K. Donor D donates 20% of its monthly designated sum to charity J and 80% to charity K, donor E donates 30% of its sum to charity J and 70% to charity K, and donor F donates 50% of its sum to charity J and 50% to charity K.

The flow of money from the three donors to the two charities can be represented by the network in Figure 5. For example, a total donation of £1000 from donor D in a given month is modelled by an input of £1000 at node D. The label of the arc connecting D to J tells you that 0.2, or 20%, of that sum, which is £200, reaches the output node J.

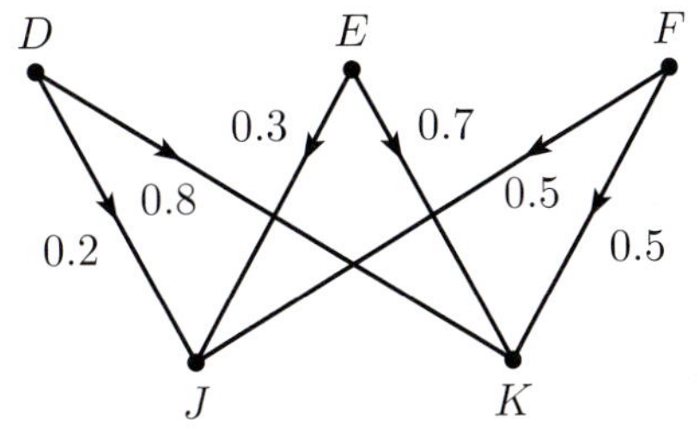

Figure 5 A network diagram representing the flow of money from three corporate donors to two charities

Activity 15 *Calculating more network outputs*

The network diagram in Figure 5 is repeated below.

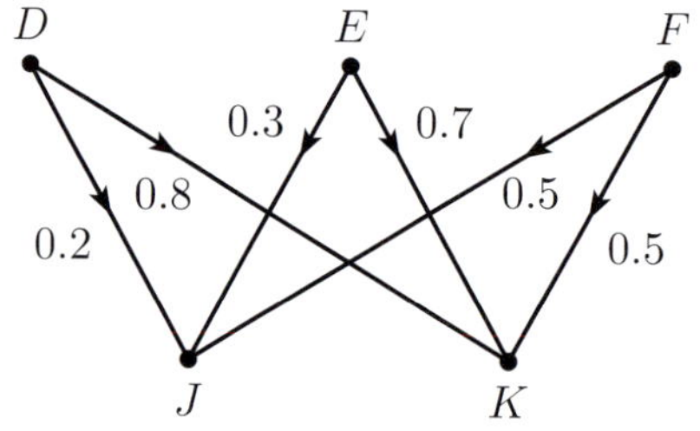

It represents the flow of money from donors D, E and F to charities J and K, as described above. How much money is output at each of nodes J and K in each of the following situations?

(a) The only input is £2000 at node D.

(b) The input at node D is £2400, the input at node E is £1700, and the input at node F is £1100.

(c) The input (in £) at node D is x, the input (in £) at node E is y, and the input (in £) at node F is z.

2.2 From networks to matrices

In this subsection you'll see how network output calculations of the kinds that you saw in Subsection 2.1 can be expressed succinctly using matrices.

Consider again the first network representing water pipes that you saw in Subsection 2.1. It is repeated in Figure 6.

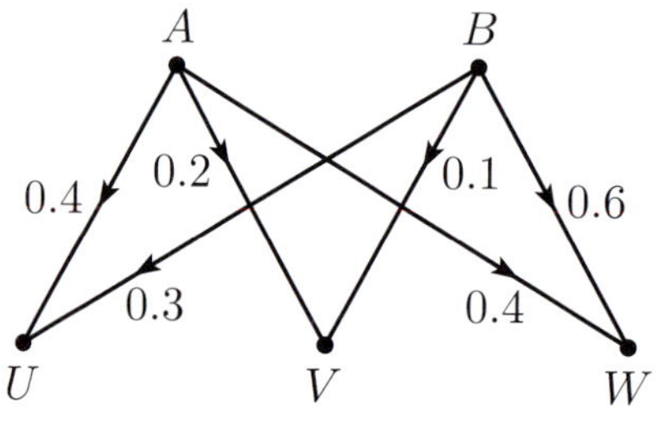

Figure 6 A network representing water pipes

The amounts of water (in litres) input at nodes A and B can be represented as a two-dimensional vector (a 2×1 matrix), whose first and second components are the inputs at A and B, respectively. For example, the vector

$$\begin{pmatrix} 0 \\ 1 \end{pmatrix}$$

represents 0 litres of water input at node A and 1 litre of water input at node B.

The outputs from the network can also be represented by a vector. For example, the vector

$$\begin{pmatrix} 0.4 \\ 0.2 \\ 0.4 \end{pmatrix}$$

represents an output of 0.4 litres of water at node U, 0.2 litres at node V, and 0.4 litres at node W.

The network itself can be represented by the matrix

$$\begin{pmatrix} 0.4 & 0.3 \\ 0.2 & 0.1 \\ 0.4 & 0.6 \end{pmatrix}.$$

Here the elements in each column are the labels of the arcs *from* a node (A or B), and the elements in each row are the labels of the arcs *to* a node (U, V or W), as shown below:

$$\begin{array}{c} \\ U \\ V \\ W \end{array} \begin{array}{c} \begin{array}{cc} A & B \end{array} \\ \begin{pmatrix} 0.4 & 0.3 \\ 0.2 & 0.1 \\ 0.4 & 0.6 \end{pmatrix} \end{array}.$$

For example, the element in the column labelled B and the row labelled U is the label of the arc from node B to node U. You'll see the reason for arranging the arc labels in this way in a moment.

You saw in Subsection 2.1 that if x litres of water is input at node A and y litres of water is input at node B, then

$0.4x + 0.3y$ litres of water is output at node U,

$0.2x + 0.1y$ litres of water is output at node V,

$0.4x + 0.6y$ litres of water is output at node W.

In other words,

if the input vector is $\begin{pmatrix} x \\ y \end{pmatrix}$, then the output vector is $\begin{pmatrix} 0.4x + 0.3y \\ 0.2x + 0.1y \\ 0.4x + 0.6y \end{pmatrix}$.

This output vector can be expressed as the product of the matrix that represents the network and the input vector:

$$\begin{pmatrix} 0.4x + 0.3y \\ 0.2x + 0.1y \\ 0.4x + 0.6y \end{pmatrix} = \begin{pmatrix} 0.4 & 0.3 \\ 0.2 & 0.1 \\ 0.4 & 0.6 \end{pmatrix} \begin{pmatrix} x \\ y \end{pmatrix}.$$

So we can say that

$$\text{if the input vector is } \begin{pmatrix} x \\ y \end{pmatrix}, \text{ then the output vector is } \begin{pmatrix} 0.4 & 0.3 \\ 0.2 & 0.1 \\ 0.4 & 0.6 \end{pmatrix} \begin{pmatrix} x \\ y \end{pmatrix}.$$

So, for any input vector, you can work out the corresponding output vector by multiplying the input vector by the matrix that represents the network. This is illustrated in the next example.

Example 3 *Using matrices to calculate network outputs*

Use the matrix that represents the network diagram in Figure 6 to calculate the outputs at nodes U, V and W if 2 litres of water is input at node A and there is no input at node B.

Solution

The matrix representing the network was found before this example.

The matrix representing the network is

$$\begin{pmatrix} 0.4 & 0.3 \\ 0.2 & 0.1 \\ 0.4 & 0.6 \end{pmatrix}.$$

The input vector (in litres) is

$$\begin{pmatrix} 2 \\ 0 \end{pmatrix}.$$

The output vector is the product of the matrix representing the network and the input vector.

Hence the output vector is

$$\begin{pmatrix} 0.4 & 0.3 \\ 0.2 & 0.1 \\ 0.4 & 0.6 \end{pmatrix} \begin{pmatrix} 2 \\ 0 \end{pmatrix} = \begin{pmatrix} 0.4 \times 2 + 0.3 \times 0 \\ 0.2 \times 2 + 0.1 \times 0 \\ 0.4 \times 2 + 0.6 \times 0 \end{pmatrix} = \begin{pmatrix} 0.8 \\ 0.4 \\ 0.8 \end{pmatrix}.$$

So the outputs at nodes U, V and W are 0.8 litres, 0.4 litres and 0.8 litres, respectively.

In general, to represent a network by a matrix in the way described above, you need a column for each input node and a row for each output node. So if the network has n input nodes and m output nodes, then the matrix has size $m \times n$.

When you find a matrix representing a network of the type considered in this section, it's useful to check that the elements in each column have sum 1. The matrix must have this property because the elements in each column are the labels on an arc *from* a particular node. You can see that the matrix representing the network in Example 3 has this property.

Activity 16 *Using matrices to calculate network outputs*

The network in Activity 14, which represents a network of water pipes, is shown again below.

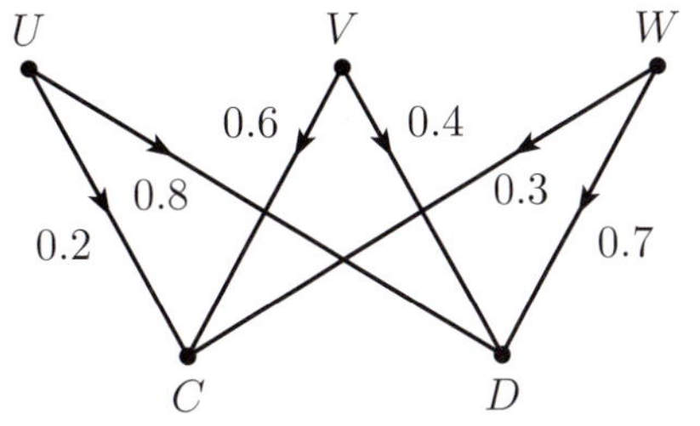

(a) Write down the matrix that represents this network.

(b) Write down a vector that represents an input of 1 litre of water at each of nodes U, V and W.

(c) Use your answer to part (a) to calculate the output vector for the input vector in part (b), and interpret your answer in terms of the amounts of water output at nodes C and D.

2.3 Combining networks

Networks can be combined by making the outputs from one network become the inputs to another network. In this section you'll see how to use matrices to calculate the outputs from a combined network using the information about the original networks.

The network diagram in Figure 4 and the one in Activity 14 are repeated in Figure 7. Both diagrams represent networks of water pipes.

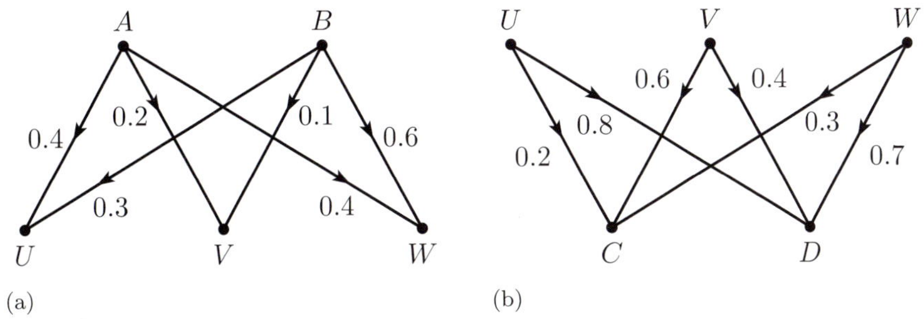

Figure 7 Two network diagrams

The first network has three output nodes, and the second network has three input nodes. We can join these nodes so that, for instance, the water output at U in the first network flows down the pipes that leave U in the second network. The combined network is shown in Figure 8.

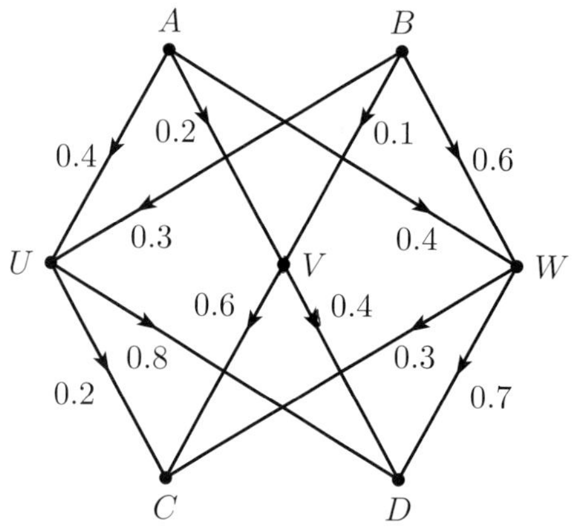

Figure 8 The combined network

The network obtained in this way can be considered as equivalent to a simpler network with arcs going directly from the input nodes A and B to the output nodes C and D, as illustrated in Figure 9. In general, we regard two networks of the type considered in this section as **equivalent** if they have the same input and output nodes, and all choices of inputs produce the same outputs in both networks.

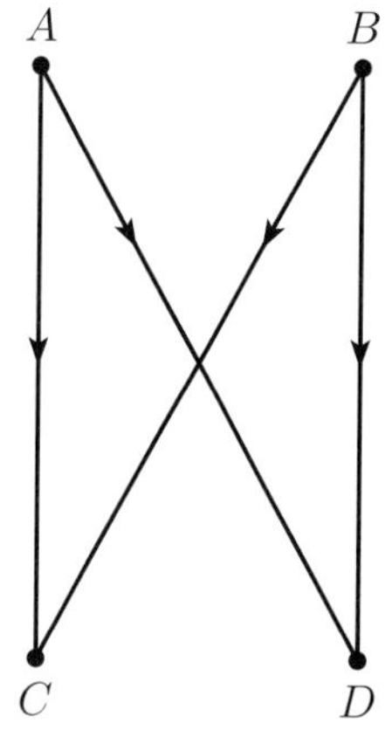

Figure 9 A network equivalent to the one in Figure 8

How can we find the labels for the arcs in the new, simpler network diagram? Here's how you can find the label for the arc from A to C, for example.

Suppose that 1 litre of water is input at node A in the combined network diagram, and there are no other inputs. Then, from the arc labels in the top half of the diagram, you know that the amounts of water passing through nodes U, V and W are 0.4, 0.2 and 0.4 litres, respectively. So, from the arc labels in the bottom half of the diagram, you can work out the following.

- Of the 0.4 litres of water that passes through node U, the amount that reaches node C is $0.2 \times 0.4 = 0.08$ litres.
- Of the 0.2 litres of water that passes through node V, the amount that reaches node C is $0.6 \times 0.2 = 0.12$ litres.
- Of the 0.4 litres of water that passes through node W, the amount that reaches node C is $0.3 \times 0.4 = 0.12$ litres.

To find the total amount of water output at node C, we add the contributions from the three different routes, which gives $0.08 + 0.12 + 0.12 = 0.32$ litres. So the proportion of the water input at node A that's output at node C is 0.32, and hence the arc from A to C should be labelled with the number 0.32.

You can find the labels for the remaining arcs in a similar way. However, these calculations are quite tedious, and it turns out that they can be simplified by using the matrices that represent the two original networks, which are

$$\begin{pmatrix} 0.4 & 0.3 \\ 0.2 & 0.1 \\ 0.4 & 0.6 \end{pmatrix} \quad \text{and} \quad \begin{pmatrix} 0.2 & 0.6 & 0.3 \\ 0.8 & 0.4 & 0.7 \end{pmatrix},$$

respectively. These matrices were found in Subsection 2.2.

Suppose that the input and output vectors (in litres) for the combined network in Figure 8 are

$$\begin{pmatrix} a \\ b \end{pmatrix} \quad \text{and} \quad \begin{pmatrix} c \\ d \end{pmatrix},$$

respectively. Suppose also that the output vector (in litres) of the top part of the network, which is also the input vector for the bottom part, is

$$\begin{pmatrix} u \\ v \\ w \end{pmatrix}.$$

Then

$$\begin{pmatrix} u \\ v \\ w \end{pmatrix} = \begin{pmatrix} 0.4 & 0.3 \\ 0.2 & 0.1 \\ 0.4 & 0.6 \end{pmatrix} \begin{pmatrix} a \\ b \end{pmatrix} \quad \text{and} \quad \begin{pmatrix} c \\ d \end{pmatrix} = \begin{pmatrix} 0.2 & 0.6 & 0.3 \\ 0.8 & 0.4 & 0.7 \end{pmatrix} \begin{pmatrix} u \\ v \\ w \end{pmatrix}.$$

Using the first of these equations to substitute in the second gives

$$\begin{pmatrix} c \\ d \end{pmatrix} = \begin{pmatrix} 0.2 & 0.6 & 0.3 \\ 0.8 & 0.4 & 0.7 \end{pmatrix} \left(\begin{pmatrix} 0.4 & 0.3 \\ 0.2 & 0.1 \\ 0.4 & 0.6 \end{pmatrix} \begin{pmatrix} a \\ b \end{pmatrix} \right).$$

You know from Section 1 that matrix multiplication is associative, so

$$\begin{pmatrix} c \\ d \end{pmatrix} = \left(\begin{pmatrix} 0.2 & 0.6 & 0.3 \\ 0.8 & 0.4 & 0.7 \end{pmatrix} \begin{pmatrix} 0.4 & 0.3 \\ 0.2 & 0.1 \\ 0.4 & 0.6 \end{pmatrix} \right) \begin{pmatrix} a \\ b \end{pmatrix};$$

that is,

$$\begin{pmatrix} c \\ d \end{pmatrix} = \begin{pmatrix} 0.32 & 0.3 \\ 0.68 & 0.7 \end{pmatrix} \begin{pmatrix} a \\ b \end{pmatrix}. \qquad (2)$$

The conclusion of this discussion is that the matrix that represents the combined network in Figure 9 is the product of the matrices that represent the original networks. So matrix multiplication provides a neat way to calculate outputs of the combined network.

Notice that the matrix representing the *top* network is the *second* matrix in the product that represents the combined network.

The labels of the simpler network that is equivalent to the combined network can be read off the product matrix. This is easier if we label the rows and the columns with the nodes as below:

$$\begin{array}{cc} & \begin{array}{cc} A & B \end{array} \\ \begin{array}{c} C \\ D \end{array} & \begin{pmatrix} 0.32 & 0.3 \\ 0.68 & 0.7 \end{pmatrix}. \end{array}$$

Notice that each column of this matrix has sum 1, as expected in view of the comment after Example 3.

The original combined network and the equivalent simpler network, with the arc labels, are shown in Figure 10.

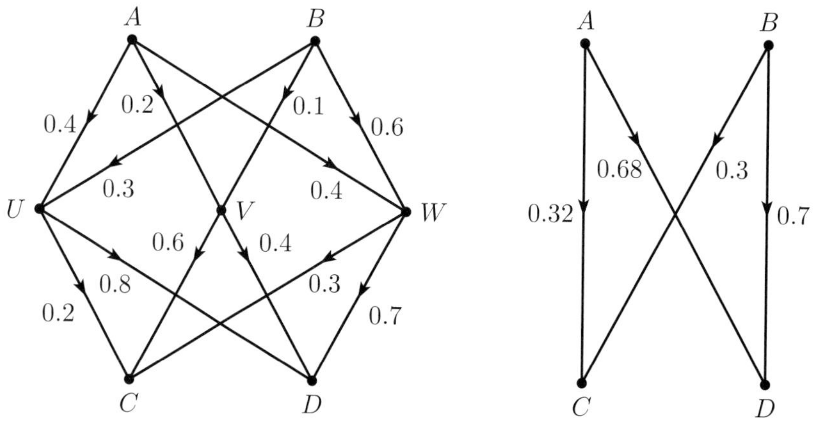

Figure 10 The two equivalent networks

Activity 17 *Finding the matrix representing a combined network*

The network below is the combination of two smaller networks, one with input nodes U and V and output nodes P and Q, and the other with input nodes P and Q and output nodes A, B, and C.

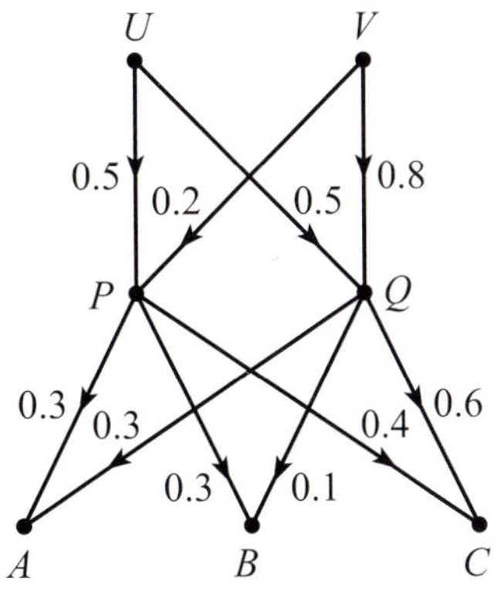

(a) Write down the matrix equation that gives the output vector $\begin{pmatrix} p \\ q \end{pmatrix}$ at nodes P and Q when the input vector is $\begin{pmatrix} u \\ v \end{pmatrix}$ at nodes U and V.

(b) Write down the matrix equation that gives the output vector $\begin{pmatrix} a \\ b \\ c \end{pmatrix}$ at nodes A, B and C when the input vector is $\begin{pmatrix} p \\ q \end{pmatrix}$ at nodes P and Q.

(c) Use your answers to parts (a) and (b) to find the matrix equation for the combined network when the input vector is $\begin{pmatrix} u \\ v \end{pmatrix}$ at nodes U and V.

(d) Draw a network with input nodes U and V and output nodes A, B and C that is equivalent to the given network.

(e) Suppose that the given network represents a network of water pipes. Determine the outputs at nodes A, B and C when 1 litre of water is input at U and 2 litres of water is input at V.

Joining networks is also useful in other contexts. For example, Figure 11 repeats the network diagram from Figure 5, which represents the flow of money from three corporate donors D, E and F to two charities J and K.

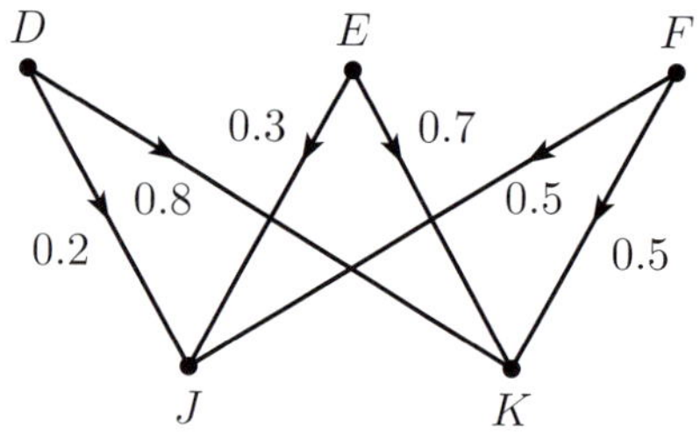

Figure 11 A network diagram representing the flow of money from three corporate donors to two charities

Suppose that donors D, E and F are the only donors to the two charities, and that the two charities spend their money in two different geographical regions, R and S, as indicated in Figure 12. So, for example, charity J spends 60% of its money in region R and 40% in region S.

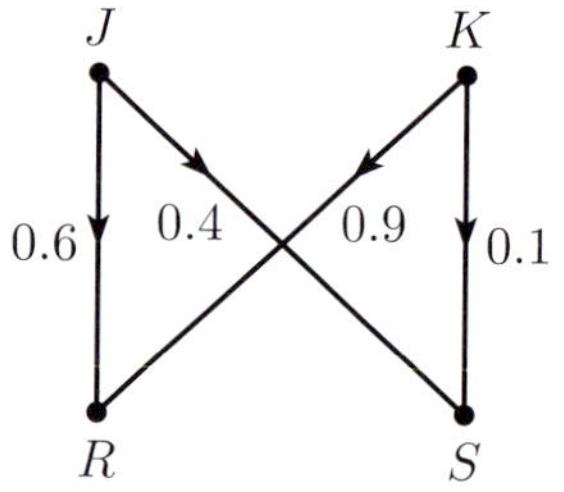

Figure 12 A network diagram representing the flow of money from two charities into two regions

In the next activity you're asked to combine the networks in Figures 11 and 12 to obtain a network that represents the flow of money from the three corporate donors to the two regions.

Activity 18 *Finding the matrix representing another combined network*

(a) Write down the matrices that represent the networks in Figures 11 and 12.

(b) Hence find the matrix that represents the network obtained by combining these two networks.

(c) Draw a simplified network that represents the combined network described in part (b).

(d) If, in a particular month, donor D donates £1700, donor E donates £1400 and donor F donates nothing, how much money is spent in region S?

In a real-life situation, networks like the one in Activity 18 would have many more input and output nodes, and the matrices representing them would be manipulated on a computer.

3 The inverse of a matrix

In Section 1 you met four matrix operations: addition, multiplication by a scalar, subtraction and matrix multiplication. No division was mentioned: it is not possible to divide a matrix by another matrix. However there is a way to manipulate matrices that in certain situations plays the role of division. To understand this method, you need to know about special matrices called *identity matrices*, which are introduced next.

3.1 Identity matrices

The number 1 has a special property among real numbers: multiplication by 1 leaves any real number unchanged. An identity matrix is a matrix that behaves like the number 1, in the sense that if a matrix $\mathbf{A}$ is multiplied by an identity matrix of an appropriate size then the result is again $\mathbf{A}$. You will meet an identity matrix in the next activity.

Activity 19 *Multiplying by an identity matrix*

Let $\mathbf{A} = \begin{pmatrix} 3 & 4 \\ -2 & 1 \end{pmatrix}$, $\mathbf{B} = \begin{pmatrix} 2 & 7 & -3 \\ 8 & 1 & 2 \end{pmatrix}$ and $\mathbf{I} = \begin{pmatrix} 1 & 0 \\ 0 & 1 \end{pmatrix}$. Evaluate each of the following products, where possible:

$$\mathbf{AI}, \quad \mathbf{IA}, \quad \mathbf{BI}, \quad \mathbf{IB}.$$

The matrix $\mathbf{I}$ in Activity 19 has the following two properties.

- If $\mathbf{A}$ is any matrix such that the product $\mathbf{AI}$ exists, then $\mathbf{AI} = \mathbf{A}$.
- If $\mathbf{A}$ is any matrix such that the product $\mathbf{IA}$ exists, then $\mathbf{IA} = \mathbf{A}$.

Any matrix $\mathbf{I}$ with these two properties is called an **identity matrix**.

For a matrix to be an identity matrix, it has to be square. To see this, suppose that $\mathbf{A}$ is an $m \times n$ matrix. Then in order for the product $\mathbf{AI}$ to be defined, $\mathbf{I}$ must have n rows. Moreover, in order for $\mathbf{AI}$ to have size $m \times n$ (which it must if it is to be equal to $\mathbf{A}$), $\mathbf{I}$ must have n columns:

$$m \times \boxed{n} \ \text{ times } \ \boxed{n} \times n \ \text{ gives } \ m \times n.$$

So $\mathbf{I}$ must have size $n \times n$.

It's straightforward to verify that the following 2×2, 3×3 and 4×4 matrices are identity matrices:

$$\begin{pmatrix} 1 & 0 \\ 0 & 1 \end{pmatrix}, \quad \begin{pmatrix} 1 & 0 & 0 \\ 0 & 1 & 0 \\ 0 & 0 & 1 \end{pmatrix}, \quad \begin{pmatrix} 1 & 0 & 0 & 0 \\ 0 & 1 & 0 & 0 \\ 0 & 0 & 1 & 0 \\ 0 & 0 & 0 & 1 \end{pmatrix}.$$

In fact, for every natural number n, the $n \times n$ matrix that has ones down the **leading diagonal** (the diagonal that starts at the top left element and ends at the bottom right element), and zeros everywhere else, is an identity matrix.

Furthermore, for every natural number n, this matrix is the *only* $n \times n$ identity matrix. To see this, let $\mathbf{I}$ be the $n \times n$ identity matrix described above, and let $\mathbf{E}$ be an $n \times n$ identity matrix. Then $\mathbf{IE} = \mathbf{I}$ and $\mathbf{IE} = \mathbf{E}$, so $\mathbf{I} = \mathbf{E}$.

When we're working with matrices, we usually reserve the letter $\mathbf{I}$ to denote the $n \times n$ identity matrix.

Activity 20 *Multiplying by an identity matrix*

Let $\mathbf{A} = \begin{pmatrix} a & b & c \\ d & e & f \end{pmatrix}$ and $\mathbf{B} = \begin{pmatrix} x \\ y \\ z \end{pmatrix}$ be general matrices, and let $\mathbf{I}$ be the 3×3 identity matrix. Verify that $\mathbf{AI} = \mathbf{A}$ and $\mathbf{IB} = \mathbf{B}$.

The facts that you've seen about identity matrices are summarised in the following box.

Identity matrices

An **identity matrix** is a square matrix $\mathbf{I}$ such that

- for any matrix $\mathbf{A}$ for which the product $\mathbf{AI}$ is defined, $\mathbf{AI} = \mathbf{A}$
- for any matrix $\mathbf{A}$ for which the product $\mathbf{IA}$ is defined, $\mathbf{IA} = \mathbf{A}$.

Each identity matrix has the form

$$\begin{pmatrix} 1 & 0 & 0 & \cdots & 0 \\ 0 & 1 & 0 & \cdots & 0 \\ 0 & 0 & 1 & \cdots & 0 \\ \vdots & \vdots & \vdots & \ddots & \vdots \\ 0 & 0 & 0 & \cdots & 1 \end{pmatrix}.$$

That is, it has ones down the leading diagonal and zeros elsewhere.

3.2 Matrix inverses

As mentioned earlier, it isn't possible to divide matrices, but it is sometimes possible to perform a manipulation with a similar effect.

Remember that, in the arithmetic of numbers, dividing by a number is the same as multiplying by its reciprocal. For example, dividing by 2 is the same as multiplying by $\frac{1}{2}$. The reciprocal of a non-zero number a is the number b such that $ab = 1$.

There is a notion in matrix arithmetic that is analogous to the notion of a reciprocal in ordinary arithmetic. Let $\mathbf{A}$ be a *square* matrix. If there is another matrix $\mathbf{B}$ of the same size with the property that

$$\mathbf{AB} = \mathbf{I} \quad \text{and} \quad \mathbf{BA} = \mathbf{I},$$

where $\mathbf{I}$ is an identity matrix, then we say that $\mathbf{A}$ is **invertible** and that $\mathbf{B}$ is an **inverse** of $\mathbf{A}$. Multiplying by an inverse of a matrix is useful in many situations where you might want to use a matrix operation analogous to division.

Activity 21 *Checking matrix inverses*

For each of the following pairs of matrices $\mathbf{A}$ and $\mathbf{B}$, check that $\mathbf{B}$ is an inverse of $\mathbf{A}$, by finding the products $\mathbf{AB}$ and $\mathbf{BA}$.

(a) $\mathbf{A} = \begin{pmatrix} 8 & 5 \\ 3 & 2 \end{pmatrix}$, $\quad \mathbf{B} = \begin{pmatrix} 2 & -5 \\ -3 & 8 \end{pmatrix}$

(b) $\mathbf{A} = \begin{pmatrix} 2 & 3 & 1 \\ -1 & 1 & 0 \\ -2 & 1 & 0 \end{pmatrix}$, $\quad \mathbf{B} = \begin{pmatrix} 0 & 1 & -1 \\ 0 & 2 & -1 \\ 1 & -8 & 5 \end{pmatrix}$

If a matrix $\mathbf{A}$ is invertible, then it has *exactly one* inverse. To see this, suppose that the matrices $\mathbf{B}$ and $\mathbf{C}$ are both inverses of the matrix $\mathbf{A}$. Then

$$\mathbf{AB} = \mathbf{I} \quad \text{and} \quad \mathbf{BA} = \mathbf{I}, \qquad \text{and also} \quad \mathbf{AC} = \mathbf{I} \quad \text{and} \quad \mathbf{CA} = \mathbf{I}.$$

Now consider the matrix $\mathbf{BAC}$. Since matrix multiplication is associative,

$$\mathbf{BAC} = (\mathbf{BA})\mathbf{C} = \mathbf{IC} = \mathbf{C},$$

and also

$$\mathbf{BAC} = \mathbf{B}(\mathbf{AC}) = \mathbf{BI} = \mathbf{B}.$$

Hence $\mathbf{B}$ and $\mathbf{C}$ are the same matrix. That is, $\mathbf{A}$ has only one inverse.

We usually write the inverse of an invertible matrix $\mathbf{A}$ as $\mathbf{A}^{-1}$. So the following facts hold.

Inverse of a matrix

If $\mathbf{A}$ is an invertible matrix, then

$$\mathbf{A}\mathbf{A}^{-1} = \mathbf{I} \quad \text{and} \quad \mathbf{A}^{-1}\mathbf{A} = \mathbf{I},$$

where $\mathbf{I}$ is the identity matrix of the same size as $\mathbf{A}$.

It follows from the definition of the inverse of a matrix that if $\mathbf{A}$ is an invertible matrix, then not only is $\mathbf{A}^{-1}$ the inverse of $\mathbf{A}$, but also $\mathbf{A}$ is the inverse of $\mathbf{A}^{-1}$. In other words, $\mathbf{A}$ and $\mathbf{A}^{-1}$ are inverses of each other.

In the rest of this subsection, you'll learn how to establish whether any particular 2×2 matrix has an inverse, and how to work out the inverse when it exists. You'll also use a computer to calculate inverses of larger matrices.

Inverses of 2 × 2 matrices

For 2×2 matrices, there is a useful formula for finding inverses.

Inverse of a 2 × 2 matrix

The inverse of the 2×2 matrix $\begin{pmatrix} a & b \\ c & d \end{pmatrix}$ is given by

$$\frac{1}{ad - bc}\begin{pmatrix} d & -b \\ -c & a \end{pmatrix}, \quad \text{provided that } ad - bc \neq 0.$$

In other words, to obtain the inverse when $ad - bc \neq 0$, swap the two elements on the leading diagonal and multiply the other two elements by -1, then multiply the resulting matrix by the scalar $1/(ad - bc)$.

You'll be asked to verify this general formula later in this subsection, but for now let's check that it works for the matrix

$$\mathbf{A} = \begin{pmatrix} 8 & 5 \\ 3 & 2 \end{pmatrix}$$

in Activity 21(a). Here $ad - bc = 8 \times 2 - 5 \times 3 = 1 \neq 0$, so the scalar $1/(ad - bc)$ is 1. You can see that the matrix

$$\mathbf{B} = \begin{pmatrix} 2 & -5 \\ -3 & 8 \end{pmatrix}$$

in Activity 21(a) is indeed obtained from $\mathbf{A}$ by swapping 8 and 2, multiplying the other two elements by -1, and multiplying the resulting matrix by the scalar 1.

Notice that the box above includes the condition $ad - bc \neq 0$. This is because the formula involves dividing by $ad - bc$, and division by 0 is not defined. If the elements of a 2×2 matrix do not satisfy this condition, then the matrix does *not* have an inverse. You'll see a proof of this fact in the next section.

In fact the quantity $ad - bc$ that appears in the formula is so important that it deserves a name, as follows.

Determinant of a 2 × 2 matrix

Let $\mathbf{A} = \begin{pmatrix} a & b \\ c & d \end{pmatrix}$. Then the number $ad - bc$ is called the **determinant** of $\mathbf{A}$, and written as $\det \mathbf{A}$.

If the determinant of $\mathbf{A}$ is not zero, then $\mathbf{A}$ is invertible.

If the determinant of $\mathbf{A}$ is zero, then $\mathbf{A}$ is not invertible.

Some texts denote $\det \mathbf{A}$ by $|\mathbf{A}|$, which looks like the notation for the modulus of a real number, but has a different meaning in this context. This notation is also used when a matrix is written out in full; for example,

$$\begin{vmatrix} 2 & 1 \\ -3 & 0 \end{vmatrix} = 2 \times 0 - 1 \times (-3).$$

When seeking the inverse of a 2×2 matrix, you should first check that the matrix is invertible by calculating the determinant. If the determinant is not zero, then you can use the rule for finding the inverse.

Example 4 *Finding inverses of 2×2 matrices*

For each of the matrices below, check whether its inverse exists, and find it if it does.

(a) $\mathbf{A} = \begin{pmatrix} 4 & 7 \\ 3 & 6 \end{pmatrix}$ (b) $\mathbf{B} = \begin{pmatrix} 3 & 4 \\ 6 & 8 \end{pmatrix}$

Solution

(a) Let $\mathbf{A} = \begin{pmatrix} 4 & 7 \\ 3 & 6 \end{pmatrix}$.

Check whether $\det \mathbf{A} = 0$.

Here $\det \mathbf{A} = 24 - 21 = 3$. Since $\det \mathbf{A} \neq 0$, the matrix has an inverse.

Use the formula to write down the inverse of $\mathbf{A}$.

The inverse is

$$\mathbf{A}^{-1} = \frac{1}{3}\begin{pmatrix} 6 & -7 \\ -3 & 4 \end{pmatrix} = \begin{pmatrix} 2 & -\frac{7}{3} \\ -1 & \frac{4}{3} \end{pmatrix}.$$

If you wish, check the answer by verifying that $\mathbf{AA}^{-1} = \mathbf{I}$ and $\mathbf{A}^{-1}\mathbf{A} = \mathbf{I}$.

Check: we have

$$\begin{pmatrix} 2 & -\frac{7}{3} \\ -1 & \frac{4}{3} \end{pmatrix} \begin{pmatrix} 4 & 7 \\ 3 & 6 \end{pmatrix} = \begin{pmatrix} 1 & 0 \\ 0 & 1 \end{pmatrix}$$

and

$$\begin{pmatrix} 4 & 7 \\ 3 & 6 \end{pmatrix} \begin{pmatrix} 2 & -\frac{7}{3} \\ -1 & \frac{4}{3} \end{pmatrix} = \begin{pmatrix} 1 & 0 \\ 0 & 1 \end{pmatrix}.$$

(b) For $\mathbf{B} = \begin{pmatrix} 3 & 4 \\ 6 & 8 \end{pmatrix}$, we have

$$\det \mathbf{B} = 3 \times 8 - 4 \times 6 = 24 - 24 = 0.$$

The determinant of $\mathbf{B}$ is zero.

Thus $\mathbf{B}$ is not invertible; that is, $\mathbf{B}^{-1}$ does not exist.

You can practise finding the inverses of 2×2 matrices in the next activity.

Activity 22 *Finding inverses of 2×2 matrices*

Determine which of the following matrices are invertible. For each matrix that is invertible, write down its inverse.

(a) $\mathbf{A} = \begin{pmatrix} 13 & 5 \\ 5 & 2 \end{pmatrix}$ (b) $\mathbf{B} = \begin{pmatrix} 3 & 0 \\ -5 & 2 \end{pmatrix}$

(c) $\mathbf{C} = \begin{pmatrix} -3 & \frac{1}{5} \\ 5 & -\frac{1}{3} \end{pmatrix}$ (d) $\mathbf{D} = \begin{pmatrix} 1.5 & 2.5 \\ 0.5 & 1.5 \end{pmatrix}$

In the next activity you're asked to verify that the formula for the inverse of a 2×2 matrix works in general.

Activity 23 *Verifying the formula for the inverse of a 2×2 matrix*

Let $\mathbf{A} = \begin{pmatrix} a & b \\ c & d \end{pmatrix}$, with $ad - bc \neq 0$, and $\mathbf{B} = \dfrac{1}{ad - bc} \begin{pmatrix} d & -b \\ -c & a \end{pmatrix}$.

Work out the products $\mathbf{AB}$ and $\mathbf{BA}$, and deduce that $\mathbf{B} = \mathbf{A}^{-1}$.

Hint: use the fact that $\mathbf{P}(k\mathbf{Q}) = k(\mathbf{PQ})$ for any matrices $\mathbf{P}$ and $\mathbf{Q}$ for which the product $\mathbf{PQ}$ is defined and for any scalar k.

A matrix that does not have an inverse is called a **non-invertible** matrix. In some texts, non-invertible matrices are called **singular** matrices, and invertible matrices are called **non-singular** matrices.

Determinants and inverses of larger matrices

You have seen that a 2×2 matrix has an inverse only if its determinant is not zero.

The determinant is defined for square matrices of any size, and it plays the same role as in the 2×2 case: any square matrix $\mathbf{A}$ such that $\det \mathbf{A} \neq 0$ has an inverse $\mathbf{A}^{-1}$. If $\det \mathbf{A} = 0$, then no inverse can be found.

It is possible to use hand calculations to work out the determinant and the inverse (when it exists) of a square matrix whose size is larger than 2×2, but the process is rather complicated, and beyond the scope of this module. However, you can use the computer algebra system to calculate determinants and inverses of larger matrices.

Activity 24 *Working with matrix inverses on a computer*

Work through Subsection 10.2 of the *Computer algebra guide*.

4 Simultaneous linear equations and matrices

In Section 3 of Unit 2, you met simultaneous linear equations and you saw two methods (substitution and elimination) for solving them. Here you'll meet a third method, which uses matrices.

4.1 Solving simultaneous linear equations in two unknowns

Consider the following simultaneous linear equations in the two unknowns x and y:

$$\begin{aligned} 0.4x + 0.2y &= 4 \\ 0.6x + 0.8y &= 11. \end{aligned} \qquad (3)$$

There's a useful way to write equations (3) as a single equation involving two-dimensional vectors (which are 2×1 matrices). To do this, you write the sides of each equation as the components of a vector, as follows:

$$\begin{pmatrix} 0.4x + 0.2y \\ 0.6x + 0.8y \end{pmatrix} = \begin{pmatrix} 4 \\ 11 \end{pmatrix}.$$

You can now write the left-hand side of this equation as the product of a 2×2 matrix and a two-dimensional vector:

$$\begin{pmatrix} 0.4 & 0.2 \\ 0.6 & 0.8 \end{pmatrix} \begin{pmatrix} x \\ y \end{pmatrix} = \begin{pmatrix} 4 \\ 11 \end{pmatrix}.$$

The 2×2 matrix that appears in this equation is called the *coefficient matrix* of the simultaneous linear equations (3). Its elements are the numbers that appear on the left-hand sides of equations (3). Any pair of simultaneous linear equations in two unknowns can be written in this form.

Matrix form of simultaneous linear equations

The simultaneous linear equations

$$\begin{aligned} ax + by &= e \\ cx + dy &= f \end{aligned}$$

can be written as the matrix equation

$$\begin{pmatrix} a & b \\ c & d \end{pmatrix} \begin{pmatrix} x \\ y \end{pmatrix} = \begin{pmatrix} e \\ f \end{pmatrix}.$$

The matrix $\begin{pmatrix} a & b \\ c & d \end{pmatrix}$ is called the **coefficient matrix**.

Note that the coefficient matrix of simultaneous linear equations depends on how the equations are written. For example, if you write equations (3) as

$$\begin{aligned} 0.6x + 0.8y &= 11 \\ 0.4x + 0.2y &= 4, \end{aligned}$$

then you obtain the coefficient matrix

$$\begin{pmatrix} 0.6 & 0.8 \\ 0.4 & 0.2 \end{pmatrix},$$

which is different from the coefficient matrix obtained above for the same equations.

Example 5 *Writing simultaneous linear equations in matrix form*

Write the following pair of equations in matrix form.

$$y = 2x - 1$$
$$3y = 2$$

Solution

Rearrange each equation so that the unknowns are on the left-hand side and in the same order in each equation, and the constant terms are on the right-hand sides.

The equations can be rewritten as follows:

$$2x + (-1)y = 1$$
$$0x + \quad 3y = 2.$$

Now read off the matrix form from the equations.

Therefore the equations can be written in matrix form as

$$\begin{pmatrix} 2 & -1 \\ 0 & 3 \end{pmatrix} \begin{pmatrix} x \\ y \end{pmatrix} = \begin{pmatrix} 1 \\ 2 \end{pmatrix}.$$

There are other possible solutions to Example 5. For example, the equations can be rewritten in the form

$$-2x + \quad y = -1$$
$$0x + 3y = 2,$$

which gives the matrix form

$$\begin{pmatrix} -2 & 1 \\ 0 & 3 \end{pmatrix} \begin{pmatrix} x \\ y \end{pmatrix} = \begin{pmatrix} -1 \\ 2 \end{pmatrix}.$$

In the next activity you can practise finding matrix forms of pairs of simultaneous linear equations.

Activity 25 *Writing simultaneous linear equations in matrix form*

Write each of the following pairs of equations in matrix form.

(a) $2x + 3y = 3$
$\quad x + 4y = -1$

(b) $2x - 6 = 0$
$\quad 3x + 6y = 15$

(c) $\frac{3}{5}x - \frac{4}{5}y = 18$
$\quad \frac{4}{5}x + \frac{3}{5}y = -1$

The matrix form of a pair of simultaneous linear equations is not just convenient shorthand: it is the initial step in a method for solving the equations. Let's see how this method works for equations (3) from the beginning of this subsection, which have the matrix form

$$\begin{pmatrix} 0.4 & 0.2 \\ 0.6 & 0.8 \end{pmatrix} \begin{pmatrix} x \\ y \end{pmatrix} = \begin{pmatrix} 4 \\ 11 \end{pmatrix}.$$

Let $\mathbf{A}$ be the coefficient matrix; that is,

$$\mathbf{A} = \begin{pmatrix} 0.4 & 0.2 \\ 0.6 & 0.8 \end{pmatrix}.$$

Then the matrix form of the equations can be written as

$$\mathbf{A} \begin{pmatrix} x \\ y \end{pmatrix} = \begin{pmatrix} 4 \\ 11 \end{pmatrix}.$$

Since $\det \mathbf{A} = 0.4 \times 0.8 - 0.2 \times 0.6 = 0.2$, the matrix $\mathbf{A}$ has an inverse $\mathbf{A}^{-1}$. If we multiply both sides of the equation above by $\mathbf{A}^{-1}$, then the result is

$$\mathbf{A}^{-1} \left(\mathbf{A} \begin{pmatrix} x \\ y \end{pmatrix} \right) = \mathbf{A}^{-1} \begin{pmatrix} 4 \\ 11 \end{pmatrix}.$$

By the associativity of matrix multiplication, this equation can be rearranged as:

$$(\mathbf{A}^{-1}\mathbf{A}) \begin{pmatrix} x \\ y \end{pmatrix} = \mathbf{A}^{-1} \begin{pmatrix} 4 \\ 11 \end{pmatrix}.$$

Since $\mathbf{A}^{-1}\mathbf{A} = \mathbf{I}$, where $\mathbf{I}$ is the 2×2 identity matrix, the equation above can be rewritten as

$$\mathbf{I} \begin{pmatrix} x \\ y \end{pmatrix} = \mathbf{A}^{-1} \begin{pmatrix} 4 \\ 11 \end{pmatrix}.$$

Since $\mathbf{I}$ is an identity matrix, our equation becomes

$$\begin{pmatrix} x \\ y \end{pmatrix} = \mathbf{A}^{-1} \begin{pmatrix} 4 \\ 11 \end{pmatrix}. \qquad (4)$$

Therefore, if we work out $\mathbf{A}^{-1}$ and calculate the product on the right-hand side of this matrix equation, then we'll find a solution to the original equations. Now

$$\mathbf{A}^{-1} = \frac{1}{0.2} \begin{pmatrix} 0.8 & -0.2 \\ -0.6 & 0.4 \end{pmatrix} = \begin{pmatrix} 4 & -1 \\ -3 & 2 \end{pmatrix}.$$

So substituting for $\mathbf{A}^{-1}$ in equation (4) gives

$$\begin{pmatrix} x \\ y \end{pmatrix} = \begin{pmatrix} 4 & -1 \\ -3 & 2 \end{pmatrix} \begin{pmatrix} 4 \\ 11 \end{pmatrix} = \begin{pmatrix} 5 \\ 10 \end{pmatrix}.$$

In other words,

$$x = 5 \quad \text{and} \quad y = 10.$$

The example above illustrates a general method for solving simultaneous linear equations, which can be summarised as follows.

Strategy:
To solve a pair of simultaneous linear equations in two unknowns using matrices

Write the simultaneous linear equations in matrix form $\mathbf{Ax} = \mathbf{b}$, where $\mathbf{A}$ is the coefficient matrix, $\mathbf{x}$ is the corresponding vector of unknowns, and $\mathbf{b}$ is the vector whose components are the corresponding right-hand sides of the equations.

If the matrix $\mathbf{A}$ is invertible, then the solution is given by

$$\mathbf{x} = \mathbf{A}^{-1}\mathbf{b}.$$

Example 6 *Using matrices to solve simultaneous linear equations*

Use matrices to solve the following pair of simultaneous linear equations.

$$x - 5y = -3$$
$$x + 3y = 13$$

Solution

Find the matrix form of the equations.

The matrix form of the equations is

$$\begin{pmatrix} 1 & -5 \\ 1 & 3 \end{pmatrix} \begin{pmatrix} x \\ y \end{pmatrix} = \begin{pmatrix} -3 \\ 13 \end{pmatrix}.$$

Find the inverse of the coefficient matrix.

The determinant of the coefficient matrix is

$$1 \times 3 - (-5) \times 1 = 8,$$

so the inverse of the coefficient matrix is

$$\frac{1}{8}\begin{pmatrix} 3 & 5 \\ -1 & 1 \end{pmatrix} = \begin{pmatrix} \frac{3}{8} & \frac{5}{8} \\ -\frac{1}{8} & \frac{1}{8} \end{pmatrix}.$$

Use the inverse matrix to solve the simultaneous equations.

Hence

$$\begin{pmatrix} x \\ y \end{pmatrix} = \frac{1}{8}\begin{pmatrix} 3 & 5 \\ -1 & 1 \end{pmatrix}\begin{pmatrix} -3 \\ 13 \end{pmatrix}$$
$$= \begin{pmatrix} 7 \\ 2 \end{pmatrix}.$$

The solution is $x = 7$, $y = 2$.

In the calculation above it is convenient to use the form of the inverse matrix with the fraction $1/8$ outside the matrix, in order to simplify the working.

As usual, you can check the answer to Example 6 by substituting back into the original equations.

You can practise the method in Example 6 in the next activity.

Activity 26 *Using matrices to solve simultaneous linear equations*

Use matrices to solve the following pairs of simultaneous linear equations. (You were asked to write these pairs of equations in matrix form in Activity 25.)

(a) $2x + 3y = 3$
$x + 4y = -1$

(b) $2x - 6 = 0$
$3x + 6y = 15$

(c) $\frac{3}{5}x - \frac{4}{5}y = 18$
$\frac{4}{5}x + \frac{3}{5}y = -1$

4.2 Non–invertible coefficient matrices

You saw in Unit 2 that a pair of simultaneous linear equations in two unknowns can have no solutions, exactly one solution or infinitely many solutions. If it has exactly one solution, then we say that it has a *unique* solution.

In this subsection you'll see that a simple way to tell whether a pair of simultaneous linear equations in two unknowns has a unique solution is to work out the determinant of its coefficient matrix.

If the determinant is non-zero, then the coefficient matrix has an inverse given by the formula that you saw in Subsection 3.2, and hence the equations have a unique solution, by what you saw in Subsection 4.1.

Let's now look at what happens when the determinant is zero. Let's start by looking at two examples.

First, consider the following pair of simultaneous equations:

$$\begin{aligned} 2x - 3y &= 5 \\ -4x + 6y &= 7. \end{aligned} \tag{5}$$

The coefficient matrix is

$$\begin{pmatrix} 2 & -3 \\ -4 & 6 \end{pmatrix},$$

which has determinant $2 \times 6 - (-3) \times (-4) = 0$. Let's look for a solution to the equations using the elimination method. If we multiply both sides of the first one of equations (5) by -2, then we get

$$\begin{aligned} -4x + 6y &= -10 \\ -4x + 6y &= \quad 7. \end{aligned}$$

These equations have identical left-hand sides, but different right-hand sides. So they have no solution: no values of x and y can make $-4x + 6y$ equal to both -10 and 7.

Now consider the following pair of simultaneous equations:

$$\begin{aligned} x + 2y &= -6 \\ -3x - 6y &= 18. \end{aligned} \tag{6}$$

The coefficient matrix is

$$\begin{pmatrix} 1 & 2 \\ -3 & -6 \end{pmatrix},$$

which has determinant $1 \times (-6) - 2 \times (-3) = 0$. Again, let's look for a solution to the equations using the elimination method. If we multiply both sides of the first of equations (6) by -3, then we get

$$\begin{aligned} -3x - 6y &= 18 \\ -3x - 6y &= 18. \end{aligned}$$

These equations are identical. So they have infinitely many solutions: all values of x and y that satisfy the first equation also satisfy the second equation.

You've now seen two cases where the coefficient matrix has determinant zero. In the first case, the simultaneous equations had no solution. In the second case, they had infinitely many solutions. In fact, whenever the coefficient matrix of a pair of simultaneous linear equations has determinant zero, one of these two possibilities must occur.

To see why, consider any two simultaneous linear equations

$$\begin{aligned} ax + by &= e \\ cx + dy &= f, \end{aligned}$$

and suppose that the determinant of the coefficient matrix is zero; that is, $ad - bc = 0$. We can assume that in each equation at least one of the coefficients is non-zero; otherwise, the equation is of the form $0x + 0y =$ constant, so the pair of equations has either no solutions or infinitely many solutions.

Suppose that in the first equation $a \neq 0$; a similar argument holds if $b \neq 0$. Since $ad - bc = 0$, we have $d = bc/a$. Hence $c \neq 0$, since if $c = 0$ then $d = 0$ also. Multiplying the first equation by c/a gives

$$cx + (bc/a)y = ce/a$$
$$cx + dy = f.$$

Since $d = bc/a$, these two equations can be rewritten as

$$cx + dy = ce/a$$
$$cx + dy = f.$$

These equations have identical left-hand sides. So they (and hence the original equations) have either no solution (if $ce/a \neq f$) or infinitely many solutions (if $ce/a = f$).

So, in conclusion, if the coefficient matrix has determinant zero, then the equations have either no solution or infinitely many solutions.

We can deduce from this fact that if a 2×2 matrix has determinant zero, then it is *not* invertible. This is a fact that was mentioned in Subsection 3.2. To see how, consider any 2×2 matrix with determinant zero. If the matrix *were* invertible, then any pair of simultaneous equations with this matrix as its coefficient matrix would have a unique solution, which cannot be the case, since you saw above that if the determinant of the coefficient matrix is zero, then the equations have no solution or infinitely many solutions.

So you have learned the following important payoffs from the matrix method for simultaneous linear equations.

For a pair of simultaneous linear equations in two unknowns:

- if the determinant of the coefficient matrix is non-zero, then this matrix is invertible and the equations have a unique solution
- if the determinant of the coefficient matrix is zero, then this matrix is not invertible and the equations have no solution or infinitely many solutions.

Thus, calculating the determinant of the coefficient matrix gives a quick test for the existence of a unique solution.

Activity 27 *Determining whether simultaneous equations have unique solutions*

Which of the following pairs of simultaneous linear equations has a unique solution? (You are not asked to find the solution.)

(a) $5x - 3y = 5$
$-x + 4y = 7$

(b) $3x - y = -6$
$-9x + 3y = 18$

4.3 Simultaneous linear equations in more than two unknowns

The matrix method that you met in Subsection 4.2 can be used to find a solution for more than two simultaneous equations in more than two unknowns.

In Subsection 5.3 of Unit 1, you learned that a *linear* equation is one in which, after you've expanded any brackets and cleared any fractions, each term is either a constant term or a number times a variable. For example, the equation

$$x + 2y - z = 3$$

is linear, but none of the equations

$$x + y^2 = 2, \quad x + yz = 1 \quad \text{and} \quad x + 2^y = 1$$

are linear. In this subsection we'll usually write a linear equation in a form similar to the linear equation above, that is, with its left-hand side as a sum of terms, each of which is a number times a variable (and with each variable appearing just once), and with the right-hand side as a number.

We'll refer to a collection of linear equations in a given set of unknowns as a **system of linear equations**. For example, here's a system of three linear equations in three unknowns, x, y and z:

$$\begin{aligned} 2x + 2y + 3z &= 2 \\ 2x + 3y + 4z &= -1 \\ 4x + 7y + 10z &= 3. \end{aligned}$$

A **solution** to a system of linear equations is an assignment of values to the unknowns that makes all the equations hold simultaneously. As in Subsection 3.1 of Unit 2, the process of finding these values is known as *solving the equations simultaneously*. We'll look only at systems where the number of unknowns is equal to the number of equations.

Numerical problems that can be modelled by systems of linear equations often arise in applications. Consider, for example, a bank that offers three types of investment to its customers: a financial product A that pays guaranteed interest of 2.5% over two years and is quite safe, and two financial products B and C that in the same time period are likely to yield interests of 5% and 6% respectively, but are more risky.

Suppose that a customer is seeking to invest £9000 over two years, and is advised to spread his investment across the three types of financial products. He is risk-averse, so he wants to invest a total of only £4000 across the two riskier products B and C. Of this £4000, he wants to invest just enough in the higher-yielding product C to ensure that his total £9000 investment is likely to yield an overall return of 4%, with the remainder of the £4000 invested in the lower-yielding product B, to spread his risks. How should he spread his investment across the three products in order to achieve his desired return?

First of all, the amounts invested in each product must add up to £9000. So, if x, y and z represent the amounts invested (in £) in products A, B and C, respectively, then

$$x + y + z = 9000.$$

Moreover, the investor wants to invest a total of £4000 in products B and C, so

$$y + z = 4000.$$

Since the interest on his investment in product A is 2.5%, the investor's expected interest payment from his investment in product A is $0.025x$. Likewise, his expected interest payments from his investments in products B and C are $0.05y$ and $0.06z$, respectively. The investor is seeking an overall return of 4%, that is, an overall interest payment of $0.04 \times £9000 = £360$. Thus the amounts invested in the three products must also satisfy the equation

$$0.025x + 0.05y + 0.06z = 360.$$

Therefore, the amounts x, y and z must simultaneously satisfy the three linear equations

$$\begin{aligned} x + \quad y + \quad z &= 9000 \\ y + \quad z &= 4000 \\ 0.025x + 0.05y + 0.06z &= 360. \end{aligned} \tag{7}$$

Such systems of linear equations can be solved efficiently by methods similar to elimination and substitution – you can probably see that equations (7) can be solved fairly easily for x, y and z by

- first subtracting the second equation from the first to give $x = 5000$
- then substituting this value for x into the third equation and solving the resulting two simultaneous equations for y and z.

Methods of this sort are covered in detail in other modules. Systems of linear equations like those above can also be solved by using matrices, and the rest of this subsection covers this approach.

Any system of n linear equations in n unknowns can be written as a single matrix equation

$$\mathbf{Ax} = \mathbf{b},$$

where $\mathbf{A}$ is an $n \times n$ coefficient matrix, $\mathbf{x}$ is an n-dimensional vector whose components are the unknowns, and $\mathbf{b}$ is an n-dimensional vector, as you saw for the case $n = 2$ in Subsection 4.2. If the coefficient matrix $\mathbf{A}$ is invertible, then the matrix equation can be solved by multiplying both sides by the inverse of $\mathbf{A}$, giving $\mathbf{x} = \mathbf{A}^{-1}\mathbf{b}$, just as in the 2×2 case.

In Section 3 you learned how to use a computer to find the inverses of matrices of size 3×3 and larger. This enables you to use the matrix method to solve systems of three or more linear equations. In the next example, this method is used to solve equations (7).

Example 7 *Solving a system of three linear equations*

Use the matrix method to solve the following system of three linear equations.

$$\begin{aligned} x + \quad y + \quad z &= 9000 \\ y + \quad z &= 4000 \\ 0.025x + 0.05y + 0.06z &= 360 \end{aligned}$$

Solution

The matrix form of the system of three equations is

$$\begin{pmatrix} 1 & 1 & 1 \\ 0 & 1 & 1 \\ 0.025 & 0.05 & 0.06 \end{pmatrix} \begin{pmatrix} x \\ y \\ z \end{pmatrix} = \begin{pmatrix} 9000 \\ 4000 \\ 360 \end{pmatrix}.$$

Use the computer algebra system to find the inverse of the coefficient matrix.

The inverse of the coefficient matrix is

$$\begin{pmatrix} 1 & -1 & 0 \\ 2.5 & 3.5 & -100 \\ -2.5 & -2.5 & 100 \end{pmatrix}.$$

Multiply each side of the matrix equation by the inverse of the coefficient matrix.

Therefore the solution is given by

$$\begin{pmatrix} x \\ y \\ z \end{pmatrix} = \begin{pmatrix} 1 & -1 & 0 \\ 2.5 & 3.5 & -100 \\ -2.5 & -2.5 & 100 \end{pmatrix} \begin{pmatrix} 9000 \\ 4000 \\ 360 \end{pmatrix} = \begin{pmatrix} 5000 \\ 500 \\ 3500 \end{pmatrix}.$$

Note that you could use the computer algebra system to solve the simultaneous equations (7) directly, that is, without first finding the inverse of the coefficient matrix, and so check the answer to Example 7.

The solution to equations (7) found in this example tells our investor that the best way to spread his investment in order to achieve a 4% return is as follows:

£5000 on investment A
£500 on investment B
£3500 on investment C.

You can practise finding inverses of matrices and using them to solve systems of more than two linear equations in the next activity.

Activity 28 *Solving systems of three linear equations*

Solve each of the following systems of simultaneous linear equations by using the computer algebra system to find the inverse of the coefficient matrix.

(a)
$$\begin{aligned} 2x + 2y + 3z &= 2 \\ 2x + 3y + 4z &= -1 \\ 4x + 7y + 10z &= 3 \end{aligned}$$

(b)
$$\begin{aligned} 3x + 2y + z &= 1 \\ 4x + 3y + z &= 2 \\ 7x + 5y + z &= 1 \end{aligned}$$

In fact using matrix inverses is not usually the most efficient way to solve systems of linear equations, but the method has theoretical importance.

For example, it can be shown that, for all $n \geq 2$,

- if the determinant of the coefficient matrix of a system of n equations is non-zero, then this matrix is invertible and the equations have a unique solution
- if the determinant of the coefficient matrix of a system of n equations is zero, then this matrix is not invertible and the equations have no solution or infinitely many solutions.

You saw the case $n = 2$ of this result in Subsection 4.2.

Example 8 *Determining whether a system of linear equations has a unique solution*

Does the system

$$\begin{aligned} x + 3y - 2z &= 12 \\ \tfrac{1}{2}x - 8y + z &= 0 \\ 7x \phantom{{}+3y} - z &= 2 \end{aligned}$$

have a unique solution?

Solution

The coefficient matrix of this system is

$$\mathbf{A} = \begin{pmatrix} 1 & 3 & -2 \\ \frac{1}{2} & -8 & 1 \\ 7 & 0 & -1 \end{pmatrix}.$$

Using the computer algebra system we find that $\det \mathbf{A} = -81.5$. Since $\det \mathbf{A} \neq 0$, the system has a unique solution.

The following activity provides further practice on the significance of the determinant of the coefficient matrix for systems of three or more linear equations.

Activity 29 *Determining whether systems of linear equations have unique solutions*

Which of the following systems of linear equations have a unique solution?

(a)
$$\begin{aligned} x + 2y + 3z &= 5 \\ 2x + 3y + z &= 6 \\ 4x + 7y + 7z &= 1 \end{aligned}$$

(b)
$$\begin{aligned} 2x + 2y + 3z &= 5 \\ 2x + 3y + z &= 6 \\ 4x + 7y + 7z &= 1 \end{aligned}$$

(c)
$$\begin{aligned} x + y - w + z &= 0 \\ x - y + w + z &= 2 \\ 2x + 3w - z &= 1 \\ 6x + 2y + 4w &= 0 \end{aligned}$$

Learning outcomes

After studying this unit, you should be able to:

- use matrix notation
- evaluate sums, differences and scalar multiples of matrices
- establish whether two matrices can be multiplied together
- evaluate a matrix product both by hand and by using the computer algebra system
- determine whether a given $n \times n$ matrix has an inverse
- work out the determinant and inverse (if it exists) of a 2×2 matrix by hand
- use the computer algebra system to work out the determinant and inverse (if it exists) of a square matrix
- write a system of linear equations in matrix form and solve it using matrix methods where possible.

Solutions to activities

Solution to Activity 1

We have $a_{33} = 3$, $a_{34} = -4$, $a_{43} = -7$, $a_{14} = -1$, $a_{44} = -9$, $a_{42} = -5$.

The matrix $\mathbf{A}$ has four rows and four columns, so its size is 4×4.

Solution to Activity 2

(a) $\mathbf{A}+\mathbf{B} = \begin{pmatrix} 2 & 1 \\ 3 & 0 \end{pmatrix} + \begin{pmatrix} 4 & 6 \\ 2 & -1 \end{pmatrix} = \begin{pmatrix} 6 & 7 \\ 5 & -1 \end{pmatrix}$

(b)
$$\mathbf{A}+\mathbf{B} = \begin{pmatrix} 2 & 0 & -4 \\ -1 & 6 & 0 \end{pmatrix} + \begin{pmatrix} 1 & 0 & 1 \\ 0 & 1 & 1 \end{pmatrix} = \begin{pmatrix} 3 & 0 & -3 \\ -1 & 7 & 1 \end{pmatrix}$$

(c)
$$\mathbf{A}+\mathbf{B} = \begin{pmatrix} a & b \\ c & d \end{pmatrix} + \begin{pmatrix} 3 & -1 \\ -1 & 2 \end{pmatrix} = \begin{pmatrix} a+3 & b-1 \\ c-1 & d+2 \end{pmatrix}$$

Solution to Activity 3

(a) $\mathbf{A}-\mathbf{B} = \begin{pmatrix} 2 & -2 \\ 1 & 0 \end{pmatrix} - \begin{pmatrix} \frac{1}{2} & 4 \\ -1 & 2 \end{pmatrix} = \begin{pmatrix} \frac{3}{2} & -6 \\ 2 & -2 \end{pmatrix}$

(b) $\mathbf{B}-\mathbf{A} = \begin{pmatrix} \frac{1}{2} & 4 \\ -1 & 2 \end{pmatrix} - \begin{pmatrix} 2 & -2 \\ 1 & 0 \end{pmatrix} = \begin{pmatrix} -\frac{3}{2} & 6 \\ -2 & 2 \end{pmatrix}$

(c)
$$\mathbf{B}-\mathbf{C} = \begin{pmatrix} \frac{1}{2} & 4 \\ -1 & 2 \end{pmatrix} - \begin{pmatrix} 0 & -x \\ x & 0 \end{pmatrix} = \begin{pmatrix} \frac{1}{2} & 4+x \\ -1-x & 2 \end{pmatrix}$$

Solution to Activity 4

(a) (i) $3\begin{pmatrix} 4 & 3 \\ 2 & 1 \end{pmatrix} = \begin{pmatrix} 12 & 9 \\ 6 & 3 \end{pmatrix}$

(ii) $\frac{1}{2}\begin{pmatrix} 4 & 6 \\ 2 & -1 \end{pmatrix} = \begin{pmatrix} 2 & 3 \\ 1 & -\frac{1}{2} \end{pmatrix}$

(iii) $\frac{2}{3}\begin{pmatrix} 6 & x \\ 3 & y \end{pmatrix} = \begin{pmatrix} 4 & \frac{2}{3}x \\ 2 & \frac{2}{3}y \end{pmatrix}$

(b) (i) $\begin{pmatrix} 4.5 & 3 \\ 2 & 1.5 \end{pmatrix} = 0.5\begin{pmatrix} 9 & 6 \\ 4 & 3 \end{pmatrix}$

(ii) $\begin{pmatrix} \frac{3}{2} \\ \frac{2}{3} \end{pmatrix} = \frac{1}{6}\begin{pmatrix} 9 \\ 4 \end{pmatrix}$

(iii) $\begin{pmatrix} 2x & 3x \\ 0 & 5x \end{pmatrix} = x\begin{pmatrix} 2 & 3 \\ 0 & 5 \end{pmatrix}$

In each of the answers in this part, there is more than one way to choose the scalar multiple. For example, another correct solution to part (i) is

$$\begin{pmatrix} 4.5 & 3 \\ 2 & 1.5 \end{pmatrix} = 0.25\begin{pmatrix} 18 & 12 \\ 8 & 6 \end{pmatrix}.$$

Solution to Activity 5

We have

$$\mathbf{A}-\mathbf{C} = \begin{pmatrix} \frac{1}{2} & 1 & 2 \\ 4 & -2 & 0 \end{pmatrix} - \begin{pmatrix} \frac{3}{2} & 0 & -1 \\ -1 & 3 & 5 \end{pmatrix} = \begin{pmatrix} -1 & 1 & 3 \\ 5 & -5 & -5 \end{pmatrix}.$$

Next,

$$2\mathbf{A} = \begin{pmatrix} 1 & 2 & 4 \\ 8 & -4 & 0 \end{pmatrix}.$$

The matrix $2\mathbf{A}$ has size 2×3 and $\mathbf{B}$ has size 2×2; hence the sum $\mathbf{B} + 2\mathbf{A}$ is not defined.

Finally,

$$\mathbf{C}-2\mathbf{A} = \begin{pmatrix} \frac{3}{2} & 0 & -1 \\ -1 & 3 & 5 \end{pmatrix} - \begin{pmatrix} 1 & 2 & 4 \\ 8 & -4 & 0 \end{pmatrix} = \begin{pmatrix} \frac{1}{2} & -2 & -5 \\ -9 & 7 & 5 \end{pmatrix}.$$

Solution to Activity 6

For each pair of matrices in this question, the sizes of the matrices are written underneath the matrices, and the middle two numbers are boxed in order to use the method described in the text.

(a) $\begin{pmatrix} 2 & 9 \\ 7 & 1 \\ -6 & 8 \end{pmatrix} \begin{pmatrix} -2 & 3 & 1 \\ 4 & 0 & 5 \end{pmatrix}$

$3 \times \boxed{2} \quad \boxed{2} \times 3$

The middle numbers are the same, so these matrices can be multiplied together. Their product is

$$\begin{pmatrix} 2 & 9 \\ 7 & 1 \\ -6 & 8 \end{pmatrix} \begin{pmatrix} -2 & 3 & 1 \\ 4 & 0 & 5 \end{pmatrix} = \begin{pmatrix} 32 & 6 & 47 \\ -10 & 21 & 12 \\ 44 & -18 & 34 \end{pmatrix}.$$

(b) $\begin{pmatrix} 1 & 2 & 3 \end{pmatrix} \begin{pmatrix} 9 \\ 8 \\ 7 \end{pmatrix}$

$1 \times \boxed{3} \quad \boxed{3} \times 1$

The middle numbers are the same, so these matrices can be multiplied together. Their product is

$$\begin{pmatrix} 1 & 2 & 3 \end{pmatrix} \begin{pmatrix} 9 \\ 8 \\ 7 \end{pmatrix} = (46).$$

(c) $\begin{pmatrix} 2 & 9 \\ 7 & 1 \\ -6 & 8 \end{pmatrix} \begin{pmatrix} 0 & 10 \\ 4 & 1 \\ -2 & 7 \end{pmatrix}$

$3 \times \boxed{2} \quad \boxed{3} \times 2$

The middle numbers are different, so these matrices cannot be multiplied together.

(d) $\begin{pmatrix} 9 \\ 8 \\ 7 \end{pmatrix} \begin{pmatrix} 1 & 2 & 3 \end{pmatrix}$

$3 \times \boxed{1} \quad \boxed{1} \times 3$

The middle numbers are the same, so these matrices can be multiplied together. Their product is

$$\begin{pmatrix} 9 \\ 8 \\ 7 \end{pmatrix} \begin{pmatrix} 1 & 2 & 3 \end{pmatrix} = \begin{pmatrix} 9 & 18 & 27 \\ 8 & 16 & 24 \\ 7 & 14 & 21 \end{pmatrix}.$$

(e) $\begin{pmatrix} 1 & 2 & 3 \\ 7 & 6 & 1 \end{pmatrix} \begin{pmatrix} 1 & 0 & 1 \\ 1 & 0 & -1 \\ 2 & 1 & 3 \end{pmatrix}$

$2 \times \boxed{3} \quad \boxed{3} \times 3$

The middle numbers are the same, so these matrices can be multiplied together. Their product is

$$\begin{pmatrix} 1 & 2 & 3 \\ 7 & 6 & 1 \end{pmatrix} \begin{pmatrix} 1 & 0 & 1 \\ 1 & 0 & -1 \\ 2 & 1 & 3 \end{pmatrix} = \begin{pmatrix} 9 & 3 & 8 \\ 15 & 1 & 4 \end{pmatrix}.$$

Solution to Activity 7

(a) (i) Here $\mathbf{A}$ has size 2×2 and $\mathbf{B}$ has size 2×1. For $\mathbf{AB}$, the size check is

$2 \times \boxed{2} \quad \boxed{2} \times 1.$

The middle numbers are the same, so the product $\mathbf{AB}$ is defined.

For $\mathbf{BA}$, the size check is

$2 \times \boxed{1} \quad \boxed{2} \times 2.$

The middle numbers are different, so the product $\mathbf{BA}$ is undefined.

(ii) Here both $\mathbf{A}$ and $\mathbf{B}$ have size 2×2, so the products $\mathbf{AB}$ and $\mathbf{BA}$ are both defined.

(b) For the matrices in part (a)(i), the product $\mathbf{BA}$ is not defined.

For the matrices in part (a)(ii),

$$\mathbf{AB} = \begin{pmatrix} 1 & 1 \\ -1 & 0 \end{pmatrix} \begin{pmatrix} 0 & 1 \\ 2 & 3 \end{pmatrix} = \begin{pmatrix} 2 & 4 \\ 0 & -1 \end{pmatrix}$$

and

$$\mathbf{BA} = \begin{pmatrix} 0 & 1 \\ 2 & 3 \end{pmatrix} \begin{pmatrix} 1 & 1 \\ -1 & 0 \end{pmatrix} = \begin{pmatrix} -1 & 0 \\ -1 & 2 \end{pmatrix}.$$

Therefore $\mathbf{AB} \neq \mathbf{BA}$.

Solution to Activity 8

(a) The size of $\mathbf{A}$ is 1×2 and the size of $\mathbf{B}$ is 2×2. The size check for the product $\mathbf{AB}$ is

$1 \times \boxed{2} \quad \boxed{2} \times 2.$

Thus $\mathbf{AB}$ has size 1×2.

The size of $\mathbf{C}$ is 2×3. The size check for the product $\mathbf{BC}$ is

$2 \times \boxed{2} \quad \boxed{2} \times 3.$

Thus $\mathbf{BC}$ has size 2×3.

We have

$$\mathbf{AB} = \begin{pmatrix} -1 & 0 \end{pmatrix} \begin{pmatrix} 1 & -1 \\ 2 & -1 \end{pmatrix} = \begin{pmatrix} -1 & 1 \end{pmatrix}, \text{ and}$$

$$\mathbf{BC} = \begin{pmatrix} 1 & -1 \\ 2 & -1 \end{pmatrix} \begin{pmatrix} 1 & 3 & 0 \\ -3 & 4 & 1 \end{pmatrix} = \begin{pmatrix} 4 & -1 & -1 \\ 5 & 2 & -1 \end{pmatrix}.$$

(b) From part (a), the size of **AB** is 1×2. Since **C** has size 2×3, the size check for $(\mathbf{AB})\mathbf{C}$ is

$$1 \times \boxed{2} \quad \boxed{2} \times 3.$$

So the product $(\mathbf{AB})\mathbf{C}$ exists and has size 1×3.

From part (a), the size of **BC** is 2×3. Since **A** has size 1×2, the size check for $\mathbf{A}(\mathbf{BC})$ is

$$1 \times \boxed{2} \quad \boxed{2} \times 3.$$

So the product $\mathbf{A}(\mathbf{BC})$ exists and has size 1×3.

Using the answers for the matrices **AB** and **BC** found in part (a) we obtain

$$(\mathbf{AB})\mathbf{C} = \begin{pmatrix} -1 & 1 \end{pmatrix} \begin{pmatrix} 1 & 3 & 0 \\ -3 & 4 & 1 \end{pmatrix} = \begin{pmatrix} -4 & 1 & 1 \end{pmatrix}$$

and

$$\mathbf{A}(\mathbf{BC}) = \begin{pmatrix} -1 & 0 \end{pmatrix} \begin{pmatrix} 4 & -1 & -1 \\ 5 & 2 & -1 \end{pmatrix} = \begin{pmatrix} -4 & 1 & 1 \end{pmatrix}.$$

We can see that $(\mathbf{AB})\mathbf{C} = \mathbf{A}(\mathbf{BC})$.

Solution to Activity 9

We have

$$2\mathbf{B} = 2\begin{pmatrix} 1 & 0 \\ 2 & 1 \end{pmatrix} = \begin{pmatrix} 2 & 0 \\ 4 & 2 \end{pmatrix},$$

so

$$\mathbf{A}(2\mathbf{B}) = \begin{pmatrix} 1 & 2 \\ 3 & 4 \end{pmatrix} \begin{pmatrix} 2 & 0 \\ 4 & 2 \end{pmatrix} = \begin{pmatrix} 10 & 4 \\ 22 & 8 \end{pmatrix}.$$

Also,

$$\mathbf{AB} = \begin{pmatrix} 1 & 2 \\ 3 & 4 \end{pmatrix} \begin{pmatrix} 1 & 0 \\ 2 & 1 \end{pmatrix} = \begin{pmatrix} 5 & 2 \\ 11 & 4 \end{pmatrix},$$

so

$$2(\mathbf{AB}) = 2\begin{pmatrix} 5 & 2 \\ 11 & 4 \end{pmatrix} = \begin{pmatrix} 10 & 4 \\ 22 & 8 \end{pmatrix}.$$

Therefore $\mathbf{A}(2\mathbf{B}) = 2(\mathbf{AB})$, as required.

Solution to Activity 10

We have

$$\mathbf{B} + \mathbf{C} = \begin{pmatrix} 1 \\ 2 \end{pmatrix} + \begin{pmatrix} -1 \\ 1 \end{pmatrix} = \begin{pmatrix} 0 \\ 3 \end{pmatrix},$$

so

$$\mathbf{A}(\mathbf{B} + \mathbf{C}) = \begin{pmatrix} 1 & -1 \\ 3 & 4 \end{pmatrix} \begin{pmatrix} 0 \\ 3 \end{pmatrix} = \begin{pmatrix} -3 \\ 12 \end{pmatrix}.$$

Also,

$$\mathbf{AB} = \begin{pmatrix} 1 & -1 \\ 3 & 4 \end{pmatrix} \begin{pmatrix} 1 \\ 2 \end{pmatrix} = \begin{pmatrix} -1 \\ 11 \end{pmatrix},$$

$$\mathbf{AC} = \begin{pmatrix} 1 & -1 \\ 3 & 4 \end{pmatrix} \begin{pmatrix} -1 \\ 1 \end{pmatrix} = \begin{pmatrix} -2 \\ 1 \end{pmatrix},$$

so

$$\mathbf{AB} + \mathbf{AC} = \begin{pmatrix} -1 \\ 11 \end{pmatrix} + \begin{pmatrix} -2 \\ 1 \end{pmatrix} = \begin{pmatrix} -3 \\ 12 \end{pmatrix}.$$

Thus $\mathbf{A}(\mathbf{B} + \mathbf{C}) = \mathbf{AB} + \mathbf{AC}$, as required.

Solution to Activity 11

(a) The product of two matrices **A** and **B** can be formed only if the number of columns of **A** is equal to the number of rows of **B**. Since **M** has two rows and three columns, the number of its rows is not equal to the number of its columns. Thus the product **MM** cannot be formed.

(b) The number of rows of **N** is equal to the number of columns of **N**, so the product **NN** is defined. We have

$$\mathbf{NN} = \begin{pmatrix} 2 & -1 \\ 0 & 2 \end{pmatrix} \begin{pmatrix} 2 & -1 \\ 0 & 2 \end{pmatrix} = \begin{pmatrix} 4 & -4 \\ 0 & 4 \end{pmatrix}.$$

(c) A matrix **A** must have the same number of rows as columns in order for the product **AA** to be defined. So only matrices that have the same number of rows as columns can be squared. (You might remember that such matrices are called *square* matrices.)

Solution to Activity 12

(a)

$$\begin{pmatrix} -1 & 4 \\ 2 & 0 \end{pmatrix}^2 = \begin{pmatrix} -1 & 4 \\ 2 & 0 \end{pmatrix} \begin{pmatrix} -1 & 4 \\ 2 & 0 \end{pmatrix} = \begin{pmatrix} 9 & -4 \\ -2 & 8 \end{pmatrix}$$

$$\begin{pmatrix} 1 & 0 & 1 \\ 0 & 2 & 0 \\ 0 & 0 & 3 \end{pmatrix}^2 = \begin{pmatrix} 1 & 0 & 1 \\ 0 & 2 & 0 \\ 0 & 0 & 3 \end{pmatrix} \begin{pmatrix} 1 & 0 & 1 \\ 0 & 2 & 0 \\ 0 & 0 & 3 \end{pmatrix} = \begin{pmatrix} 1 & 0 & 4 \\ 0 & 4 & 0 \\ 0 & 0 & 9 \end{pmatrix}$$

(b) From the text preceding this activity, we know that $\mathbf{A}^3 = \begin{pmatrix} 20 & 7 \\ 21 & 6 \end{pmatrix}$. Therefore

$$\begin{aligned} \mathbf{A}^6 &= \mathbf{A}^3\mathbf{A}^3 = (\mathbf{A}^3)^2 \\ &= \begin{pmatrix} 20 & 7 \\ 21 & 6 \end{pmatrix}^2 = \begin{pmatrix} 20 & 7 \\ 21 & 6 \end{pmatrix}\begin{pmatrix} 20 & 7 \\ 21 & 6 \end{pmatrix} \\ &= \begin{pmatrix} 547 & 182 \\ 546 & 183 \end{pmatrix}. \end{aligned}$$

Solution to Activity 14

(a) Inputting 1 litre of water at node U gives 0.2 litres of water output at node C and 0.8 litres of water output at node D.

Inputting 2 litres of water at node V gives 2×0.6 litres of water output at node C and 2×0.4 litres of water output at node D.

So, in total, we have $0.2 + 2 \times 0.6 = 1.4$ litres of water output at node C, and $0.8 + 2 \times 0.4 = 1.6$ litres of water output at node D.

(b) Using a method similar to that of part (a) gives $0.2x + 0.6y + 0.3z$ litres of water output at node C and $0.8x + 0.4y + 0.7z$ litres of water output at node D.

Solution to Activity 15

(a) If £2000 is input at node D, then the output at node J is $0.2 \times £2000 = £400$.

Similarly, the output at node K is $0.8 \times £2000 = £1600$.

(b) If the input at node D is £2400, at node E is £1700, and at node F is £1100, then the output (in £) at node J is

$$0.2 \times 2400 + 0.3 \times 1700 + 0.5 \times 1100 = 1540,$$

and at node K is

$$0.8 \times 2400 + 0.7 \times 1700 + 0.5 \times 1100 = 3660.$$

(c) If the input (in £) at node D is x, at node E is y, and at node F is z, then the output (in £) at node J is

$$0.2x + 0.3y + 0.5z,$$

and at node K is

$$0.8x + 0.7y + 0.5z.$$

Solution to Activity 16

(a) Since there are three input nodes and two output nodes, the network can be represented by a 2×3 matrix.

The number that labels the arc from the first input node (U) to the first output node (C) is 0.2, and this goes in the first row and first column. The number that labels the arc from the second input node (V) to the first output node (C) is 0.6, and this goes in the first row and second column.

Proceeding similarly for all the inputs and outputs of the network gives the following matrix:

$$\begin{array}{c} \\ C \\ D \end{array} \begin{array}{c} \quad U \quad\; V \quad\; W \\ \begin{pmatrix} 0.2 & 0.6 & 0.3 \\ 0.8 & 0.4 & 0.7 \end{pmatrix}. \end{array}$$

(b) The vector that represents an input of 1 litre of water at nodes U, V and W is

$$\begin{pmatrix} 1 \\ 1 \\ 1 \end{pmatrix}.$$

(c) The output vector for the input vector found in part (b) is

$$\begin{aligned} &\begin{pmatrix} 0.2 & 0.6 & 0.3 \\ 0.8 & 0.4 & 0.7 \end{pmatrix}\begin{pmatrix} 1 \\ 1 \\ 1 \end{pmatrix} \\ &= \begin{pmatrix} 0.2 \times 1 + 0.6 \times 1 + 0.3 \times 1 \\ 0.8 \times 1 + 0.4 \times 1 + 0.7 \times 1 \end{pmatrix} = \begin{pmatrix} 1.1 \\ 1.9 \end{pmatrix}. \end{aligned}$$

This vector represents an output of 1.1 litres of water at node C and 1.9 litres of water at node D.

Solution to Activity 17

(a) The matrix that represents the network with nodes U, V, P and Q is

$$\begin{array}{c} \\ P \\ Q \end{array}\begin{array}{c} \begin{array}{cc} U & V \end{array} \\ \begin{pmatrix} 0.5 & 0.2 \\ 0.5 & 0.8 \end{pmatrix} \end{array}.$$

The matrix equation giving the output $\begin{pmatrix} p \\ q \end{pmatrix}$ for the input $\begin{pmatrix} u \\ v \end{pmatrix}$ is

$$\begin{pmatrix} p \\ q \end{pmatrix} = \begin{pmatrix} 0.5 & 0.2 \\ 0.5 & 0.8 \end{pmatrix}\begin{pmatrix} u \\ v \end{pmatrix}.$$

(b) The matrix that represents the network with nodes P, Q, A, B and C is

$$\begin{array}{c} \\ A \\ B \\ C \end{array}\begin{array}{c} \begin{array}{cc} P & Q \end{array} \\ \begin{pmatrix} 0.3 & 0.3 \\ 0.3 & 0.1 \\ 0.4 & 0.6 \end{pmatrix} \end{array}.$$

The matrix equation giving the output $\begin{pmatrix} a \\ b \\ c \end{pmatrix}$ for the input $\begin{pmatrix} p \\ q \end{pmatrix}$ is

$$\begin{pmatrix} a \\ b \\ c \end{pmatrix} = \begin{pmatrix} 0.3 & 0.3 \\ 0.3 & 0.1 \\ 0.4 & 0.6 \end{pmatrix}\begin{pmatrix} p \\ q \end{pmatrix}.$$

(c) The matrix equation giving the output $\begin{pmatrix} a \\ b \\ c \end{pmatrix}$ for an input $\begin{pmatrix} u \\ v \end{pmatrix}$ in the combined network is obtained by using parts (a) and (b) as follows.

$$\begin{aligned} \begin{pmatrix} a \\ b \\ c \end{pmatrix} &= \begin{pmatrix} 0.3 & 0.3 \\ 0.3 & 0.1 \\ 0.4 & 0.6 \end{pmatrix}\begin{pmatrix} p \\ q \end{pmatrix} \\ &= \begin{pmatrix} 0.3 & 0.3 \\ 0.3 & 0.1 \\ 0.4 & 0.6 \end{pmatrix}\left(\begin{pmatrix} 0.5 & 0.2 \\ 0.5 & 0.8 \end{pmatrix}\begin{pmatrix} u \\ v \end{pmatrix}\right) \\ &= \left(\begin{pmatrix} 0.3 & 0.3 \\ 0.3 & 0.1 \\ 0.4 & 0.6 \end{pmatrix}\begin{pmatrix} 0.5 & 0.2 \\ 0.5 & 0.8 \end{pmatrix}\right)\begin{pmatrix} u \\ v \end{pmatrix} \\ &= \begin{pmatrix} 0.3 & 0.3 \\ 0.2 & 0.14 \\ 0.5 & 0.56 \end{pmatrix}\begin{pmatrix} u \\ v \end{pmatrix} \end{aligned}$$

(d) Labelling the matrix that represents the combined network gives

$$\begin{array}{c} \\ A \\ B \\ C \end{array}\begin{array}{c} \begin{array}{cc} U & V \end{array} \\ \begin{pmatrix} 0.3 & 0.3 \\ 0.2 & 0.14 \\ 0.5 & 0.56 \end{pmatrix} \end{array}.$$

A simplified network equivalent to the original one is given below.

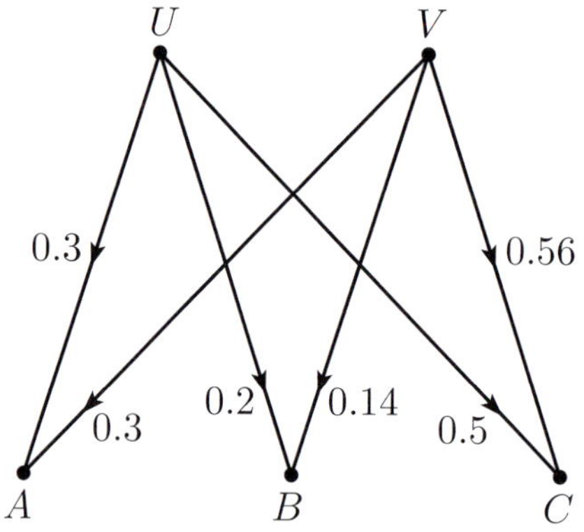

(e) An input of 1 litre at node U and 2 litres at node V is represented by the input vector $\begin{pmatrix} 1 \\ 2 \end{pmatrix}$. Using the equation for the combined network found in part (c), the corresponding output is given by

$$\begin{pmatrix} a \\ b \\ c \end{pmatrix} = \begin{pmatrix} 0.3 & 0.3 \\ 0.2 & 0.14 \\ 0.5 & 0.56 \end{pmatrix}\begin{pmatrix} 1 \\ 2 \end{pmatrix} = \begin{pmatrix} 0.9 \\ 0.48 \\ 1.62 \end{pmatrix}.$$

Therefore the output is 0.9 litres at node A, 0.48 litres at node B and 1.62 litres at node C.

Solution to Activity 18

(a) The matrices (with row and column headings) are

$$\begin{array}{c} \\ J \\ K \end{array}\begin{array}{c} \begin{array}{ccc} D & E & F \end{array} \\ \begin{pmatrix} 0.2 & 0.3 & 0.5 \\ 0.8 & 0.7 & 0.5 \end{pmatrix} \end{array} \quad \text{and} \quad \begin{array}{c} \\ R \\ S \end{array}\begin{array}{c} \begin{array}{cc} J & K \end{array} \\ \begin{pmatrix} 0.6 & 0.9 \\ 0.4 & 0.1 \end{pmatrix} \end{array},$$

respectively.

(b) The matrix that represents the combined network is

$$\begin{pmatrix} 0.6 & 0.9 \\ 0.4 & 0.1 \end{pmatrix}\begin{pmatrix} 0.2 & 0.3 & 0.5 \\ 0.8 & 0.7 & 0.5 \end{pmatrix} = \begin{pmatrix} 0.84 & 0.81 & 0.75 \\ 0.16 & 0.19 & 0.25 \end{pmatrix}.$$

(c) The simplified network is shown below.

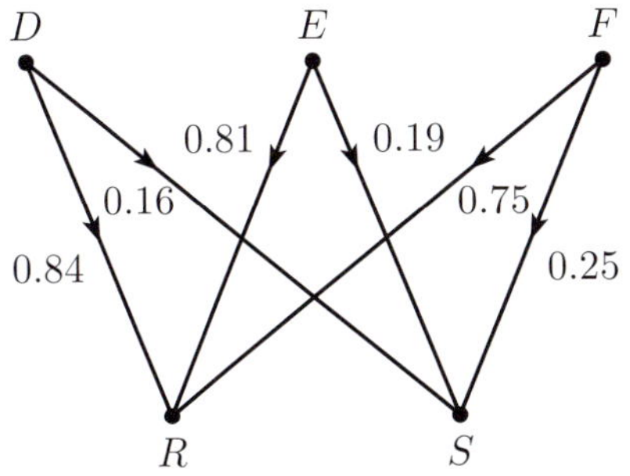

(d) If the input vector (in £) to the combined network is

$$\begin{pmatrix}1700\\1400\\0\end{pmatrix},$$

then the output vector (in £) is

$$\begin{pmatrix}0.84 & 0.81 & 0.75\\0.16 & 0.19 & 0.25\end{pmatrix}\begin{pmatrix}1700\\1400\\0\end{pmatrix}=\begin{pmatrix}2562\\538\end{pmatrix}.$$

So £538 is spent in region S.

Solution to Activity 19

The products $\mathbf{AI}$, $\mathbf{IA}$ and $\mathbf{IB}$ are defined, and they give the following results:

$$\mathbf{AI}=\begin{pmatrix}3 & 4\\-2 & 1\end{pmatrix}\begin{pmatrix}1 & 0\\0 & 1\end{pmatrix}=\begin{pmatrix}3 & 4\\-2 & 1\end{pmatrix}=\mathbf{A},$$

$$\mathbf{IA}=\begin{pmatrix}1 & 0\\0 & 1\end{pmatrix}\begin{pmatrix}3 & 4\\-2 & 1\end{pmatrix}=\begin{pmatrix}3 & 4\\-2 & 1\end{pmatrix}=\mathbf{A},$$

$$\mathbf{IB}=\begin{pmatrix}1 & 0\\0 & 1\end{pmatrix}\begin{pmatrix}2 & 7 & -3\\8 & 1 & 2\end{pmatrix}=\begin{pmatrix}2 & 7 & -3\\8 & 1 & 2\end{pmatrix}=\mathbf{B}.$$

The product $\mathbf{BI}$ is not defined because $\mathbf{B}$ has 3 columns and $\mathbf{I}$ has 2 rows.

Solution to Activity 20

We have

$$\mathbf{AI}=\begin{pmatrix}a & b & c\\d & e & f\end{pmatrix}\begin{pmatrix}1 & 0 & 0\\0 & 1 & 0\\0 & 0 & 1\end{pmatrix}$$
$$=\begin{pmatrix}a & b & c\\d & e & f\end{pmatrix}.$$

Also,

$$\mathbf{IB}=\begin{pmatrix}1 & 0 & 0\\0 & 1 & 0\\0 & 0 & 1\end{pmatrix}\begin{pmatrix}x\\y\\z\end{pmatrix}=\begin{pmatrix}x\\y\\z\end{pmatrix}.$$

Therefore $\mathbf{AI}=\mathbf{A}$ and $\mathbf{IB}=\mathbf{I}$, as required.

Solution to Activity 21

(a) $$\mathbf{AB}=\begin{pmatrix}8 & 5\\3 & 2\end{pmatrix}\begin{pmatrix}2 & -5\\-3 & 8\end{pmatrix}=\begin{pmatrix}1 & 0\\0 & 1\end{pmatrix},$$

$$\mathbf{BA}=\begin{pmatrix}2 & -5\\-3 & 8\end{pmatrix}\begin{pmatrix}8 & 5\\3 & 2\end{pmatrix}=\begin{pmatrix}1 & 0\\0 & 1\end{pmatrix}.$$

(b) $$\mathbf{CD}=\begin{pmatrix}2 & 3 & 1\\-1 & 1 & 0\\-2 & 1 & 0\end{pmatrix}\begin{pmatrix}0 & 1 & -1\\0 & 2 & -1\\1 & -8 & 5\end{pmatrix}$$
$$=\begin{pmatrix}1 & 0 & 0\\0 & 1 & 0\\0 & 0 & 1\end{pmatrix},$$

$$\mathbf{DC}=\begin{pmatrix}0 & 1 & -1\\0 & 2 & -1\\1 & -8 & 5\end{pmatrix}\begin{pmatrix}2 & 3 & 1\\-1 & 1 & 0\\-2 & 1 & 0\end{pmatrix}$$
$$=\begin{pmatrix}1 & 0 & 0\\0 & 1 & 0\\0 & 0 & 1\end{pmatrix}.$$

Solution to Activity 22

(a) We have $\det\mathbf{A}=13\times 2-5\times 5=1\neq 0$.
Thus $\mathbf{A}$ is invertible and

$$\mathbf{A}^{-1}=\begin{pmatrix}2 & -5\\-5 & 13\end{pmatrix}.$$

(b) We have $\det\mathbf{B}=3\times 2-0\times(-5)=6\neq 0$.
Thus $\mathbf{B}$ is invertible and

$$\mathbf{B}^{-1}=\frac{1}{6}\begin{pmatrix}2 & 0\\5 & 3\end{pmatrix}=\begin{pmatrix}\frac{1}{3} & 0\\\frac{5}{6} & \frac{1}{2}\end{pmatrix}.$$

(Either of these two forms of the inverse is acceptable as your final answer.)

(c) We have $\det\mathbf{C}=-3\times(-\frac{1}{3})-\frac{1}{5}\times 5=0$.
Thus $\mathbf{C}$ is not invertible.

(d) We have $\det\mathbf{D}=1.5\times 1.5-2.5\times 0.5=1\neq 0$.
Thus $\mathbf{D}$ is invertible and

$$\mathbf{D}^{-1}=\begin{pmatrix}1.5 & -2.5\\-0.5 & 1.5\end{pmatrix}.$$

Solution to Activity 23

Using the property of matrix products given in the hint, we have

$$\begin{aligned}\mathbf{AB} &= \begin{pmatrix} a & b \\ c & d \end{pmatrix}\left(\frac{1}{ad-bc}\begin{pmatrix} d & -b \\ -c & a \end{pmatrix}\right)\\ &= \frac{1}{ad-bc}\begin{pmatrix} a & b \\ c & d \end{pmatrix}\begin{pmatrix} d & -b \\ -c & a \end{pmatrix}\\ &= \frac{1}{ad-bc}\begin{pmatrix} ad-bc & 0 \\ 0 & ad-bc \end{pmatrix}\\ &= \frac{1}{ad-bc}\times(ad-bc)\begin{pmatrix} 1 & 0 \\ 0 & 1 \end{pmatrix}\\ &= \begin{pmatrix} 1 & 0 \\ 0 & 1 \end{pmatrix} = \mathbf{I}.\end{aligned}$$

Moreover,

$$\begin{aligned}\mathbf{BA} &= \left(\frac{1}{ad-bc}\begin{pmatrix} d & -b \\ -c & a \end{pmatrix}\right)\begin{pmatrix} a & b \\ c & d \end{pmatrix}\\ &= \frac{1}{ad-bc}\begin{pmatrix} ad-bc & 0 \\ 0 & ad-bc \end{pmatrix} = \mathbf{I}.\end{aligned}$$

Hence $\mathbf{AB} = \mathbf{BA} = \mathbf{I}$, and it follows that $\mathbf{B} = \mathbf{A}^{-1}$.

Solution to Activity 25

(a) The coefficient matrix of the equations

$$\begin{aligned}2x + 3y &= 3\\ x + 4y &= -1\end{aligned}$$

is $\begin{pmatrix} 2 & 3 \\ 1 & 4 \end{pmatrix}$. Thus, the matrix form of the equations is

$$\begin{pmatrix} 2 & 3 \\ 1 & 4 \end{pmatrix}\begin{pmatrix} x \\ y \end{pmatrix} = \begin{pmatrix} 3 \\ -1 \end{pmatrix}.$$

(b) First rearrange the first equation as

$$2x = 6.$$

The matrix form of the equations

$$\begin{aligned}2x \qquad &= 6\\ 3x + 6y &= 15\end{aligned}$$

is

$$\begin{pmatrix} 2 & 0 \\ 3 & 6 \end{pmatrix}\begin{pmatrix} x \\ y \end{pmatrix} = \begin{pmatrix} 6 \\ 15 \end{pmatrix}.$$

(c) The matrix form of the equations

$$\begin{aligned}\tfrac{3}{5}x - \tfrac{4}{5}y &= 18\\ \tfrac{4}{5}x + \tfrac{3}{5}y &= -1\end{aligned}$$

is

$$\begin{pmatrix} \frac{3}{5} & -\frac{4}{5} \\ \frac{4}{5} & \frac{3}{5} \end{pmatrix}\begin{pmatrix} x \\ y \end{pmatrix} = \begin{pmatrix} 18 \\ -1 \end{pmatrix}.$$

Solution to Activity 26

(a) Activity 25(a) gives the matrix form of the equations as

$$\begin{pmatrix} 2 & 3 \\ 1 & 4 \end{pmatrix}\begin{pmatrix} x \\ y \end{pmatrix} = \begin{pmatrix} 3 \\ -1 \end{pmatrix}.$$

The determinant of the coefficient matrix is

$$2 \times 4 - 3 \times 1 = 5,$$

so the inverse of the coefficient matrix is

$$\frac{1}{5}\begin{pmatrix} 4 & -3 \\ -1 & 2 \end{pmatrix} = \begin{pmatrix} \frac{4}{5} & -\frac{3}{5} \\ -\frac{1}{5} & \frac{2}{5} \end{pmatrix}.$$

Hence

$$\begin{aligned}\begin{pmatrix} x \\ y \end{pmatrix} &= \frac{1}{5}\begin{pmatrix} 4 & -3 \\ -1 & 2 \end{pmatrix}\begin{pmatrix} 3 \\ -1 \end{pmatrix}\\ &= \begin{pmatrix} 3 \\ -1 \end{pmatrix}.\end{aligned}$$

The solution is $x = 3$, $y = -1$.

(b) Activity 25(b) gives the matrix form of the equations as

$$\begin{pmatrix} 2 & 0 \\ 3 & 6 \end{pmatrix}\begin{pmatrix} x \\ y \end{pmatrix} = \begin{pmatrix} 6 \\ 15 \end{pmatrix}.$$

The determinant of the coefficient matrix is

$$2 \times 6 - 0 \times 3 = 12,$$

so the inverse of the coefficient matrix is

$$\frac{1}{12}\begin{pmatrix} 6 & 0 \\ -3 & 2 \end{pmatrix} = \begin{pmatrix} \frac{1}{2} & 0 \\ -\frac{1}{4} & \frac{1}{6} \end{pmatrix}.$$

Hence

$$\begin{pmatrix} x \\ y \end{pmatrix} = \frac{1}{12}\begin{pmatrix} 6 & 0 \\ -3 & 2 \end{pmatrix}\begin{pmatrix} 6 \\ 15 \end{pmatrix} = \begin{pmatrix} 3 \\ 1 \end{pmatrix}.$$

The solution is $x = 3$, $y = 1$.

(c) Activity 25(c) gives the matrix form of the equations as

$$\begin{pmatrix} \frac{3}{5} & -\frac{4}{5} \\ \frac{4}{5} & \frac{3}{5} \end{pmatrix}\begin{pmatrix} x \\ y \end{pmatrix} = \begin{pmatrix} 18 \\ -1 \end{pmatrix}.$$

The determinant of the coefficient matrix is

$$\tfrac{3}{5} \times \tfrac{3}{5} - \left(-\tfrac{4}{5}\right) \times \tfrac{4}{5} = 1,$$

so the inverse of the coefficient matrix is

$$\frac{1}{1}\begin{pmatrix} \frac{3}{5} & \frac{4}{5} \\ -\frac{4}{5} & \frac{3}{5} \end{pmatrix} = \begin{pmatrix} \frac{3}{5} & \frac{4}{5} \\ -\frac{4}{5} & \frac{3}{5} \end{pmatrix}.$$

Hence

$$\begin{pmatrix} x \\ y \end{pmatrix} = \begin{pmatrix} \frac{3}{5} & \frac{4}{5} \\ -\frac{4}{5} & \frac{3}{5} \end{pmatrix} \begin{pmatrix} 18 \\ -1 \end{pmatrix}$$

$$= \begin{pmatrix} \frac{50}{5} \\ -\frac{75}{5} \end{pmatrix} = \begin{pmatrix} 10 \\ -15 \end{pmatrix}.$$

The solution is $x = 10$, $y = -15$.

Solution to Activity 27

(a) The matrix form of the equations is

$$\begin{pmatrix} 5 & -3 \\ -1 & 4 \end{pmatrix} \begin{pmatrix} x \\ y \end{pmatrix} = \begin{pmatrix} 5 \\ 7 \end{pmatrix}.$$

The determinant of the coefficient matrix is

$$5 \times 4 - (-3) \times (-1) = 17.$$

This is non-zero, so the coefficient matrix is invertible and the equations have a unique solution.

(b) The matrix form of the equations is

$$\begin{pmatrix} 3 & -1 \\ -9 & 3 \end{pmatrix} \begin{pmatrix} x \\ y \end{pmatrix} = \begin{pmatrix} -6 \\ 18 \end{pmatrix}.$$

The determinant of the coefficient matrix is

$$3 \times 3 - (-1) \times (-9) = 0.$$

This is zero, so the coefficient matrix is non-invertible and the equations do not have a unique solution.

Solution to Activity 28

(a) The matrix form of the equations is

$$\begin{pmatrix} 2 & 2 & 3 \\ 2 & 3 & 4 \\ 4 & 7 & 10 \end{pmatrix} \begin{pmatrix} x \\ y \\ z \end{pmatrix} = \begin{pmatrix} 2 \\ -1 \\ 3 \end{pmatrix}.$$

We have

$$\begin{pmatrix} 2 & 2 & 3 \\ 2 & 3 & 4 \\ 4 & 7 & 10 \end{pmatrix}^{-1} = \begin{pmatrix} 1 & \frac{1}{2} & -\frac{1}{2} \\ -2 & 4 & -1 \\ 1 & -3 & 1 \end{pmatrix}.$$

Therefore

$$\begin{pmatrix} x \\ y \\ z \end{pmatrix} = \begin{pmatrix} 1 & \frac{1}{2} & -\frac{1}{2} \\ -2 & 4 & -1 \\ 1 & -3 & 1 \end{pmatrix} \begin{pmatrix} 2 \\ -1 \\ 3 \end{pmatrix} = \begin{pmatrix} 0 \\ -11 \\ 8 \end{pmatrix}.$$

The solution is $x = 0$, $y = -11$, $z = 8$.

(b) The matrix form of the equations is

$$\begin{pmatrix} 3 & 2 & 1 \\ 4 & 3 & 1 \\ 7 & 5 & 1 \end{pmatrix} \begin{pmatrix} x \\ y \\ z \end{pmatrix} = \begin{pmatrix} 1 \\ 2 \\ 1 \end{pmatrix}.$$

We have

$$\begin{pmatrix} 3 & 2 & 1 \\ 4 & 3 & 1 \\ 7 & 5 & 1 \end{pmatrix}^{-1} = \begin{pmatrix} 2 & -3 & 1 \\ -3 & 4 & -1 \\ 1 & 1 & -1 \end{pmatrix}.$$

Therefore

$$\begin{pmatrix} x \\ y \\ z \end{pmatrix} = \begin{pmatrix} 2 & -3 & 1 \\ -3 & 4 & -1 \\ 1 & 1 & -1 \end{pmatrix} \begin{pmatrix} 1 \\ 2 \\ 1 \end{pmatrix} = \begin{pmatrix} -3 \\ 4 \\ 2 \end{pmatrix}.$$

The solution is $x = -3$, $y = 4$, $z = 2$.

(Details of how to use the CAS to find the inverses of the matrices are given in the *Computer algebra guide*, in the section 'Computer methods for CAS activities in Books A–D'.)

Solution to Activity 29

(a) The matrix form of the equations is

$$\begin{pmatrix} 1 & 2 & 3 \\ 2 & 3 & 1 \\ 4 & 7 & 7 \end{pmatrix} \begin{pmatrix} x \\ y \\ z \end{pmatrix} = \begin{pmatrix} 5 \\ 6 \\ 1 \end{pmatrix}.$$

The determinant of the coefficient matrix is zero, so the coefficient matrix is non-invertible and the equations do not have a unique solution.

(b) The matrix form of the equations is

$$\begin{pmatrix} 2 & 2 & 3 \\ 2 & 3 & 1 \\ 4 & 7 & 7 \end{pmatrix} \begin{pmatrix} x \\ y \\ z \end{pmatrix} = \begin{pmatrix} 5 \\ 6 \\ 1 \end{pmatrix}.$$

The determinant of the coefficient matrix is 14. This is non-zero, so the coefficient matrix is invertible and the equations have a unique solution.

(c) The matrix form of the equations is

$$\begin{pmatrix} 1 & 1 & -1 & 1 \\ 1 & -1 & 1 & 1 \\ 2 & 0 & 3 & -1 \\ 6 & 2 & 4 & 0 \end{pmatrix} \begin{pmatrix} x \\ y \\ w \\ z \end{pmatrix} = \begin{pmatrix} 0 \\ 2 \\ 1 \\ 0 \end{pmatrix}.$$

The determinant of the coefficient matrix is zero, so the coefficient matrix is non-invertible and the equations do not have a unique solution.

(Details of how to use the CAS to find the determinants are given in the *Computer algebra guide*, in the section 'Computer methods for CAS activities in Books A–D'.)

Acknowledgements

Grateful acknowledgement is made to the following sources:

Page 219: Taken from: http://en.wikipedia.org/wiki/File:Ms_sf_2.jpg. This file is licensed under the Creative Commons Attribution-Share Alike Licence http://creativecommons.org/licenses/by-sa/3.0/

Page 221: Courtesy Rob Cottingham. www.robcottingham.ca

Page 238: © Unknown

Every effort has been made to contact copyright holders. If any have been inadvertently overlooked the publishers will be pleased to make the necessary arrangements at the first opportunity.

Index